Innovation Communities

Klaus Fichter • Severin Beucker
Editors

Innovation Communities

Teamworking of Key Persons - A Success Factor in Radical Innovation

Springer

Editors
Klaus Fichter
Borderstep Institute for Innovation
and Sustainability
Clayallee 323
14169 Berlin
Germany
fichter@borderstep.de

Severin Beucker
Borderstep Institute for Innovation
and Sustainability
Clayallee 323
14169 Berlin
Germany
beucker@borderstep.de

This publication is based on the results of the research and implementation project "Innovation Communities: Networks of promotors as a success factor for radical innovation (InnoCo)" (FKZ 01FM07075).
InnoCo was part of the research programme on "Innovation Strategies Beyond Traditional Management" and was funded by the German Federal Ministry of Education and Research and the European Social Fund.

ISBN 978-3-642-22127-9 e-ISBN 978-3-642-22128-6
DOI 10.1007/978-3-642-22128-6
Springer Heidelberg Dordrecht London New York

Library of Congress Control Number: 2011944015

Printed on acid-free paper

Springer is a part of Springer Science+Business Media (www.springer.com)

Preface

Stimulated by Schumpeter's works on the central role of the entrepreneur in innovation processes, there has been more than half a century's worth of research on the role of key persons in innovation. Since the introduction of the term "champion" by Schon in 1963, and the "promotor"[1] concept by Witte (1973), there has been little doubt that the human factor plays a decisive part in innovation. A wide array of studies shows that transformational leaders are still a key factor for innovation success today, but that their roles and contributions to the effectiveness and efficiency of innovation projects need to be re-examined in the light of changing framework conditions of innovation. The increased dynamics and complexity of innovation processes requires disparate information assets to be levelled and various interrelated knowledge and resource bases inside and outside the organization to be integrated. Self-organizing networks have hence become the dominant innovators for complex technological solutions and systems in the twenty-first century.

The Innoco Project

Taking new challenges like increased complexity, dynamics and division of labour in innovation processes into account, this book presents the new concept of "innovation communities" and the results from empirical and practical work on innovation communities, carried out in the framework of the research and implementation project "*Innovation Communities: Networks of promotors as a success factor for radical innovation (InnoCo)*". InnoCo was part of the research programme on "Innovation Strategies Beyond Traditional Management", and was

[1] Throughout the book, we employ the specific German term "promotor", derived from the Latin verb "promovere", to distinguish it from the more generic term "promoter".

funded by the German Federal Ministry of Education and Research and the European Social Fund of the European Union from 2008 to 2011.

Partners of the project alliance InnoCo are the Borderstep Institute for Innovation and Sustainability, Berlin (a non-profit institute coordinating the project), the Department of Marketing and Innovation Management at the Brandenburg Technical University of Cottbus, the Institute of Ergonomics and Technology Management (IAT) at the University of Stuttgart, and the InTraCoM GmbH (LLC), Stuttgart. Practice partners include the innovation network association Connected Living, the carbon nanotubes innovation alliance Inno.CNT, the industry partners io-li-tec GmbH & Co. KG, FutureCarbon, and more than ten further companies and institutions.

Goals and Guiding Questions

In the framework of the InnoCo project, an *innovation community* is defined as an informal network of likeminded individuals (often from various different companies and organizations) that team up in a project-like fashion and commonly promote a specific innovation, either on one or several levels of an innovation system, hence acting as universal or specialised "promotors". Innovation communities therefore are characterised as promotor networks or as informal personal networks of innovators.

The InnoCo project aimed at achieving the following goals:

- Reaching an understanding of the role and relevance of innovation communities for the development and implementation process of radical, "breakthrough" innovations.
- Developing concepts for the establishment and development of innovation communities, and rendering them convertible to various industries and fields of application.
- Identifying and developing productive methods and tools that support the establishment and development of innovation communities.
- Processing and evaluating research results, developing methods and tools for training purposes, and broad transfer and utilization procedures.

For these purposes, the research and implementation project InnoCo project comprised three main activities:

1. *Empirical research*, including large-scale surveys of innovation networks as well as multiple case studies on the role of innovation communities in radical innovation projects.
2. *Practical projects* in collaboration with companies, intermediaries and research institutions, which initiated and developed innovation communities in three practice fields.

3. *Development of methods and tools* to support the establishment and development of innovation communities and their contribution to the overall success of the innovation project.

Within its scope of activities, the InnoCo project focused on five key questions:

1. *Success factor*: What is the role and relevance of transformational leaders like promotors and champions in innovation networks, and what is their impact on innovation success?
2. *Performance*: How exactly do innovation communities contribute to the success of innovation projects?
3. *Evolution*: How does the size and structure of innovation communities change over time?
4. *Quality*: Which features make innovation communities successful?
5. *Methods*: Which measures and methods constitute and improve the performance of innovation communities?

This publication presents the results of the InnoCo project, containing findings and experiences that are valuable for the following groups in particular:

- Entrepreneurs
- Innovation, R&D and network managers
- Innovation and strategy consultants
- Innovation and start-up intermediaries
- Innovation researchers
- Government officials and politicians responsible for R&D and innovation programmes and funding

Structure and Contents of the Book

The book is structured into six chapters:

Chapter 1 introduces the concept of "Innovation Communities" and, in the light of changing innovation framework conditions, explains why it is helpful in successfully managing radical innovation projects, thus providing the basic conceptual background of the book.

Based on the conceptual framework developed in Chap. 1, the following *Chap. 2* presents empirical findings regarding the role and relevance of transformational leaders like "promotors" and "champions" in innovation networks, and their effect on innovation success. Based on research hypotheses and measures for promotor roles, group identity, group interaction, project performance and degree of innovation, this chapter presents the findings of the innovation network survey. The hypothesized relationships were tested using a comprehensive sample of 107 networks of German companies and research institutions.

While Chap. 2 indicates that innovation communities have a significantly positive effect on project performance and can therefore be considered a success factor for radical innovations, some questions remain in terms of the composition (How does the size and structure of innovation communities change over time?), the performance (How exactly do innovation communities contribute to the success of innovation projects?), and the quality of innovation communities (Which features make them successful?). Against this background, Chap. 3 presents the findings from five case studies to address the issues articulated above.

At the core of Chap. 4 is the question whether and how innovation communities can be actively supported and managed, for instance by influencing their configuration or providing the structural or methodological background. While the case studies in the preceding Chap. 3 give insight into five innovation communities and the effects and interactions observed between the members of those communities, the three practice fields presented in Chap. 4 have a different scope. They were selected and accompanied to collect experience in the practical field of innovation community management. All three practice fields focus on radical innovations that take place in various sectors of high-tech industry (information and communication technology, renewable energies, material sciences).

Based on the empirical findings of a survey of innovation networks (Chap. 2), the results of the five case studies (Chap. 3), and the practical experience acquired in the development of innovation communities in the preceding Chap. 4, Chap. 5 presents hands-on methods for establishing and developing innovation communities. This chapter merges empirical and practical knowledge as to which methods and measures can be employed by innovation and network managers for the practical development of innovation communities. These findings are summarized in a "toolbox". Within the InnoCo project, special attention was paid to two methodological approaches that are as methodologically complex as they are important for the development of innovation communities. These are systemic innovation coaching for innovation communities, and Delphi-based scenario development for innovation networks. These methods are presented in detail in Chap. 5.

Chap. 6, finally, summarizes key insights from the InnoCo project that have been presented in detail throughout the book. It also discusses implications for managerial practice and future implementations of the innovation community concept. InnoCo has demonstrated the relevance of cross-organizational and cross-level networks of promotors for the success of radical innovation projects, and has generated many important new insights into the evolution and management of innovation communities. Nevertheless, there are important research questions that are still left unanswered and call for further investigations. The final section of Chap. 6 outlines limitations of present research and pathways for future research.

Berlin Klaus Fichter and Severin Beucker

Acknowledgements

This publication is a communal achievement! The concept of "innovation communities", the central topic of the book, also applies to the team that, within the framework of the InnoCo project, jointly developed, examined and tested a new concept of innovation networks across various organizational and systemic levels. We would like to thank our project partners for this excellent example of collaboration: Prof. Daniel Baier and Dr. Alexandra Rese, BTU Cottbus; Stefanie Bunzel, University of Stuttgart; Dr. Dierk-Oliver Kiehne, intracom GmbH, Stuttgart; and Dr. Willy Bierter, Product-Life Institute, Geneva. We are most obliged to our industry partners, too: the working group "Connected Energy", part of the innovation network association Connected Living e.V.; the FutureCarbon GmbH and other companies within the innovation alliance for carbon nanotubes Inno. CNT; the io-li-tec GmbH & Co. KG, Solaronix SA, and other members of the innovation network for dye-sensitized solar cells. Furthermore we would like to express our gratitude to our colleague Linda Bergset, who was part of the Borderstep research team. Financial support from the German Federal Ministry of Education and Research BMBF and the European Social Fund of the European Union is gratefully acknowledged; the InnoCo project would not have been possible otherwise. We would especially like to express our gratitude to Prof. Ingeborg Bootz who was in charge of the project on the part of the project executing agency, PT-DLR (German Aerospace Center), within the promotional activity field "Innovation Beyond Traditional Management".

We would further like to thank our Borderstep colleagues Anke Posthumus-Mendner for her organizational support and administrative supervision, and Anwen Roberts for editorial services, proofreading and translation; the production of the book at hand would not have been possible without them.

Berlin — Klaus Fichter and Severin Beucker

Contents

Author Bibliographies

Dr Daniel Baier is full professor of Marketing and Innovation Management at the Brandenburg University Technology Cottbus, Germany. He received his PhD and venia legendi in marketing-oriented product development at the University of Karlsruhe. His works have appeared in: the Journal of Econometrics, the Annals of Operations Research, Zeitschrift für betriebswirtschaftliche Forschung (zfbf), Zeitschrift für Betriebswirtschaftslehre, Marketing ZFP, R&D Management, the International Journal of Technology Intelligence and Planning, as well as the Journal of Retailing and Consumer Services.

His current research focuses on the market-oriented development of innovative products and services, on innovation management in networks, data analysis, statistics, and operations research. He teaches bachelor and master students as well as PhD students and executives classes in marketing and innovation management in Economics and Industrial Engineering courses.

daniel.baier@tu-cottbus.de
www.tu-cottbus.de/fakultaet3/de/marketing/

Linda Bergset works as a researcher at the Borderstep Institute for Innovation and Sustainability in Berlin with special focus on sustainable entrepreneurship and innovation. One particular emphasis of her research lies in funding and resource acquisition strategies for sustainable start-ups and innovation projects.

bergset@borderstep.de
www.borderstep.de

Dr Severin Beucker is cofounder of the Borderstep Institute for Innovation and Sustainability, based in Berlin, Germany. His research focuses on the analysis of innovation and cooperation networks in the field of sustainable technologies and green future markets. In addition, he has a strong focus on technology management, especially the assessment of resource productivity and the cost-effectiveness of new technologies.

He studied environmental engineering at the Technical University of Berlin. He worked for the Institute for Human Factors and Technology Management (IAT) at the University of Stuttgart and later for its cooperation partner, the Fraunhofer Institute for Industrial Engineering, IAO.

He has managed various joint research projects and industry-funded projects, and directed task forces for the development of emergent research fields, such as nanotechnology and biotechnology, for the Fraunhofer Society and the German Ministry of Education and Research. He completed his doctoral thesis at the Faculty of Mechanical Engineering at the University of Stuttgart in 2005. He has been teaching cooperation and network management at the Carl von Ossietzky University of Oldenburg since 2010.

beucker@borderstep.de
www.borderstep.de

Dr Willy Bierter is co-director of the Product-Life Institute in Geneva, Switzerland, and research fellow and senior consultant at the Borderstep Institute for Innovation and Sustainability. His research and collaborative work in the fields of environment, sustainability, green future markets and technology, eco-entrepreneurship, and mass customization go back several decades. As a coach or co-leader, he has worked with more than 100 enterprises on developing sustainable business innovation and business transformation projects as well as eco-design systems, processes, products, and services. He was a lecturer in environmental science at the Swiss Polytechnic University ETH in Zurich, and director of the working group "New Welfare Models" at the Wuppertal Institute for Climate, Environment and Energy. He is the author and co-author of a number of books and research papers.

bierter@borderstep.de
www.borderstep.de and www.product-life.org

Stefanie Bunzel (Dipl.-Ing.) studied Mechanical Engineering at the University of Stuttgart and has been working at the Fraunhofer Institute for Industrial Engineering, IAO, as a research associate since 2007. She works in numerous industrial and research projects in the business fields Technology and Innovation Management with a focus on technology analysis, technology assessment and social network analysis. One main focus of her research is in the field of nanotechnology, with the aim of identifying the innovation potential of nano-based solutions, i.e. creating added value to a product by improved material attributes or new functions, at an early stage. She supports a considerable number of small and large companies by providing methodical approaches to systematic technology and innovation management.

Stefanie.Bunzel@iao.fraunhofer.de
www.innovation.iao.fraunhofer.de

Dr Klaus Fichter is the founder and director of the Borderstep Institute for Innovation and Sustainability, based in Berlin, Germany. He is professor of Innovation Management and Sustainability at the Carl von Ossietzky University of Oldenburg, member of the board of directors of the Oldenburg Center for Sustainability Economics and Management, CENTOS, and head of master programmes in innovation management and sustainable entrepreneurship.

The emphasis of his research is on evolutionary and interaction economics, innovation management, eco-entrepreneurship and sustainable innovations. His particular focus is on co-creation, innovation communities, the entrepreneurial role within the innovation process (entrepreneurship, interpreneurship), and on strategic and interactive methods.

fichter@borderstep.de
www.borderstep.de and www.innovation.uni-oldenburg.de

Dr Dierk-Oliver Kiehne founded InTraCoM GmbH in 2000 and was its CEO until 2010, when he split the company into one consulting and one software company, InTraCoM and IPR-Systems. In the same year, he became the director of the Berlin-based IPR-Systems corporation. Both companies focus on innovation management, consulting, and software solutions. In addition to innovation management, they offer solutions for the research, analysis and evaluation of intellectual property (IP) issues.

Dr Kiehne's research focus is on innovation management methods and tool development, and on the application and continuous improvement of innovation and quality management methods in various industrial projects.

Kiehne@ipr-systems.de
www.intracomgroup.de and www.ipr-systems.de

Dr Alexandra Rese is assistant professor in the department of Marketing and Innovation Management at the Brandenburg University of Technology (TU) in Cottbus, Germany. She received her PhD in sociology and entrepreneurship from the University of Karlsruhe while working at the Fraunhofer Institute for Systems and Innovation Research, ISI, in Karlsruhe. Her work has been published in journals like R&D Management and International Journal of Technology Intelligence and Planning. Her current research focuses on innovative and entrepreneurial teams, and on the skills required in entrepreneurship, as well as conflicts, information management, and computer-based methods for new product development. She also teaches business administration to students of the TU Cottbus.

rese@tu-cottbus.de
www.tu-cottbus.de/fakultaet3/de/marketing/

Chapter 1
Innovation Communities: A New Concept for New Challenges

Klaus Fichter

Abstract This chapter introduces the concept of "innovation communities" and explains why it is helpful for successfully managing radical innovation projects in the light of changing innovation framework conditions, providing the basic conceptual background of the book.

Section 1.1 describes new challenges to innovation processes through increased dynamics, complexity and increased division of labour. Because of this fundamental change in framework conditions, self-organizing networks have become the dominant innovators for complex technological and systems solutions.

Practical experience and empirical research in innovation management have shown that cooperation and efficient networking are important success factors in (nearly) all innovation processes. At the same time, key persons like entrepreneurs and their relevance for initiating and implementing of innovations have been a major focus of start-up policy makers, venture capitalists, innovation incubators, and innovation researchers. However, so far, only little empirical evidence exists on the cooperation of key persons across organizational borders and across different levels of innovation systems. Against this background, Sect. 1.2 describes shortfalls of existing theories and introduces conceptual sources for cross-organizational and cross-level networks of transformational leaders.

Section 1.3 further establishes the new concept of "innovation communities". On the basis of extended promotor theory and the concept of three-level innovation systems, the term "innovation communities" is here used in the sense of promotor networks, or informal personal networks of innovators.

K. Fichter and S. Beucker (eds.), *Innovation Communities*,
DOI 10.1007/978-3-642-22128-6_1, © Springer-Verlag Berlin Heidelberg 2012

Practical experience as well as empirical research in innovation management has shown that cooperation and efficient networking are key success factors and enablers in (nearly) all innovation processes. For this reason, developing and maintaining innovation networks and regional innovation clusters has been a key concern of R&D policy, innovation intermediaries and innovative companies for the last couple of decades. Innovative cooperation also has been investigated with particular interest in innovation research for quite some years.[1] At the same time, key persons like entrepreneurs and their relevance for initiation and implementation of innovations have been a major focus of start-up policy makers, venture capitalists, innovation incubators as well as innovation researchers. However, so far, only little empirical evidence exists about the cooperation of key persons across organizational borders as well as their role and significance for the innovation process.

Stimulated by Schumpeter's works on the central role of the entrepreneur in innovation processes (1947, 1993), there has been more than half a century's worth of research on the role of key persons in innovation. Since the introduction of the term "champion" by Schon in 1963, and the "promotor"[2] concept by Witte (1973), there has been little doubt that the human factor plays an important role in innovation (Rothwell 1994). A vast array of studies shows that transformational leaders, here distinguished as "champions" and "promotors" are still a key factor for innovation success today, but that their role and contribution to the effectiveness and efficiency of innovation projects needs to be reexamined in the light of changing framework conditions of innovation (Fichter 2009a; Sand and Rese 2011).

1.1 New Challenges: Change in Innovation Framework Conditions

In the last decades, the conditions of innovation have changed significantly. The development and implementation of process, product, service, and system innovations are taking place in a field of increased dynamics, complexity and division of labor.

1.1.1 Increased Dynamics in Innovation Processes

The acceleration and increased variability of technological change, market structures[3] and innovation processes is determined by the idea of "dynamisation".

[1] For a more detailed discussion of the state of the art in innovation research see Sect. 2.1.

[2] Throughout the book, we employ the specific German term "promotor", derived from the Latin verb "promovere", to distinguish it from the more generic term "promoter".

[3] According to Meffert and Burmann (2000, 180) the dynamics of a market can be described with the following figures: rate of innovation, changes in demand preference, changes in competition, technological transformations as well as the growth potential of a given market.

The increased dynamics of innovation is the result of two main causes: Firstly, the increasing power and efficiency of information and communication technologies, and their increasing use in the economic and innovation processes. Greater availability of information is accompanied by the intensified creation of knowledge and by a temporal and spatial segmentation of innovation processes and projects. Secondly, the dynamisation of innovation is due to the liberalization of global trading and increasing international competition for innovators. This increases the pressure to create powerful national and regional centres of innovation, to specialize within the global competitive innovation market, and to further accelerate development and market periods.

1.1.2 *Increased Complexity: Self-Organizing Networks as Dominant Innovators*

In addition to the increased dynamics, innovation is now equally subject to increased complexity.[4] This refers both to the amount of decision-relevant facts and variables in the corporate environment (element complexity) and their mutual reliance (relation complexity), and to the increased number of actors and unequal distribution of resources like knowledge and financial capital that need to be coordinated and adjoined in innovation projects.

This rise in complexity is itself a reflection of increased division of labour and specialization on a national and international level. Central causes and driving forces include the rapid development of computing in information and communication technologies (ICT) and the internationalization of competition. Shorter product lifespans and increasing development costs require new products to be introduced in as many markets simultaneously as possible. New ICT reduces transaction costs and simplifies network structures, e.g. for large corporate networks and in innovation management comprising distributed locations. Increased division of labour has its benefits, as flexible organizations are more agile and more quickly adaptable to turbulent and changing conditions in the innovation market and environment. Even so, this only increases the demand for coordination of complex networks and the integration of fragmented information sources or disparate interests.

The increased dynamics and complexity of innovation processes also requires disparate information assets to be leveled and various interrelated knowledge bases inside and outside the organization to be integrated (Staber 2004). For this reason, self-organizing networks have become dominant innovators for complex technological and systems solutions (Kash and Rycoft 2000). An analysis of ten leading

[4] Complexity is defined in varying ways in the pertinent literature. Younger system theory describes it as "the degree of sophistication, connectivity und implications of a decision-based field of activity" (Willke 2000, 22).

journals from the field of technology and innovation management clearly indicates that collective constructs of innovation like "teams", "networks" or "communities" have received considerable attention since the early 1990s. From 1990 to 1995, some 175 articles focused on one or several of these three groups. This increased to 288 articles between 1996 and 2001, and to 425 articles between 2002 and 2007 (Fichter 2009a, 358).

1.1.3 Increased Division of Labour: Managing Open Innovation

The increased dynamics and complexity of innovation notably impacts the interaction of corporate and non-corporate innovation activities as well as labour division within the innovation project itself. Based on studies conducted in research-based industries like IT, pharmaceuticals and biotechnology, Chesbrough (2003) notes a fundamental change in the way innovative ideas and inventions are dealt with and how they are successfully marketed. He characterizes this change as a fundamental paradigmatic shift in the way technological information is processed and utilized, from the formerly dominant paradigm of "closed innovation" to one of "open innovation" (Chesbrough 2003).

Within the "closed innovation" paradigm, innovation activities generally follow the view that innovations require firm control and rely primarily on the company's research, development and marketing: "If you want something done right, you've got to do it yourself" (Chesbrough 2003, XX). The conception, evolution and commercialization of innovative ideas largely take place within corporate boundaries. This paradigm of a closed innovation process, which in the past has been particularly dominant in R&D-intensive industries, is now increasingly being eroded, due to the increasing momentum, complexity and competition in innovation since the 1980s described above. The new perspective, however, is based on a modified "informational landscape" and primarily implies the opening of the innovation process to incorporate external ideas and R&D capacities, as well as more transparency in marketing innovative technologies and products.

The "open innovation" paradigm affords external ideas or marketing formats the same relevance as in-house innovation. This approach to innovation can be understood as a concept of labour division in response to changing conditions. The growing importance of networking beyond organizational boundaries refers to the increasing necessity of letting ideas flow out of the corporation, to find better sites for their monetization, and flow back into the corporation in the shape of new submissions and business models (Chesbrough 2003). In this sense, the "open innovation" paradigm treats R&D itself as an open system (Chesbrough 2006, 1) and highlights the importance of shared processes, connecting outside-in and inside-out motions by working within alliances of complementary companies (Gassmann and Enkel 2006). There is currently a broad awareness for open innovation and its relevance to corporate research. The implications and trends that support open innovation and their economic implications are actively discussed from strategic, organizational, behavioral, informational, legal and business perspectives (Enkel et al. 2009, 311).

1.2 Key Persons in the Innovation Process: Conceptual Sources and Deficits

To date, the theoretical and empirical work on "informal transformational leaders" (Howell and Higgins 1990), namely "champions" and "promotors," has focused on in-house innovation management and accordingly, largely been limited to an intra-organizational perspective. However, the growing need for inter-organizational cooperation to achieve innovation success calls for a closer look at this concept.

In the following, we will draw on relevant theoretical and conceptual foundations in order to develop a broader concept of the key players in collective innovation processes, with special regard to the increasing importance of open innovation and innovation network management.

On the basis of our research, we consider four sources as especially relevant to this endeavour. First, interaction economics which allows a conceptualization of the role of key players from an evolutionary perspective; second, innovation system theory which helps in grasping the relevant levels and institutional settings of innovation, third, champion research which establishes the importance of key persons in innovation projects, and finally, fourth, promotor theory which can be considered one of the most elaborate concepts of key persons and their role in overcoming innovation barriers.

1.2.1 Interaction Economics: Key Actors from an Evolutionary Perspective

The recent theoretical field of interaction economics (Antes and Fichter 2011, 262) is a micro-economic approach that sees the social interaction between actors as a central "location" of self-organization as well as decision-making and change in economic processes, and therefore makes this aspect its main objective for conceptualization. Therefore the focus of interaction economics is, like evolutionary economics, in the broadest sense the change in economic systems. Rather than describe an internal status, it concentrates on transitional processes and the emergence of new ideas both in terms of spreading and of the impact of novelty. Therein, the interaction between individuals within the system is conceptualized as the central "place of change" (Fichter 2009b).

Here, interaction economics can draw on different fruitful theoretical sources and branches of academia. The three most important ones are:

- Evolutionary economics that models changes in economic systems and, with the concepts of contingency and of path dependency, provides valuable theoretical framework (Lehmann-Waffenschmidt and Reichel 2000; Lehmann-Waffenschmidt 2009).

- Interactive innovation theories, with their historical foundations going back to Schumpeter's concept of "Creative Response" (1947), and today including a variety of concepts for interactive modes of innovation (lead user approach, boundary spanning, open innovation, innovation communities etc.).
- The "Structuration Theory" as developed by Giddens (1984) and its reception in business administration research, allowing a focus on institutional aspects as an integral part of "rule-resource complexes" (structures) and a perspective on institutions that not only treats them as limiting or enabling, but in any case formalized interaction, but also conceptualizes the changes in institutional structures driven by individuals as well.

The Role of Social Interaction

For an explanation of innovation based on interaction economics, the main question is what function social interaction assumes in a given innovation process. There are three basic features of social interaction which can be distinguished (Fichter 2005, 249):

1. Based on the fact that goods are unevenly distributed between economic actors and within markets, due to division of labour and heterogeneous resources and equipment, a demand for external resources such as raw materials, semi-finished goods, services, rights or financial capital arises. With a decreasing vertical range of production and increasing division of labour, the need for interaction increases. In the economic context, interaction not only allows goods and services to be exchanged, but can also, especially in innovation processes, reduce market-dependent and technological insecurities, e.g. by integrating clients into the development process or by seeking R&D partnerships with leading technological and academic institutions at an early stage.
2. Through the increased dynamics and complexity of the market and current technological development, uncertainty about future development is growing and increasing the demand for interaction needed to integrate knowledge and establish directional security.
3. The third basic function of interaction lies in the fact that it allows the resolution of conflicts that arise from differences in stakeholders' interests or from competing targets. Here it can generally be assumed that the degree of interaction required increases with the conflict potential of the projects. Certainly, insecurity, risk and conflict potentials are among the more dominant characteristics of innovation challenges (Fig. 1.1).

In the light of the circumstances detailed above, it becomes clear that interaction processes relate to the acquisition not only of physical goods, services, rights or financial capital, i.e. the exchange of allocative resources, but also of competition-

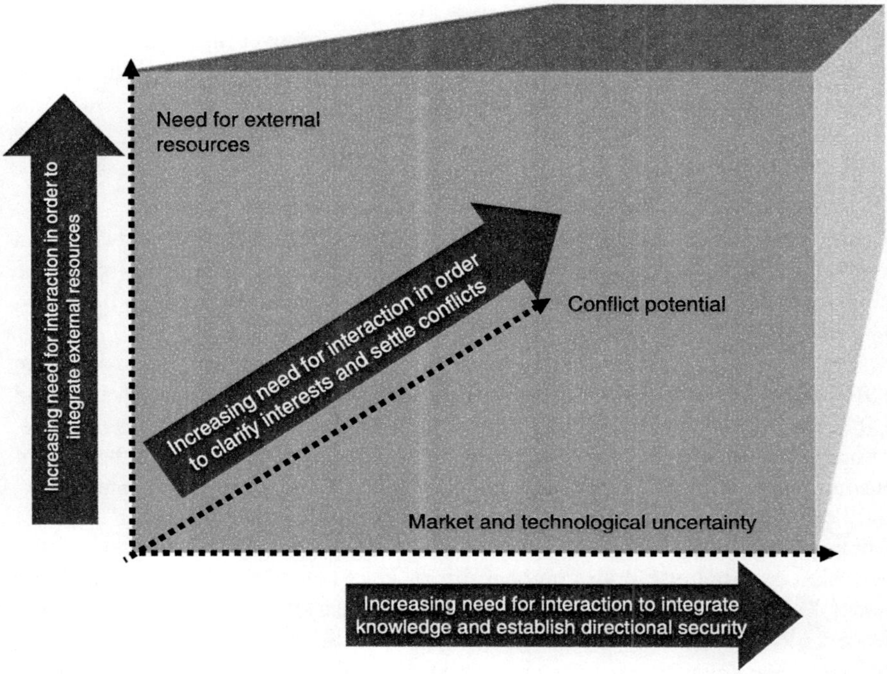

Fig. 1.1 Basic features of interaction in the economic context (Source: Author)

and innovation-related data and information and the exchange of values, views and perceptions. Functional network associations in innovative partnerships enable, for example, the development of a framework of mutual understanding which the collective integration and generation of knowledge would not work without.

Against this background, "interaction" will in this work be defined as follows:

> "Interaction is a process in which at least two partners take on a sequence of interdependent actions and reactions" (Lüthje 2000, 73). The interaction process . . . is based on communication and co-presence, is shaped by the interests and power relations of the stakeholders, is directed at obtaining resources, resolving conflict and reducing uncertainty in the economic context, and can arise when participants align their verbal and nonverbal actions with each other. (Author)

Key Player Concept

The conceptualization of key actors presents an important explanative model in interaction economics. The "key player" concept builds on the assumption in interaction economics that, with respect to their heterogeneous characteristics, e.g., ethical dispositions, preferences, strategic behaviour, participants in economic processes can be differentiated as diverse types of actors, such as different consumer types, different corporate types, etc. Applying this assumed heterogeneity of players to the progress of innovation and diffusion over time, the question of which stakeholders particularly influence the course of events moves into focus. This is where the key player approach grasps.

> Key players are those individuals, organizations or networks that are seen to have a significantly greater influence on the development and implementation of new solutions within a focal innovation or diffusion process. (Author)

Key players can be encountered in all groups of stakeholders involved in an innovation and diffusion process, as well as in politics and administration, in the supply and demand end of the market as well as intermediaries, or in civil society, e.g. in associations, interest groups or media.

The focus on key players in fact has a long tradition within economic sciences, particularly in the area of innovation and diffusion research. Within innovation and diffusion research, a number of key player concepts are to be found. This includes, among others, the "promotor" concept going back to Witte (1973) from the field of business innovation research (Hausschild and Gemünden 1999), the "gatekeeper" and "champion" models (Hauschildt and Schewe 1999), the "lead user" concept (Hippel 1988, 2005), and the concept of "innovation communities" (Fichter 2009a). Diffusion research also draws on concepts of key players, notably those "opinion leaders" and "change agents" found to be particularly relevant to the diffusion process by Rogers (2003). Another rich source is entrepreneurship research, which deals with the key player per se in innovation: with the entrepreneurs themselves or their entrepreneurial syndicate (Fichter 2005).

Based on these key player concepts, a "key person" can be defined as follows:

> A key person is an individual within a focal innovation or diffusion process who is observed to have a significantly higher influence on the creation and diffusion of novel processes and innovative solutions than other participants. (Author)

Key persons can generally impact the course of an innovation enterprise in two different ways, on the one hand by supporting the development and implementation of innovative solutions, on the other hand by preventing or inhibiting them.

1.2.2 Innovation Systems

Most research on collaboration in the pertinent literature focuses on formal aspects of inter-organizational collaboration, usually subsuming personal relationships under this dominant concept (Macdonald and Piekkari 2005, 444). Also, in open innovation research, the majority of analysis has been on the corporate and inter-organizational level (West et al. 2006, 287), paying less heed to the role of personal networks. It could therefore be concluded that innovation network research is of little help when explaining the role of "transformational leader" networks. However, the literature on innovation networks presents yet another important conceptual element, which can assist in framing the nature of innovation communities, namely the conceptualization of "innovation systems".

While most research on innovation networks has focused on inter-organizational cooperation, there have been growing numbers of research activities conceptualizing these networks as part of an "innovation system" since the 1980s (Freeman 1987). Most of them focus on national innovation systems which can be defined as "market and non-market institutions in a country that influence the direction and speed of innovation and technology diffusion [...] But innovation systems also exist at other levels, e.g. as world-wide, regional or local networks of firms and clusters of industries" (OECD 1999, 23). Drawing on the interactive paradigm of innovation (Lundvall 1988) and the idea of "inter-organizational collectivities" (Van de Ven et al. 1976), more recently, scholars (Lynn et al. 1996) have developed an "innovation community framework" for studying relevant organizations and relationships as structural systems. They propose the term "innovation community" to refer to the organizations significantly involved in the commercialization of a new technology or innovative product. Thus they conceptualize "innovation communities" as inter-organizational networks, "embedded in a dense web of social and economic relationships" (Lynn et al. 1996, 98). Technology is perceived to be at the centre of the innovation community, which is suggested to consist of a "substructure" and a "superstructure".

"Organizations in the substructure produce either the innovation or its technological complementaries. [...] Superstructure organizations provide collective goods to their members, often specializing in coordinating flows of information or coordinating the activities of substructure organizations" (Lynn et al. 1996, 98). Many superstructure organizations are thus "linking organizations," such as professional societies, occupational or industry associations, government organizations, as well as non-corporate R&D institutions. Superstructure organizations promote innovation and technology integration in different ways, linking diverse bodies of knowledge, competence and skills.

The framework of Lynn et al. (1996) has helped the understanding of the way in which inter-organizational innovation communities function (Lynn 1998), allowing for different levels of innovation systems to be differentiated. Above the individual and single company level, there is an organizational and network level directly involved in "producing" innovations (substructure) and a third level (superstructure) that links to further innovation organizations and intermediates. Thus, the relevant point here is that innovation systems consist of different interconnected levels of innovation protagonists (Vanhaverbeke and Cloodt 2006, 276). This supports the notion that interaction takes place not only across organizational boundaries, but also across different levels of the innovation system.

Nevertheless, one main shortfall of the innovation community framework of Lynn et al. (1996) is that it exclusively examines inter-organizational aspects, omitting the role of interpersonal networks within the innovation system. Thus, there is a conceptual gap in what might be called the "microstructure" of an innovation system. Under these circumstances, it is helpful to turn to "promotor" theory and "champion" research in order to conceptualize the interpersonal micro-structure within innovation systems.

1.2.3 *Promotor Theory and Champion Research*

Promotor[5] theory is based on the notion that the success of innovation processes depends on overcoming certain barriers; it requires promotors who commit enthusiastically to specific innovation projects and help overcome those barriers. Witte (1973, 15) defined promotors as 'individuals who actively and intensively support the innovation process'. With regard to barriers, Witte (1977) differentiates between two kinds of specialisation, the 'power promotor' and the 'expert promotor', and assumes that there is a correspondence between specific barriers and specialised roles. The 'power promotor' contributes through hierarchical power and the 'expert promotor' contributes through expert knowledge (Witte 1973, 17). Another assumption of promotor theory is that the innovation process will be more successful if both types of specialised promotors work closely together (Hauschildt and Kirchmann 1997, 68).

Witte's original two-centre theory of power and knowledge has been extended since its introduction in the 1970s. In the 1980s, Hauschildt and Chakrabarti (1988, 385 f) described a third barrier that can hinder economic progress: administrative barriers. For this reason, they introduced the role of a 'process promotor', who actively arbitrates between the technical and the economic world by means of organizational knowledge (Hauschildt 1999, 174). Gemünden and Walter (1995) developed a fourth type of specialised promotor: 'relationship promotors' actively encourage an innovation process by means of innovation-related business relationships inside and between the organization and its external partners. The defining characteristic of relationship promotors is their extensive network competence, i.e. powerful relationships with other parties (Table 1.1).

Table 1.1 Barriers, power bases and promotor roles in innovation processes

Barrier type	Power base	Promotor role
Knowledge	Specialized knowledge	Expert promotor
Ignorance, opposition, resources	Hierarchical potential, control of resources	Power promotor
Administrative	Organizational know-how, communication skills	Process promotor
Cooperation, dependency	Networking skills, potential for interaction	Relationship promotor

Source: Author

[5] Throughout the book, we employ the specific German term "promotor", derived from the Latin verb "promovere", to distinguish it from the more generic term "promoter".

Promotor theory stresses that the different specialised promotor roles do not have to be played by different individuals, but can also be combined in one person, the 'universal promotor'.

Promotor theory offers a consistent and elaborate base for describing and explaining the role of transformational leaders in innovation processes; its conceptual focus on a single organization is, however, too limited in scope (Fichter 2005). For this reason, the original theory has to be extended, by adding two new assumptions:

1. *Cross-boundary cooperation of promotors*: In cases of open innovation and complex technologies, the innovation process will be successful only if universal or specialized promotors from cooperating organizations work closely together.
2. *Promotor roles on all levels of innovation systems*: Promotor roles are not limited to those organizations involved in "producing" innovations, but can also be played by innovation intermediates (Howells 2006) or individuals from organizations of the superstructure of the innovation system (Lynn et al. 1996; Lynn 1998).

The new construct of "innovation communities" will draw on the concept of multi-level innovation systems, because it helps to clarify and configure cross-boundary relationships and allows systematic connections to the research on "superstructures" of regional, national and international innovation systems (Lynn et al. 1996; Lynn 1998).

In contrast to promotor theory, the Anglo-Saxon research has so far mainly been focused on the "champion" concept introduced by Schon (1963). Although Rost et al. (2007) propose that champions and promotors differ with respect to the kind of knowledge they possess, a closer look at the various specifications of the term "champion" reveals that innovation champions can in fact be conceptualized as "universal promotors", as they combine two or more of the promotor's roles in one (Sand and Rese 2011, 235).

1.2.4 Need for a New Concept

Against this background, a new construct for transformational leaders in collective innovation procedures seems necessary. This should systematically involve the role of different types of transformational leaders and comprise all relevant levels of an innovation system. For that reason, the conceptual framework presented in the following will be based on two founding theories: Firstly, promotor theory and secondly, the concept of multi-level innovation systems.

Promotor theory has been selected because it presents a consistent explanatory framework for working with the construct of innovation barriers, and provides the concept of power bases, which are essential to overcoming innovation barriers (Hauschildt 1999, 171). Furthermore, the promotor model offers the possibility of describing the case of the "universal promotor," which equates to the "champion" model, as well as the chance to explain the division of labor between different types of specialized promotors. There is extensive empirical proof for both the specialized and the universal cases (Rost et al. 2007), but the former is especially relevant when it comes to mastering complex technologies and finding system solutions (Hauschildt 1999, 174).

1.3 A New Concept: Innovation Communities

An important development in innovation studies in the past decade has been the acknowledgement of the role of communities outside and beyond corporate boundaries in creating, shaping and disseminating technological and social innovations. Innovation Communities and their role in the innovation process fit in here and offer the opportunity to extend the company-based concept of open innovation (West and Lakhani 2008, 223).

1.3.1 Multi-level Innovation Systems

Drawing on the idea of multi-level innovation systems, the conceptual framework of innovation communities will use three distinct system levels where collaboration of promotors can take place.

1. *The company level*: this has been the 'classical' level of analysis of champion research and original promotor theory.
2. *The value chain level* of innovating actors, comprising cooperation or networks of companies and/or individuals who 'produce' an innovation, either in their functional role as supplier, manufacturer or user (von Hippel 1988). This 'substructure' of an innovation system (Lynn et al. 1996) could also be called the 'value chain of innovation'.
3. *The level of framing and interlinking organizations*. This 'superstructure' level hosts organizations that enable other actors to innovate (Winch and Courtney 2007, 751). Intermediation comprises a broad variety of functions, such as foresight, roadmapping, information scanning, matchmaking, brokering, testing, etc. (Howells 2006). Many superstructural organizations are thus framing and linking organizations, like technology brokers, industry associations, government organizations or national R&D organizations.

1.3.2 Innovation Communities as Promotor Networks

On the basis of extended promotor theory and the concept of three-level innovation system, the term 'innovation communities' shall be defined as follows:

> An innovation community is an informal network of likeminded individuals, acting as universal or specialised promotors, often from more than one company and different organizations that team up in a project related fashion, and commonly promote a specific innovation, either on one or across different levels of an innovation system. (Author)

Innovation communities therefore are characterised as promotor networks or as informal personal networks of innovators. They can be differentiated from scientific communities, which follow specific research topics (R&D communities), or communities that follow specific professional interests, by their declared and primary aim to support the breakthrough of a specific innovation. Innovation communities should, for this reason, not be confused with 'Communities of Practice' (Lave and Wenger 1991; Wenger et al. 2002; Amin and Roberts 2008) but are a specific form of communities that are related to concrete innovation projects (Fig. 1.2).

It can be concluded that 'innovation communities', as defined here, can be differentiated from other forms of social networks in innovation processes by three key criteria:

1. The community is always related to a specific innovation idea or project.
2. All community members play a promotor role in this process.
3. The community members collaborate closely and informally, and they perceive themselves as a 'team', a 'group' or a similar entity, with a feeling of group identity.

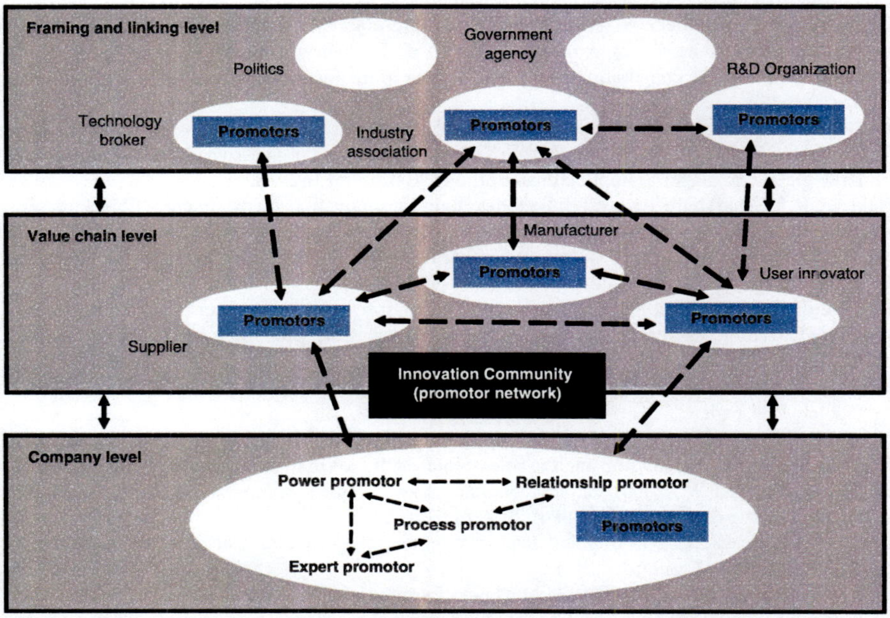

Fig. 1.2 The innovation community as promotor network (Source: Author)

References

Amin A, Roberts J (2008) Knowing in action: beyond communities of practice. Res Pol 37(2):353–369

Antes R, Fichter K (2011) Die Bedeutung von Intermediären und institutionellen Arrangements für einen nachhaltigen Konsum: Theoretische und empirische Einsichten auf Basis interaktionsökonomischer und institutionentheoretischer Grundlagen. In: Antoni-Komar I, Lehmann-Waffenschmidt M, Pfriem R, Welsch H (eds) Wenke2: Wege zum nachhaltigen Konsum. Metropolis, Marburg, pp 159–216

Chesbrough H (2003) Open innovation: the new imperative for creating and profiting from technology. Harvard Business School Press, Boston

Chesbrough H (2006) Open innovation: a new paradigm for understanding industrial innovation. In: Chesbrough H, Vanhaverbeke W, West J (eds) Open innovation: researching a new paradigm. Oxford University Press, Oxford, pp 1–12

Enkel E, Gassmann O, Chesbrough H (2009) Open R&D and open innovation: exploring the phenomenon. R&D Manage 39(4):311–316

Fichter K (2005) Innovation communities: Die Rolle von Promotorennetzwerken bei Nachhaltigkeitsinnovationen. In: Pfriem R, Antes R, Fichter K, Müller M, Paech N, Seuring S, Siebenhüner B (eds) Innovationen für Nachhaltige Entwicklung. Deutscher Universitätsverlag, Wiesbaden, pp 287–300

Fichter K (2009a) Innovation communities: the role of networks of promotors in open innovation. R&D Manage 39(4):357–371

Fichter K (2009b) Interaktionsökonomik [Electronic version]. Borderstep, Berlin. Retrieved 10 Apr 2011 from http://www.borderstep.de/details.php?menue=70&le=en

Freeman C (1987) Technology policy and economic performance: lessons from Japan. Printer Publisher, London

Gassmann O, Enkel E (2006) Open Innovation: Externe Hebeleffekte in der Innovation erzielen. Zeitschrift Führung+Organisation 3:132–138

Gemünden HG, Walter A (1995) Beziehungspromotoren, Schlüsselpersonen für interorganisationale Innovationsprozesse. Zeitschrift für Betriebswirtschaft 65:971–986

Giddens A (1984) The constitution of society: outline of the theory of structuration. Polity Press, Cambrigde

Hauschildt J (1999) Promotors and champions in innovations – development of a research paradigm. In: Brockhoff K, Chakrabarti AK, Hauschildt J (eds) The dynamics of innovation: strategic and managerial implications. Springer, Berlin, pp 167–182

Hauschildt J, Chakrabarti AK (1988) Arbeitsteilung im Innovationsmanagement – Forschungsergebnisse, Kriterien und Modell. Zeitschrift für Organisation 57:378–388

Hauschildt J, Kirchmann E (1997) Arbeitsteilung im Innovationsmanagement: Zur Existenz und Effizienz von Prozesspromotoren. Zeitschrift Führung und Organisation 66:68–73

Hauschildt J, Schewe G (1999) Gatekeeper und Prozesspromotoren. In: Hauschildt J, Gemünden HG, Hauschildt J, Gemünden HG (eds) Promotoren, Champions der Innovation, 2nd edn. Gabler, Wiesbaden, pp S. 159–176

Hausschild J, Gemünden HG (eds) (1999) Promotoren, Champions der Innovation, 2nd edn. Gabler, Wiesbaden

Howell JM, Higgins CA (1990) Champions of technological innovation. Am Sci Quart 45:317–331

Howells J (2006) Intermediation and the role of intermediaries in innovation. Res Pol 35:715–728

Kash DE, Rycoft RW (2000) Patterns of innovating complex technologies: a framework for adaptive network strategies. Res Pol 29(7–8):819–831

Lave J, Wenger EC (1991) Situated learning: legitimate peripheral participation. Cambridge University Press, Cambridge, MA

Lehmann-Waffenschmidt M (2009) Gibt es eine Evolution in der Wirtschaft? Zur Diagnose und komparativ-evolutorischen Analyse des wirtschaftlichen Wandels. In: Antoni-Komar I et al (eds) Neue Konzepte der Ökonomik: Unternehmen zwischen Nachhaltigkeit, Kultur und Ethik. Metropolis, Marburg, pp 369–396

Lehmann-Waffenschmidt M, Reichel M (2000) Kontingenz, Pfadabhängigkeit und Lock-In als handlungsbeeinflussende Faktoren der Unternehmenspolitik. In: Beschorner T, Pfriem R (eds) Evolutorische Ökonomik und Theorie der Unternehmung. Metropolis, Marburg, pp 337–376
Lundvall B-A (1988) Innovation as an interactive process: from user-producer interaction to the national system of innovation. In: Dosi G et al (eds) Technical change and economic theory. Printer Publishers, London, pp 349–369
Lüthje C (2000) Kundenorientierung im Innovationsprozess. Eine Untersuchung der Kunden-Hersteller-Interaktion in Konsumgütermärkten. Deutscher Universitätverlag, Wiesbaden
Lynn LH (1998) The commercialization of the transistor radio in Japan: the functioning of an innovation community. IEEE Trans Eng Manage 45:220–229
Lynn LH, Reddy NM, Aram JD (1996) Linking technology and institutions: the innovation community framework. Res Pol 25:91–106
Macdonald S, Piekkari R (2005) Out of control: personal networks in European collaboration. R&D Manage 35(4):441–453
Meffert H, Burmann Ch (2000) Strategische Flexibilität und Strategiewechsel in turbulenten Märkten. In: Häflinger GE, Meier JD (eds) Aktuelle Tendenzen im Innovationsmanagement. Physica-Verlag, Heidelberg, pp 173–215
OECD (1999) Managing international innovation systems. OECD, Paris
Rogers EM (2003) Diffusion of innovations, 5th edn. Free Press, New York
Rost K, Hölzle K, Gemünden HG (2007) Pros and cons of role specialisation for economic process. Schmalenbach Bus Rev 59(4):340–363
Rothwell R (1994) Industrial innovation: success, strategy, trends. In: Dodgson M, Rothwell R (eds) The handbook of industrial innovation. Edward Elgar Publishing, Hants, pp 33–53
Sand N, Rese A (2011) Comparing promotor networks in radical and incremental innovation projects: roles and interaction as success factors. In: Baier D, Gaul W, Rese A, Würth R (eds) The entrepreneurship – innovation – marketing – interface. Proceedings of the 3rd symposium in Cottbus. Swiridoff, Künzelsau, pp 233–250
Schon DA (1963) Champions for radical new inventions. Harvard Bus Rev 41(2):77–86
Schumpeter JA (1947) The creative response in economic history. J Econ Hist 7:149–159
Schumpeter JA (1993) Theorie der wirtschaftlichen Entwicklung (8. Auflage, unveränderter Nachdruck der 1934 erschienenen 4. Auflage, 1. Auflage 1911). Duncker Humblot, Berlin
Staber U (2004) Networking beyond organizational boundaries: the case of project organizations. Creativity Innovat Manage 13:30–40
Van de Ven AH, Delbecq AL, Koenig R (1976) Determinants of coordination modes within organizations. Am Sociol Rev 41:322–338
Vanhaverbeke W, Cloodt M (2006) Open innovation in value networks. In: Chesbrough H, Vanhaverbeke W, West J (eds) Open innovation, researching a new paradigm. Oxford University Press, New York, pp 258–281
von Hippel E (1988) The sources of innovation. Oxford, New York
von Hippel E (2005) Democratizing innovation. MIT Press, Cambridge, MA
Wenger E, McDermott R, Snyder WM (2002) Cultivating communities of practice, a guide to managing knowledge. Harvard Business School Press, Boston, MA
West J, Lakhani KR (2008) Getting clear about communities in open innovation. Ind Innovat 15(2):223–231
West J, Vanhaverbeke W, Chesbrough H (2006) Open innovation: a research agenda. In: Chesbrough H, Vanhaverbeke W, West J (eds) Open innovation, researching a new paradigm. Oxford University Press, New York, pp 285–307
Willke H (2000) Systemtheorie I: Grundlagen, 6th edn. Lucius & Lucius, Stuttgart
Winch G, Courtney R (2007) The organization of innovation brokers: an international study. Tech Anal Strat Manage 19(6):747–763
Witte E (1973) Organisation für Innovationsentscheidungen: Das Promotorenmodell. Schwart, Göttingen
Witte E (1977) Power and innovation: a two center theory. Int Stud Manage Organ 7:47–70

Chapter 2
Networks of Promotors as Success Factor for Radical Innovations: Theoretical Considerations and Empirical Findings

Alexandra Rese and Daniel Baier

Abstract Based on the conceptual framework developed in the previous chapter, this chapter presents empirical findings regarding the role and relevance of transformational leaders like "promotors" and "champions" in innovation networks, and their effect on innovation success.

Section 2.1 gives an introduction to state-of-the-art innovation network research, success factor research and open innovation research, as well as linking key findings from these areas to the research agenda of Innovation Communities. Section 2.2 develops the theoretical background and terminology for studying the role of innovators and extends well-known studies on key persons in innovation projects by using a cross-company, network-oriented perspective. The chapter also develops concepts of group interaction and identity and operationalizes the Innovation Community concept for a large-scale survey. In Sect. 2.3, the research hypotheses on innovation communities are developed, and measures for promotor roles, group identity, group interaction, project performance and degree of innovativeness are introduced.

Section 2.4 presents the findings of the innovation networks survey. The hypothesized relationships were tested using a comprehensive sample of 107 German networks of companies and research institutions. The results support the proposition that group interaction and group identity factors are significantly higher in successful innovation networks. One important finding of the survey is that innovation communities have a considerable (positive) effect on project performance and overall innovation success, especially in radical innovation projects. These "promotor networks" can be considered to be an important success factor for radical innovation projects. The implications of these findings are discussed in Sect. 2.5.

K. Fichter and S. Beucker (eds.), *Innovation Communities*,
DOI 10.1007/978-3-642-22128-6_2,

2.1 Introduction

For quite some time, cooperations and networks have been central topics of innovation research (Powell et al. 1996; Ritter and Gemünden 2003; Dhanaraj and Parkhe 2006). In turbulent markets and complex or fast-changing environments, small and medium enterprises in particular see the formation of cooperations and alliances as one possibility of obtaining advantages through synergies (van Dijk et al. 1997; Rosenfeld 1996; Human and Provan 1997; Baier et al. 2006; Rese and Baier 2011). Empirically already well investigated are network formation and design (e.g. Thorgren et al. 2009), management (Anand and Khanna 2000; Schreiner et al. 2009), suitability of partners (Un et al. 2010), success factors regarding innovation management (Baier et al. 2006; Rese and Baier 2011), dynamics (Nakamura et al. 1996), stability (Leblebici and Salancik 1982; Teusler 2008), and the innovation network environment (e.g., Baum et al. 2000). With respect to theory, current research considerably relies on sociological and economical network and cooperation analysis. For example, the examination of formal and informal knowledge networks in R&D draws on social network analysis (Cantner and Graf 2006; Allen et al. 2007), whereas collaboration skills and communication patterns are investigated using statistical methods (Ritter 1999; Thoms 2003; Schreiner et al. 2009).

Regarding the question what in fact determines an innovation project's success, research so far has shown that collaboration on new product development (NPD) does not necessarily improve commercial performance. According to LaBahn et al. (1996), integrating external partners can protract development cycle periods. Regarding another factor relevant to success, Ledwith and Coughlan (2005) found that the process of integrating external partners is difficult to manage and requires a range of skills. These results have yielded this survey's main focus on interpersonal relations in innovation networks.

Innovation research has revealed that the involvement of key actors can be vital to innovation success. With regard to the individual level within one company, the initiators and key actors of innovation processes have already been extensively examined empirically (cf. Rost et al. 2007; Gemünden et al. 2007; Lichtenthaler and Ernst 2009; Markham et al. 2010). Their function has been analyzed in the context of the "champion" approach, the term "champion" having first been suggested by Schon (1963) and relating to Witte's "promotor" model (1973). However, when it comes to innovation networks, the (informal) role of innovative actors in contributing to an innovation project, their (informal) interactions, and their identification with the network have not been sufficiently investigated empirically so far (Cf. Thoms 2003) for first conceptualizations of promotors' roles in innovation networks, Gerybadze (2003) for a conceptual discussion of group dynamics in innovation networks, Fichter and Beucker (2008) and Fichter (2009) for further conceptual discussions and first case studies).

The Open Innovation literature emphasizes the interactive nature of innovation, and that the integration of lead users, suppliers, competitors, customers or universities is just as relevant to the success of an innovation as the presence of the innovators themselves (Laursen and Salter 2006; Un et al. 2010). The focus of studies has so far been on the alliances that companies can form, and their effect on innovative output (Laursen and Salter 2006). Laursen and Salter (2006, 132) also point out that these innovators "tend to band together in teams and coalitions".

Drawing on the discussion of "communities of practice" (e.g., Cook and Brown 1999), they recommend that individual innovators should operate as a group by developing "a shared understanding and common ways of working together" (Laursen and Salter 2006, 136). Empirical research has shown that the success of collaborative product development projects depends on the way in which partners contribute to the joint project (and whether they contribute as expected), that they communicate frequently and develop a substantial degree of mutual trust. In addition, the presence of a so-called "collaboration champion" can enhance the success. This champion is described as someone in a senior management position who is in charge of collaboration, can provide resources and is determined to overcome all obstacles (Bruce et al. 1995; Littler et al. 1995).

The comparatively new concept of "innovation communities" (or ICs, see Chap. 1) now focuses on the relationship between partners from different companies or institutions in the innovation process. The term Innovation Community is generally understood in this context as an informal group or network of like-minded people from different organizations supporting a specific innovation project (Gerybadze 2003; Fichter 2009, Chap. 1). The concept was first introduced into German literature by Gerybadze (2003) who emphasized the aspects group dynamics, communication and common understanding for the functioning of such a community. Fichter (2009, 360) proposed the concept of innovation communities in the context of open innovation, defining them as "networks of innovators". This concept was expanded by the idea that IC members take on specific roles in the innovation process. In first case studies, Fichter could show that the ICs examined sported high degrees of interaction and community identity (Fichter 2009). As the main focus here is on the quality of identity and interactions within the group, in the following, we will use the terms *group interaction* and *group identity*. The Innovation Community concept is similar in some respects to the "Communities of Practice" concept, even though the latter was developed for the field of learning, information and knowledge management in particular. But three elements which are thought to be crucial for these communities of practice – shared interests, a shared community and a shared practice (Lave and Wenger 1991) – are assumed to be important for Innovation Communities as well. By focusing on individuals and groups that initiate and implement a specific innovation project, the IC concept can help to understand why specific innovation projects are successful and others are not. A deeper understanding of innovation communities could help to reinforce innovation capacities, and contribute to a successful organization of the innovation process, for instance regarding the selection of suitable partners. In addition, a better understanding of informal relations in R&D could lead to a better understanding of the innovation process, more efficient information exchange and use of new ideas.

Given these findings and considerations, in the following section, the theoretical background and terminology regarding informal roles in innovation networks, the formation of group identity and the group interaction of key actors are discussed. Empirical measurement concepts are evaluated and a heron definition of an innovation community is derived. In the following, a special focus lies on the expected impact of an innovation community on project performance and consequently, on the innovation network's success. Additionally, further hypotheses concerning innovation networks are derived.

After describing the research methodology and measures used in this study, we present our results drawing on a large-scale sample of German networks of small and medium enterprises and research institutions cooperating in NPD. The networks are described, for instance, in terms of the industry sector, innovation phase or geographical distribution of project partners. Then the key actors of the respective networks, their experience with innovation projects, their position within their firms, the way in which they became mutually acquainted, and their communication channels and frequency (e.g., of meetings) are presented. Their informal roles are compared to empirical studies from within companies and analyzed with regard to network performance, while differences in group-based interaction and identity, as characteristics of an IC, are examined with respect to project performance. Based on concepts of role constellations, group identity and group interaction, the quantitative definition of an innovation community is tested, in particular regarding project performance. Conclusively, the following central issues are presented below:

- Does the presence of an innovation community have a positive effect on network performance?
- Do group interaction and group identity have a positive effect on network performance?
- Is the degree of novelty or inventiveness of the innovation project decisive for the presence of an innovation community?
- In which role constellation do innovation communities have a significant influence on the success of innovation projects?

Finally, the results and their implications for innovation management are discussed. Prior to this publication, selected results of the empirical findings were already presented at the Product Development Management Conference (Rese et al. 2009; 2010a), the QMOD-ICQSS[1] conference (Rese et al. 2010b) and the EIM[2] Symposium (Sand and Rese 2011). Journal publications are currently in review.

2.2 Theoretical Background and Terminology

The Innovation Community concept, as introduced in Chap. 1, is based on the vision of a group of key actors from one or several organizations who promote an innovation project. To determine what a key actor promoting an innovation project

[1] Quality Management and Organisational Development Conference (QMOD) – International Conference on Quality and Service Sciences (ICQSS).

[2] Symposium on Entrepreneurship – Innovation – Marketing (EIM) Interface.

is, the concepts of "promotor" and "champion" theory can be employed. As for the "group", organizational theory defines it as "a collection of individuals who are interdependent in their tasks, who share responsibility for outcomes, who see themselves and are seen by others as an intact social entity, embedded in one or more larger social systems (for example, business unit or the corporation), and who manage their relationships across organizational boundaries" (Cohen and Bailey 1997, 141). In order to transfer this definition to the innovation community concept, two dimensions of inter-group relations seem to be relevant – group identity and group interaction. Despite the informal setup prevailing, an innovation community is not just seen as a loose network of key actors, but as a "real" group with close ties between members (cf. Gerybadze 2003; Fichter 2009). In the following, the role concept and the two dimensions of intergroup relations are explained in detail.

2.2.1 *Analysis of Innovator Roles with a Focus on Networks*

Innovation research has found extensive empirical proof that enthusiastic and committed individuals play an important role in promoting innovations. Roles are defined as behavioural instructions assigned to an individual that result from the interaction with others (e.g. Katz and Kahn 1966). In relation to others, role players assume a position which is attached to a status (Linton 1936). Therefore, apart from expected duties, these roles will also include expected rights, e.g. resource disposability (Turner 2002; Markham et al. 2010).

In general, role-based models have been examined intensively in innovation research (for an overview, see Gemünden et al. 2007; Rost et al. 2007). With its "champion" concept, the Anglo-Saxon research has focused on one single person as being responsible for the innovation project, beginning with Schon (1963). The focus lies on generalists who play multiple roles in innovation processes. According to Howell et al. (2005, 641), champion behaviour consists of the following three facets: "expressing enthusiasm and confidence about the success of the innovation, persisting under adversity, and getting the right people involved". For within companies, several champion roles have been proposed by different authors (Rothwell et al. 1974; Howell and Higgins 1990; Burgelman 1983; Kelley and Lee 2010; Markham et al. 2010). The roles were identified with the help of a paired comparison approach of successful and unsuccessful companies and projects. An underlying theoretical concept was not tested (Hauschildt 1999). In contrast, research in German-speaking countries focused on more than one person promoting a single innovation project together, and identified four roles (for an overview Gemünden et al. 2007; Rost et al. 2007). The underlying theoretical concept developed by Witte (1977) centres on different barriers which promotors help to overcome by improving project performance (Hauschildt 1999). Nevertheless, both streams of research point to the following key roles (see Table 2.1).

Technical innovators or *expert promotors* are the 'idea givers or 'inventors'. They make major technical contributions in the development and/or design of the

Table 2.1 Champion and promotor roles in literature

Activities	Champion concept				Promotor concept
	Rothwell et al. (1974)	Burgelman (1983)	Markham et al. (2010)	Kelley and Lee (2010)	Kirchmann (1994), Walter and Gemünden (2000), Hauschildt and Kirchmann (2001)
Power base: hierarchical potential, control of resources					
Providing organizational resources, making decisions, fixing targets, persuading others to support the team, blocking opposition, motivating others		Organizational champion	Sponsor	Sponsor	Power promotor
Power base: knowledge specialty					
Fostering new ideas, initiating innovative processes (selecting project, developing opportunity case, seeking support of sponsor)	Product champion	Product champion	Champion		
Possession of technical knowledge, knowing alternatives/internal mechanisms and the potential and limitations of the new product or process, evaluating new and complex problems, developing proposals for problem solution and implementing problem solutions, initiating and promoting technical learning processes	Technical innovator				Expert promotor
Power base: organizational know-how, communication potential					
Collecting, sorting through, translating, channelling, and passing information to others					Process promotor
Coordinating business and financial activities, planning, scheduling, monitoring and controlling (e.g. schedules), supervising technical work	Business innovator			Supervisor	
Power base: soft skills, network know-how, potential for interaction					
Searching for and bringing together appropriate partners					Relationship promotor
Solving conflicts, engaging in activities to build relations within the partners					
Achieving negotiation results					
Coordinating activities of the partners					

Source regarding activities: Chakrabarti and Hauschildt (1989), Walter and Gemünden (2000), Thoms (2003), Papies (2006), Roberts (2007), Markham et al. (2010)

innovation. Their knowledge includes the technical alternatives, internal mechanisms, and the potential and limitations of the new product or process (Chakrabarti and Hauschildt 1989).

The *sponsor, organizational champion* or *power promotor* provides guidance, supports the acquisition of resources, and protects the project team from organizational interference. He is usually a member of the middle-level management (Burgelman 1983). His activities are more of a political nature. He keeps in touch with the top management, providing information and keeping them enthusiastic about the project. Therefore he selects the projects he wants to get involved in carefully to match the company's strategy and objectives. He starts the realization of the innovation project deciding on budget, capacity and personnel (Chakrabarti and Hauschildt 1989).

The *(product) champion* identifies promising ideas, either his own or from others, and turns them into new innovation projects. Similar to the *process promotor*, a champion will become "a salesman of the idea", connecting power promotors and expert promotors and acting as a translator of the technical language of innovation into a general language (Chakrabarti and Hauschildt 1989). He has the necessary organizational know-how, inter-organizational networks and communication skills to bring the people necessary for the innovation project together. These activities are often taken on by group leaders, due to their deep involvement in the definition process (Burgelman 1983). In addition, Kirchmann (1994) who jointly analyzed product innovation in the German plant construction and engineering industry from the perspective of both manufactures and users, refers to the process promotor as *gatekeeper* acting as a 'gate' for the R&D group regarding external information flows.[3] He performs the tasks of collecting, sorting through, translating, channelling, and passing information to others, in case of an innovation community to the other members (Roberts and Fusfeld 1981; Whelan et al. 2010). Because the process promotor also develops a plan of action (Chakrabarti and Hauschildt 1989) in the present study, besides being a gatekeeper for information, the process promotor is defined as a person in an innovation network who is also responsible for the progress of the project. Rost et al. (2007) refer to the process promotor as project leader. This managerial *supervisor* role (also termed "*business innovator*") includes planning, scheduling, monitoring and evaluating project progress and controlling project decision making (Roberts 2007; Kelley and Lee 2010).

Finally, the *relationship promotor* performs activities supporting the development and maintenance of relationships, especially with external partners (Walter and Gemünden 2000). The relationship promotor has strong personal ties inside and especially outside the organization. His activities include information exchange with the actors involved, the search for appropriate partners, the coordination of

[3] The term 'gatekeeper', describing a key actor who mediates the information flow "between his organisational colleagues and the world outside" (Allen 1970), was first introduced by Allen and Cohen in 1969. Allen started his research with a study on the use of information in R&D proposal competitions (Allen 1964).

activities of the partners, the solution of conflicts and the achievement of negotiation results. Regarding networks, Woodside (1994) introduced the *network champion* who serves as a "marriage broker" or deal maker, building new relationships among firms at multiple levels to interact in NPD.

2.2.2 Group Identity and Group Interaction

Researchers have proposed that promotor or champion roles in the innovation process are linked. For example, Markham et al. (2010) found that within companies, the role players must interact with each other and coordinate their efforts in order to move innovation projects to the formal product development level. Regarding innovation projects that are developed and promoted across company boundaries, Gerybadze (2003) proposed and Fichter (2009) could show in first case studies that interaction of role players and development of a specific group identity become even more important.

Because the innovation community concept implies that the existence of a group of promoting key actors depends on a (developing) innovation project, a stable structure and group norms – which are the informal rules groups adopt to regulate and regularize group members' behaviour (Feldman 1984) – establish gradually (Gerybadze 2003). This is different from groups in organizations, in which the norms, tasks, and roles of the individual members are much more predefined by corporate law, corporate governance, and in particular the formal organizational structure, including role descriptions and managerial authority (Child 1972). Norms defining group members' behaviour are formed through interaction and are closely related to group identity. Because group membership is highly valued, members will assign these norms to themselves and conduct themselves according to the norms of the group (Postmes and Spears 2000).

We assume that innovation communities also show the characteristics of organization groups and that these characteristics are even more important. Gerybadze (2003, 146) had pointed to the exceptionally strong ties between members of an innovation community and their strong commitment to the informal community. According to the definition of innovation communities of Fichter (2009, 360), "the community members collaborate closely and informally, and they perceive themselves as a 'team', a 'group' or a similar entity, with a feeling of group identity". Therefore it is important for innovation community members to develop a specific self-concept as a group, so that they can reliably distinguish themselves from non-members on the basis of their common experience working together on an innovation project. At the same time, intra-group processing must be included and valued when group members work together (for intra-group interactions between group members see Hackman 1987).

The formation of group identity is explained by social identity theory developed by Tajfel and Turner (Tajfel 1978, 1979; Turner 1978, 1985; Tajfel and Turner 1979). Social identity theory was initially formulated to account for the psychological basis of intergroup discrimination (Tajfel et al. 1971). The basic phenomenon observed is that individuals tend to classify themselves and others as members of social groups. Group identity is formed based on group membership and defined by Tajfel (1972, 292) as the "individual's knowledge that he belongs to certain groups together with some emotional and value significance to him and this group membership". Based on three underlying intra-psychological processes – social categorization, social comparison and social identification (see Ellemers et al. 2003, 2004) – perceived group membership distinguishes between "in-groups", which are the product of membership, and "out-groups". Typically, members of in-groups and their actions are evaluated more favourably than members of out-groups.

Group membership is determined by group boundaries (Alderfer and Smith 1982). According to Gerybadze (2003, 152), boundary and interface issues become particularly crucial for innovation communities because members of an innovation community are formally assigned to different organizations and they themselves interact with different groups. In addition, while they form a collective unit and share the same goals, the goals may be contrary to the goals of the organizations which they originally come from, and they may encounter significant opposition there. These partially contradictory intergroup relationships can result in substantial conflicts. Gerybadze (2003) proposed that the conditions affecting the quality of group boundaries introduced by Friedlander (1987) are important for the group dynamics of innovation communities. Friedlander (1987) lists the following dimensions: boundary clarity and permeability, the degree of cohesion between group members, the degree of match between group members' functional identity and local language, and the climate within the group. We renamed two dimensions: We called the degree of match between group members' functional identity and local language "common language and understanding". According to Friedlander (1987), an organizational climate should promote innovation, cooperation, and mutual support. We therefore defined this dimension as a cooperative working atmosphere, supporting intergroup relationship and the sustainment of group membership. The dimensions and their (assumed or investigated) effect on the development of a group identity, such as group membership, are discussed in Table 2.2.

Since groups across organizational borders are supposed to be less structured than groups within organizations (Gerybadze 2003), and like virtual teams (Conner 2003), have to deal with increased uncertainty and ambiguity due to the disperse setup, additional clarity and direction must be provided. In addition, research has shown that in NPD projects which are highly uncertain and require identifying and sequencing activities (Kim and Wilemon 2003), commitment to the task must be strong and constant (Chiocchio and Essiembre 2009).

Group identity develops while group members interact with each other. Gerybadze (2003, 153) emphasizes the close communication of innovation community members, supporting the development of a common understanding with respect to the innovation project. Interacting behaviours of group members are distinguished – based on the work of Bales (1950) – either in tasks or in socio-emotional behaviours (Gladstein 1984; McGrath 1984). Empirical studies

Table 2.2 Dimensions of group identity in innovation communities and their effect on group membership

Dimension of group identity (Friedlander 1987)	Definition	Effect on group membership (theoretical considerations or empirical findings)
Boundary clarity	Degree to which innovation communities are independent of other groups (Friedlander 1987)	Defining group membership, distinguishing members from non-members (Ashforth and Mael 1989)
Boundary permeability	Degree of barrier potential of innovation communities to information flow restricting or increasing the inward and outward flow of communication (Alderfer 1977; Agazarian 1989) and the "possibility for individual mobility across groups" (Ellemers et al. 1988, 497)	A (non-significant) tendency towards a stronger group identification in high status groups (Ellemers et al. 1988, 505)
Cohesion	Desire of the innovation community members to remain in the innovation community and to commit energy and resources to the common innovation project (Festinger et al. 1950)	Maintaining group membership (Casey-Campbell and Martens 2009)
Common language and understanding	Development of a common understanding by pursuing common goals and tasks and a separate language to match the requirements of the innovation community's work (Friedlander 1987)	Defining group membership, distinguishing members from non-members (Zenger and Lawrence 1989; Ashforth and Mael 1996)
Cooperative working atmosphere	Shared perceptions of innovation community's policies, practices, and procedures (Reichers and Schneider 1990) promoting innovation, cooperation, and mutual support (Friedlander 1987)	Maintaining group membership (Ashforth 1985)

Source: See references in the table

conceptualize these as a multifaceted construct, embracing elements of both task-related and social interaction within groups (Gladstein 1984; Stewart and Barrick 2000; Högl and Gemünden 2001). Research on groups in organizations supports a direct (positive) relationship between group interaction among group members and group performance. Cooperative behaviour enables the group to perform more effectively, due to the coordination of group efforts towards a common goal (Gladstein 1984). In addition, the efficient use of group resources can be promoted (Podsakoff et al. 2000). Empirical studies of workgroups have consistently shown that selected intra-group processes have a positive influence on project performance

(Pinto and Pinto 1990; Stewart and Barrick 2000; Högl and Gemünden 2001; Högl et al. 2004). All six teamwork quality facets of the construct by Högl and Gemünden (2001) – communication, coordination, balance of member contributions, mutual support, effort, cohesion were significantly positively related to project performance. For alliances Schreiner et al. (2009) could empirically prove that communication, coordination (and bonding) had a positive effect on goal fulfilment including firm performance. Högl et al. (2007) could show that the dispersion of group members increased the positive impact of teamwork quality on group performance. Therefore, besides group identity, the interaction of innovation community members seems to be another important characteristic of innovation communities.

2.2.3 *Operationalizing the Innovation Community*

Using the given term definitions and the theory-based assumptions on the positive effects of a group of promotors and champions as well as the concepts of group identity and group interaction for innovation projects, we introduce the following operationalization for the term innovation community: An innovation community is a group of at least three individuals that promote a common innovation project. At least three of the individuals must occupy at least one promotor role. The group must display a high quality of group dynamics, i.e. a high quality of group interaction as well as a high quality of group identity.

2.3 Research Hypotheses and Measures

In the following, the research hypotheses and measures for defining promotor roles, group interaction, group identity and (innovation) project performance are discussed in detail.

2.3.1 *Research Hypotheses*

First we want to test general hypotheses concerning the role distribution of key actors in innovation networks. Regarding NPD within companies, empirical findings of German research support the assumption of promotor theory that separate promotors are necessary to achieve a high level of innovation success. For example, Hauschildt and Kirchmann (2001) found that a troika of power, expert, and process promotors was best suited to support the innovation process, followed by the dyad of power and expert promotor, and by a sole expert promotor. Nevertheless, Folkerts and Hauschildt (2002, 16) showed that despite role exclusivity being prevalent, in fact role combinations occurred frequently. According to Rost et al. (2007) specialized role types have more benefit in low task complexity environments (e.g. incremental innovation), while universal promotors are more qualified to reduce the barriers against radical innovation due to their general knowledge. These universal promotors who combine two or more roles show parallels to the champion concept (Rost et al. 2007). Regarding innovation networks relatively little is known about the roles actors play in the innovation process. Rost et al. (2007) found that persons with a general knowledge are better suited when it comes to decision-making between diverse parties. We therefore want to test the following hypotheses:

Hypothesis 1: In innovation networks, the champion role is more frequent than the promotor role.
Hypothesis 2: In innovation networks, the champion role has a positive effect on project performance.

Because we assume that in general, champions are more frequent in innovation networks, we also propose that this role is more frequent in innovation communities.

Hypothesis 3: In innovation communities, more champions than promotors are present.

Regarding the other two characteristics of innovation communities, group identity and group interaction, we believe that they are also important for innovation networks:

Hypothesis 4: Group identity has a positive effect on the performance of innovation networks.
Hypothesis 5: Group interaction has a positive effect on the performance of innovation networks.

Drawing on conclusions of Fichter (2009, 368) that innovation communities are in particular important in the early phases of the innovation process, we propose that innovation communities can also be found more frequently in such phases:

Hypothesis 6: In the early stages of the innovation process, innovation communities are more frequent than in advanced stages.

According to Gerybadze (2003, 155), face-to-face communication and geographical proximity is important for innovation communities, e.g. to solve conflicts and to develop a common understanding. We therefore want to test the following two hypotheses:

Hypothesis 7: Innovation communities are more frequent in innovation networks in which the partners are close by.
Hypothesis 8: Innovation community members communicate most frequently face-to-face.

Finally, we propose that innovation networks in which an innovation community is present will be more successful:

Hypothesis 9: Innovation networks with an innovation community are more successful than innovation networks without an innovation community.

2.3.2 *Measures*

A questionnaire was developed with scale items derived from an extensive review of the literature on promotors, champions, groups, networks and cooperation, and adapted where necessary. All constructs were developed as multidimensional concepts and, apart from the promotor scales, measured with a seven-point scale ranging from 1 = "strongly disagree" to 7 = "strongly agree". A preliminary version of the questionnaire was given to two network managers for feedback and comments. These suggestions were included, helping to reduce ambiguities or difficulties in responding to the scale items and to ensure clarity. In the following, the measures and items are discussed.

Measures for Promotor Roles

To measure the four distinct and ideal "promotor roles" (power promotor, expert promotor, process promotor, relationship promotor) identified in a number of empirical studies, two different approaches exist in literature. Several authors use rating scales (e.g. Walter and Gemünden 2000; Folkerts 2001; Papies 2006; Gemünden et al. 2007). The role items describe specific types of behaviour of persons occupying these roles to overcome different barriers. They also include the power base – the foundations of the role owners' influence. Another approach is to ask respondents to list colleagues present during the innovation process who fit the description of promotors. If a person exhibits a certain role, this person is assigned the value 1 (e.g. Hauschildt and Kirchmann 2001; Rost et al. 2007).

To assign a role to each of the innovation network members, we used item scales with up to ten items for each of the four promotor roles. Response options for eight innovation network members and space for further entries was provided. According to the literature the size of an optimal team ranges between five and nine people for working on a complex problem (Qurashi 1993; Högl 2005; Stewart 2006). The innovation communities of the case studies described by Fichter (2009) had consisted of 6 up to 12 individual promotors involved in the implementation phase. We used nominal scale values instead of rating scales for reasons of clarity. The value 1 indicates if an innovation community member fulfils the aspect of a promotor role, and the value 0 if not. A role was classified as occupied if at least half of the items of a role were ticked. An innovation network member can occupy more than one role in "personal union", also, a role can be shared by several innovation network members. The items of the four promotor roles are listed in the Appendix of this chapter. The items describing the promotor roles were for the main part adapted from Papies (2006), Gemünden et al. (2007) and actitivites listed by Thoms (2003). In addition, items of the expert promotor role based on items used to describe the innovator role by Hirst and Mann (2004). To describe the process and relationship promotor role, additional items of Walter (1998) and role items of Hirst and Mann (2004) were used.

Measures for Group Identity

For measuring group identity in innovation networks, item scales were developed for the dimensions which Friedlander (1987) proposed to affect boundary transactions. We added the dimension "cognitive similarity", because social identity theory implies that group identity may develop if people become aware of each other's (similar) cognitions (McGarty et al. 1995).

Boundary clarity was measured using six items that describe a closed group culture with clear boundaries between in-groups and out-groups. Items reflected the perception of being one of the group, an insider (Gefen and Ridings 2003), the external perception of the group as a social entity, the fit of the members in the group (Workman 2005), the degree to which group membership is clear (i.e. one knows who is in the group and who is not) and stability over time (Wageman et al. 2005).

Boundary permeability was assessed by asking network managers to report if the innovation community had many information-intensive links among groups. The five items focused on the intensity of interaction with other groups, networks, or partners (Workman 2005), access to important decision makers or contacts with technical information relevant to the project (Hirst and Mann 2004), and the openness for new group members.

The seven items measuring the cohesion of members of the innovation community included the commitment to the innovation project, interpersonal attraction and integration of group members, and pride of being part of the group (Mullen and Copper 1994; Högl and Gemünden 2001).

The measurement scale for common language and understanding was newly developed based on Friedlander (1987). The task purpose – i.e. the innovation idea or project in the case of the innovation network – is essential for developing functional identity. Therefore, one item addresses the importance of innovation for the group members. Two items deal with the "local" language of the group, assessing whether all group members used the same terminology and there were no conflicts due to conceptual misunderstandings.

The predominant cooperative working atmosphere is characterized by Anderson and West (1998) as one of non-threatening trust and support. In such a climate, all members of a work group are able to propose new ideas and problem solutions. The four group identity items are mainly based on Howard et al. (2005).

The item scale measuring the cognitive similarity of innovation community members was newly developed, but is based on theoretical considerations regarding shared cognition in groups. Social identity theory proposes that different forms of cognitive similarity between members of a group can be precursors to the development of a shared identity (McGarty et al. 1995). Cognition refers to knowledge structures which may be similar or beliefs within the group that there is a shared understanding (Swaab et al. 2007). Four items are used to measure perceived similarity: in education and professional training, in occupational roles, functions and tasks, in shared beliefs and opinions, and in a shared understanding.

Measures for Group Interaction

Group interaction within the innovation networks was measured using Högl and Gemünden's comprehensive concept (2001). For the purpose of the investigation, 'communication',' coordination', 'balance of member contributions', 'mutual support', and 'effort' were selected as collaborative team process dimensions. The 'cohesion' dimension which Högl and Gemünden (2001) also used as group interaction facet was assigned to the concept of group identity. Scales were adapted from Högl and Gemünden (2001) for the research context. Two items addressing effective information flow were added to the communication construct, based on Hirst and Mann (2004).

Measures for Project Performance

For measuring project performance, a multi-dimensional self-assessment construct was used (Dawes 1999). Objective performance data (e.g. financial data) was expected to be frequently unavailable in early stages of the innovation projects (Calantone et al. 2003). Two dimensions were taken into account: the comparison of the original innovation project goals relative to the adherence to budget and schedule (*efficiency*), and the achievement of set goals and/or expectations, especially with respect to the quality of the outcome (*effectiveness*) (Högl and Gemünden 2001). Because innovation communities are supposed to be rather small, we assumed that in contrast to innovation projects in larger companies (Blindenbach-Driessen et al. 2010), innovation community members are well informed in terms of these two dimensions of innovation projects. The items were formulated in such a way as to capture the actual stage of the innovation project.

Measures for Project Innovativeness

In correspondence with the literature, the degree of project novelty or *innovativeness* was measured with the help of a multi-factor construct (e.g. Garcia and Calantone 2002). Three different dimensions of product innovativeness (technology, market, organization) were thereby assessed from an ex ante point of view (see the Appendix for more detail). Items were adapted from Hauschildt and Schlaak (2001).

2.4 Research Results

First the sample and the quality of the measures are described, then descriptive results are presented with respect to characteristics of the networks and the key actors. Hypotheses are tested and the effect of roles, group interaction and group identity on project performance is analyzed. Finally, it is investigated if the presence of an innovation community has an effect on project performance. For this purpose, the quantitative definition of an innovation community (see above) was used, taking active innovation community members and a high quality of group identity and group interaction into account.

2.4.1 Sample

The sample is based on a previous study on innovation networks in Germany (see e.g. Baier et al. 2006; Rese and Baier 2011). This sample was updated by excluding inactive networks and searching for newly formed networks in highly innovative branches of industry (nanotechnology, solar energy, information and communication technology). Websites in particular (e.g. funding programmes of different German federal and federal state ministries, industry-related websites, and websites of industry associations) were evaluated. Available experts – the other authors of this book – also checked the list of networks and gave hints on missing ones. This yielded a complete sample of 767 identified networks. The focus was on networks with a high proportion of small and medium enterprises (SME), in order to avoid an overly large impact on the network by one dominant (bigger) enterprise. The so-called network managers received a questionnaire and were asked to respond in November 2008. These were centred on a specific innovation project in which they themselves were personally heavily involved. In the data collection, only one person per network was involved. The networks of the previous study had been rather small, giving individual members more insights e.g. into the social processes within the network. The data collection resulted in a total of 107 questionnaires. Additionally, 50 recipients had replied that no innovation was being developed in the network, that they had not started product development yet, that they felt they could not answer the questions, or that they simply did not want to participate. These non-response reasons yielded an effective sample size of 717, thus providing a response rate of 14.9%. With usual sample sizes of about 10% when questioning companies, the resulting response rate is quite high, but could not match the response rate of 43.5% of the previous study (Rese and Baier 2011). Taking into account that the questionnaire of this study was much more complex, this is a satisfying result.

2.4.2 Assessing the Measures

To verify the quality of the constructs, they were tested using exploratory and confirmatory factor analysis (Gerbing and Hamilton 1996). Some items had to be dismissed because of their low item to total correlation or low reliability (<0.4). A

Table 2.3 Summary statistics of group identity, group interaction and project performance facets

Factor	(Initial) number of items	Mean[a] (standard deviation)	Cronbach's α	Variance extracted (%)	AVE[b]	Composite reliability[c]
Group identity						
Boundary clarity	6	5.43 (1.01)	0.813	52.19	0.427	0.815
Boundary permeability	(5)4	5.68 (0.98)	0.783	61.04	0.524	0.803
Cohesion	(7)6	5.25 (1.06)	0.870	61.75	0.530	0.867
Common language and understanding	3	5.93 (0.90)	0.682	61.47	0.494	0.724
Cooperative working atmosphere	4	6.03 (0.75)	0.760	59.25	0.516	0.786
Cognitive similarity	(4)3	4.68 (1.31)	0.769	69.97	0.584	0.799
Group interaction						
Communication	(9)6	5.50 (1.05)	0.891	65.77	0.585	0.893
Coordination	2	5.97 (0.89)	0.683	75.97		
Balance of member contributions	3	5.19 (1.02)	0.632	59.65	0.366	0.633
Mutual support	5	5.79 (0.93)	0.886	69.06	0.650	0.897
Effort	3	5.46 (0.98)	0.777	69.59	0.625	0.819
Network performance						
Effectiveness	5	5.40 (0.99)	0.846	63.29	0.528	0.847
Efficiency	5	5.64 (0.99)	0.864	65.05	0.579	0.869

Source: Authors

$N = 105$

[a]Scale: 1 = strongly disagree, 7 = strongly agree

[b]Average variance extracted (AVE) = $\Sigma\lambda_i^2$ varF/($\Sigma\lambda_i^2$ varF + Σ variance (ε_i)), with λ = unstandardized factor loading, varF = factor variance, ε = indicator measurement error

[c]Composite reliability = $(\Sigma\lambda_i)^2$ varF/($(\Sigma\lambda_i)^2$ varF + Σ variance (ε_i)), with λ = unstandardized factor loading, varF = factor variance, ε = indicator measurement error

few missing values were replaced by mean values. The results in Table 2.3 indicate that the measures used in this study are fairly robust in terms of their composite reliability and convergent validity.

All Cronbach's α values are higher than 0.6, which is proposed as acceptable by Murphy and Davidshofer (2001). Even the guidelines of Nunnally (1978) for exploratory research that propose a value of 0.7 are mostly exceeded. The composite reliability of the constructs is also higher than Bagozzi and Yi's (1988) minimum value of 0.60, supporting the reliability of the constructs. The average variance extracted (AVE) mostly exceeds 0.5, which is the recommendation of Fornell and Larcker (1981), suggesting that there is some evidence for the discriminative validity of the constructs. Altogether, the testing of the measurement scales suggests a good quality of the constructs. Apart from these satisfactory results, the calculated variance inflation factors indicated that the group identity and group interaction dimensions are highly correlated (Rese et al. 2010a). In general, multicollinearity is a problem in success factor analysis (see the similar results of Rese and Baier 2011).

All items corresponding to a construct were added to yield a total score for that construct. The mean value for the cognitive similarity of group members is lowest with 4.68, while the highest mean value was for the cooperative working atmosphere with 6.03 on a scale from 1 = "strongly disagree" to 7 = "strongly agree".

2.4.3 Descriptive Results

Network Characteristics

Most of the innovation projects were assigned to areas that are typically very research- and innovation-intensive, e.g. new process and production technologies, energy, nano- and materials technology, or healthcare. The sample's concentration on potentially high innovative or radical projects is also reflected by the network partners coming from research- and innovation-intensive industries (see Table 2.4). In other studies in Germany, these industries also proved to be cooperation-intensive (e.g. Eggers and Kinkel 2002; Frietsch 2007).

In comparison to the previous study, the projects had advanced in the innovation process, often reaching the testing and validation stage or even the launch stage. Another third of the networks was in the development stage, while only few of them were in an early stage (see Table 2.5). Within the innovation networks, on average, 4.9 organizations were identified as the most important project partners. Altogether, the networks were much smaller than in the previous study, with no network consisting of more than ten partners (see Table 2.5). This might be the result of

Table 2.4 Descriptive statistics of the networks (focus area of innovation project, sector of industry of network partners)

Focus area of innovation project		Sector of industry of network partners	
Process and production technologies	24.0%	Medical technology, measurement and control technology/optics	25.7%
Energy	13.7%	Mechanical engineering	24.9%
Nano & materials technology	13.7%	Chemical industry	10.4%
Healthcare	11.0%	Aerospace technology	4.1%
Climate	8.2%	Pharmaceutical industry	3.7%
Food & agriculture, biotechnology	8.2%	Devices for generation and distribution of electricity	2.9%
ICT	6.8%	Office machines, computer equipment and facilities	2.1%
Transport	2.1%	Broadcast & television technology, telecommunications	2.1%
Safety	2.1%	Rail industry & rail vehicle construction	1.2%
Others	10.3%	Others	15.4%

Source: Authors

Table 2.5 Descriptive statistics of the networks (stage, geographical distribution, number of partners, percentage of SME)

Stage of the network	Present study	Rese and Baier (2011)	Number of partners	Present study	Rese and Baier (2011)
Building a business case	8.8%	47.3%	Less than 4	26.7%	4.1%
Development stage	34.3%	17.7%	4–10	73.3%	56.8%
Testing & validating	38.2%	28.8%	More than 10	—	39.1%
Launch	18.6%	6.2%			
Geographical distribution of the partners	Present study	Rese and Baier (2011)	Percentage of SME	Present study	Rese and Baier (2011)
Local	7.8%	3.4%	0%	6.9%	0.8%
Regional	23.3%	40.3%	1% up to <50%	15.6%	28.3%
National	48.9%	46.6%	>50% up to 99%	44.2%	46.4%
International	20.0%	9.7%	100%	33.3%	14.9%
			Average	67.8%	61.8%

Source: Authors

the progress of the networks in the innovation process, which had mostly left the very early stage of the innovation process and now concentrate on a specific innovation project with selected partners.

In both studies, on average, about two-third of the network partner firms were small and medium enterprises. Both questionnaires included the current definition of the European Commission for SME (European Commission 2006). More than one-third of innovation networks (36.1%) were dominated by one partner contributing to more than 50% to the innovation project. In both studies, small networks with up to four partners tended to be dominated by one partner organization (present study: 54.0%), whereas larger networks shared contributions more equally (present study: 81.1%). Regarding the geographical distribution the network partners are more nationally and internationally dispersed than in the previous study.

The networks were grouped, according to their confirmation of the project performance construct (see Appendix for the items) by means of a one-dimensional cluster analysis, into 61 "successful" and 44 "less successful" networks (two networks could not be classified due to missing data). Another clustering was completed regarding project innovativeness, resulting in 50 "radical" and 57 "incremental" innovation networks (two of which could not be classified due to missing data).

Key Actors in the Networks

The members of the innovation networks mainly got together through their own business networks (former joint project activities, business contacts), while private networks played a negligible role (see Table 2.6). Moreover, the active efforts of

Table 2.6 Descriptive statistics of the key actors (contact opportunities, frequency of meetings, communication channels)

Opportunities of key actors to get to know each other (multiple answers)		Frequency of meetings	
Former joint projects	66.0%	Every week	12.6%
Active effort of a key actor	55.7%	Every month	30.1%
Former business contact	50.0%	All 3 months	41.7%
Task forces, trade shows, conferences	29.2%	Twice a year	14.6%
Joint education/study	5.7%	Once a year	1.0%
Knew each other privately	5.7%		
Others	7.5%		
Communication channels (multiple answers)			
Telephone	95.2%	Face-to-face meetings (two persons)	70.5%
E-mail	92.4%	Task forces	56.2%
Telephone conferencing	23.8%	Meetings with the whole group	56.2%
Internet forums	5.7%	Workshops including externals	39.0%
Video conferencing	2.9%	Congress/conferences	36.2%
Blogs	1.0%	Others	2.8%

Source: Authors

one of the network members were important to bring the innovation network members together. To a lesser extent, visits of trade shows, task forces and conferences played a role in contacting possible members. In most innovation networks, the members maintained intensive business contacts and held joint project meetings every 3 months or even more frequently. Within their innovation network, members mainly communicated through conventional communication channels like telephone or e-mail, however, face-to-face communication was used frequently as well. In contrast, new communication platforms like blogs, video conferences or Internet forums hardly played a role.

Altogether, 539 people were involved in the 101 innovation networks (6 networks had been excluded due to lack of data). Of these 539 people, 352 could be classified as "role owners" according to our definition of a role as being occupied if at least half of the items were confirmed. In total, 773 roles were identified, already indicating that the majority of the role owners held more than one position. Even so, our definition excluded 187 people who did not occupy any one of the four defined promotor roles. On average, about five key actors (mean value: 5.34) took part in an innovation network.

Taking project performance or project innovativeness into account, innovation networks did not differ significantly in size. Three of these five members were role owners (mean value: 3.49). About two of the innovation network members (mean value: 1.9, $n = 186$) were identified as initiators of the network, at the time of the empirical investigation occupying 414 of the identified roles.

In more than half of the innovation networks (59.1%), all network members came from different enterprises. More than half of the key actors could be found on high hierarchical positions within their organizations (53.5%). 41.1% belonged to middle management levels, and only a small share of 5.4% to a lower hierarchical level. In addition, most of the key actors (80.3%) had rather high or very high experience with innovation projects. Only a small part (a total of 11.0%) had gathered rather low experience with innovation projects.

2.4.4 *Hypothesis Testing*

In the following, the hypotheses developed in Sect. 2.3.1 will be tested. We start with the hypotheses regarding innovation networks. Then the hypotheses on innovation communities are investigated.

Promotor Roles in the Networks

Regarding the role distribution, the results for innovation networks are quite similar to those within firms (see Table 2.7). The expert promotor role was most frequently found. As in the study of Rost et al. (2006), process and relationship promotors both occur similarly frequently. While promotor roles were more often occupied exclusively within companies, in innovation networks, the process and relationship roles in particular are mainly found in combination with other roles.

Altogether, innovation networks consisted of combined promotor role owners (champions) more often than of specialized promotor role owners. Therefore we have already found support for hypothesis 1.

In the networks, the roles of expert promotor (82.1%) and process promotor (8.1%) were most frequently occupied, followed by the relationship promotor (83.2%) and the power promotor (80.2%). An exclusive expert promotor could be found in about half of the networks (49.5%), an exclusive power promotor in a quarter of the networks (26.7%) (see Table 2.8). While the expert promotor is most frequently found exclusively in innovation networks, whithin firms this is the case with the power promotor. Despite the smaller numbers of exclusive process and relationship promotors, both roles could be found in nearly all networks, but were often occupied only once (one process promotor in 51.5% of the networks, one relationship promotor in 43.6% of the networks).

Before the other hypotheses are tested, some descriptive analysis can provide first insights into the distribution of roles in the innovation networks: Regarding multiple role owners, innovation networks are no different than firms. More than a third of the role owners occupy two or more roles. In contrast to firms, network members are more often rated as having no role than as having an exclusive promotor role (see Table 2.9). Within innovation networks, champions with 2, 3, or 4 roles are quite equally distributed, while, within firms, the number of champions decreases with an increasing number of roles.

Table 2.7 Comparison of role distribution within firms and innovation networks regarding identified roles

Promotor roles	Innovation networks			Within firms			
	Present study			Folkerts and Hauschildt (2002)			Rost et al. (2006)
	Exclusive role	Role in combination	In total	Exclusive role	Role in combination	In total	In total
Power promotor	18.1% (n = 34)	81.9% (n = 154)	24.3% (n = 188)	86.2% (n = 25)	13.8% (n = 4)	25.0% (n = 29)	23.5% (n = 308)
Expert promotor	32.0% (n = 85)	68.0% (n = 181)	34.4% (n = 266)	40.9% (n = 18)	59.1% (n = 26)	37.9% (n = 44)	32.1% (n = 422)
Process promotor	5.6% (n = 9)	94.4% (n = 152)	20.8% (n = 161)	38.3% (n = 18)	61.7% (n = 29)	37.1% (n = 47)	22.7% (n = 298)
Relationship promotor	6.3% (n = 10)	93.7% (n = 148)	20.4% (n = 158)				21.7% (n = 285)
n (identified roles)			773			120	1,313

Source: Authors; Folkerts and Hauschildt (2002), Rost et al. (2006)

Table 2.8 Distribution of the four promotor roles in innovation networks

Promotor roles	Role not occupied	Exclusive role	Role in combination	In total (n = 101)
Power promotor	19.8% (n =20)	25.7% (n = 26)	54.5% (n = 55)	80.2% (n = 81)
Expert promotor	10.9% (n = 11)	49.5% (n = 50)	39.6% (n = 40)	89.1%(n = 90)
Process promotor	11.9% (n = 12)	6.9% (n = 7)	81.2% (n = 82)	88.1% (n = 89)%
Relationship promotor	16.8% (n = 17)	8.4% (n = 9)	74.3% (n = 75)	83.2% (n = 84)

Source: Authors

Table 2.9 Distribution of multiple promotor roles in innovation networks

	Within firm (Rost et al. 2007)		Innovation networks			Radical innovation networks		
Roles	Persons	(%)	Persons	(%)	Per network	Persons	(%)	Per network
No role	47	5.5	187	34.7	2.8% (n = 3)	105	39.0	6.0% (n = 3)
Promotor	445	51.7	138	25.6	71.3% (n = 72)	54	20.1	66.0% (n = 33)
Champion	326	39.9	214	39.7	91.1% (n = 92)	110	40.9	86.0% (n = 43)
With two roles	168	20.5	77	14.3	51.5% (n = 52)	36	13.4	46.0% (n = 23)
With three roles	85	10.4	67	12.4	51.5% (n = 52)	36	13.4	52.0% (n = 26)
With four roles	73	8.9	70	13.0	45.4% (n = 46)	38	14.1	44.0% (n = 22)
n	818		539		101	269		50

Source: Authors; Rost et al. (2007)

In general, nearly all of the innovation projects were supported by multiple role owners (91.1%). There was no network in which each of the four roles was exclusively occupied by a different network member. In three networks (2.8%), no role was occupied according to our role definition (see Table 2.9). Radical innovation networks did not differ very much from the total sample of innovation networks regarding the distribution of roles (for more details on radical and incremental networks see Sand and Rese 2011). The percentage of "no role" owners was slightly higher, while the percentage of exclusive role owners was slightly smaller. At the time of the empirical investigation, the initiators of the networks to an even greater extent occupied champion roles (n = 126, 67.7%). A comparatively small fraction occupied an exclusive promotor role (n = 34, 18.3%), or had no role at all (n = 26, 14.0%).

Promotor Roles and Project Performance

With respect to project performance, it was examined whether the occupation of at least one exclusive promotor role (regardless of the specific promotor role) or one combined (champion) role had any positive effects. Regarding the four promotor roles and their effect on project performance, three different possibilities of role occupation (role not occupied, role at least once exclusively occupied and role occupied only in combination) were examined. In addition, it was analyzed whether an increase in numbers of a role was positively related to project performance. The results show that only networks with at least one champion present, occupying all four roles, were significantly more successful than innovation networks with no such champion. The results are even stronger regarding the presence of a universal champion in radical innovation networks (see Table 2.10). Therefore, hypothesis 2 is supported.

Regarding the exclusive occupation of roles or combination with other roles, only the power promotor role showed significant results (see Table 2.11). Project performance is highest when this role is occupied in combination with other roles.

Project performance significantly decreases with the number of exclusive role owners (see Table 2.12). On the other hand, with increasing numbers of champions

Table 2.10 Promotors, champions and project performance

Roles	Project performance[a]					
	All innovation networks (n = 101)			Radical innovation networks (n = 50)		
	Role not occupied	Role at least once occupied	F	Role not occupied	Role at least once occupied	F
Exclusive role	5.7 (0.92)	5.4 (0.90)	1.518	5.4 (0.96)	5.6 (0.88)	0.749
Champion with two roles	5.4 (0.87)	5.6 (0.94)	1.066	5.4 (0.92)	5.8 (0.85)	2.602
Champion with three roles	5.6 (0.84)	5.4 (0.97)	1.087	5.5 (0.91)	5.6 (0.92)	0.045
Champion with four roles	5.3 (0.94)	5.8 (0.82)	6.247*	5.2 (0.86)	6.0 (0.79)	10.205**

Source: Authors
**Significant with $p < 0.01$; *Significant with $p < 0.05$
[a]Scale: 1 = "strongly disagree" to 7 = "strongly agree"; mean value (standard deviation)

Table 2.11 Exclusive or combined promotor roles and network performance

Roles	Project performance[a]			
	Role not occupied	Role (at least once) exclusively occupied	Role occupied only in combination	F
Power promotor	5.2 (0.98)	5.3 (0.83)	5.7 (0.89)	3.129*
Expert promotor	5.3 (0.94)	5.4 (0.98)	5.7 (0.79)	1.577
Process promotor	5.4 (0.89)	5.5 (1.10)	5.5 (0.91)	0.054
Relationship promotor	5.4 (0.89)	5.3 (1.14)	5.6 (0.89)	0.659

Source: Authors
*Significant with $p < 0.05$
[a]Scale: 1 = "strongly disagree" to 7 = "strongly agree"; mean value (standard deviation)

Table 2.12 Number of roles and project performance

	Project performance
Exclusive role	−0.244*
Champion (with two roles and more)	0.197*

Source: Authors
Correlations of mean-centered variables
*Significant with $p < 0.05$

(with two roles and more), network performance also increases significantly. The results indicate that more than one key actor is important for an innovation community and its performance. The correlation analysis supports our operationalization of an innovation community with three individuals occupying at least one promotor role, even if a regression analysis with quadratic terms showed no significant results.

Group Interaction and Group Identity and Project Performance

Comparing successful and less successful innovation networks, the results show a clear picture. In successful networks, all group identity and group interaction dimensions are significantly higher than in less successful networks (see Table 2.13).

The dimensions are also higher in radical innovation networks than in incremental innovation networks, but these results are less significant. Group identity and group interaction can therefore be described as "success factors" for innovation networks, so that hypothesis 4 and hypothesis 5 are supported.

Presence of an Innovation Community in Networks

The respondents should rate the level of contribution of the group of innovators to the success of the innovation project on a scale of 1 = "low" to 7 = "very high". In almost half of the networks (46.6%) the group of innovators contributed to a very high level to project success (point 7 on the rating scale), followed by another 39.8% with a high level of contribution (point 6 on the rating scale). The importance of the group of innovators was significantly higher ($p = 0.00$) in successful networks (very high: 66.7%, high: 37.1%) than in less successful networks (very high: 18.6%, high 52.2%). The results were similar but less significant ($p < 0.05$) for radical innovation networks (mean value: 6.47) and incremental innovation networks (mean value: 6.11). Altogether, the respondents indicated that the group of innovators is an important success factor for innovation networks.

For a quantitative definition of innovation communities roles, group identity and group interaction of the key actors were taken into account. We speak of the existence of an innovation community if the following quantitative conditions are fulfilled: Presence of at least three cooperating promotors who occupy at least one promotor role each, and presence of high group identity and interaction within the innovation community. Each of the group identity and group interaction dimensions had to be five or more on a scale of 1 = "strongly disagree" to 7 = "strongly agree".

Altogether, 23 of the 107 networks fulfil these criteria and can be classified as innovation networks with an innovation community.[4] Regarding innovativeness,

---

[4] At first, 20 innovation communities were identified according to our definition. After replacing missing values, three further networks could be classified as having an innovation community.

Table 2.13 Group identity and group interaction in innovation networks classified according to performance and innovativeness

	Successful networks (n = 61)	Less successful networks (n = 44)	Radical innovation networks (n = 50)	Incremental innovation networks (n = 55)
Group identity[a]				
Boundary clarity	5.75***	4.99	5.68*	5.21
Boundary permeability	5.85*	5.44	5.86+	5.52
Cohesion	5.70***	4.62	5.56**	4.96
Common language and understanding	6.21***	5.53	6.06	5.81
Cooperative working atmosphere	6.36***	5.57	6.20*	5.87
Cognitive similarity	5.01*	4.40	5.03*	4.50
Group interaction[a]				
Communication	5.90***	4.94	5.71*	5.31
Coordination	6.31***	5.49	6.13+	5.82
Balance of member contributions	5.51***	4.74	5.44*	4.96
Mutual support	6.21***	5.20	5.92	5.67
Effort	5.94***	4.80	5.66+	5.28

Source: Authors

***Significant with $p = 0.00$; **Significant with $p < 0.01$; *Significant with $p < 0.05$; +Significant with $p < 0.1$

[a]Scale: 1 = "strongly disagree" to 7 = "strongly agree"; mean value

the majority of these 23 networks are radical innovative networks (n = 17; 73.9%). On average, more than six persons (mean value: 6.61) participated in a network with an innovation community. Of those six, an average of five (mean value: 4.83) occupied at least one role according to the role definition. Therefore, even in a network with an innovation community, one key actor can still be found to have no role. Five of the innovation networks with an innovation community consisted only of champions, while the other 18 networks had between one and five exclusive role owners. In almost all networks (95.7%), at least one champion was present. In one of the 23 networks, only promotors participated. Altogether, an average of about three champions and one promotor took part in an innovation community (see Table 2.14). We therefore found evidence for hypothesis 3. In radical innovation networks with an innovation community, even more champions occurred (mean value: 3.88), especially the champion with four roles (mean value: 1.41).

The innovation communities in radical innovation networks were slightly bigger (mean value: 6.82) with about five role players (value: 5.06).

Regarding the proposed relationships of innovation communities and stage of the innovation process, the geographical proximity of the partners, and communication channels, hypotheses 6, 7 and 8 could not be confirmed (cf. Table 2.15). Innovation communities were not more frequent in innovation networks in the early

Table 2.14 Role constellation in innovation networks with an innovation community

Roles	Present in innovation networks with an innovation community (n = 23)	Present in innovation networks without an innovation community (n = 77)	Mean value: innovation networks with an innovation community (n = 23)	Mean value: innovation networks without an innovation community (n = 77)
Promotor (exclusive role)	78.3% (n = 18)	68.8% (n = 53)	1.52	1.27
Champion	95.7% (n = 22)	89.6% (n = 69)	3.30***	1.77
with two roles	69.6% (n = 16)+	45.5% (n = 35)	1.30**	0.77
with three roles	60.9% (n = 14)	49.6% (n = 38)	0.91+	0.67
with four roles	60.9% (n = 14)	45.5% (n = 35)	1.09*	0.77

Source: Authors
***Significant with $p = 0.00$; **Significant with $p < 0.01$; *Significant with $p < 0.05$; +Significant with $p < 0.1$

stages of innovation. One reason might be that it simply takes time to develop group identity and group interaction. To some extent, the partner organizations of innovation networks with an innovation community were more often locally or regionally dispersed, but the differences were not significant. Nevertheless, it seems to be beneficial for the formation of an innovation community if not all members come from different partner firms. In 60% of innovation communities, at least some of the members were from the same partner organization. This was only the case in 37.5% of the innovation networks without an innovation community. Also, the assumption of Gerybadze (2003, 155) that innovation community members communicate exclusively face-to-face could not be confirmed (cf. Table 2.15). Meetings of the whole group were more frequent in innovation community networks, but not significantly. Instead, modern technical possibilities to communicate as a group were used significantly more often, e.g. telephone conferences or Internet forums. In addition, congresses or conferences served as meeting points.

Innovation networks with an innovation community displayed a significantly higher project performance in terms of effectiveness. The presence of an innovation community particularly seems to help to achieve set goals or expectations regarding the quality of the outcome and, less significantly, adherence to budget and schedule. If radical and incremental innovation networks are compared regarding project performance and its two dimensions effectiveness and efficiency, the results show that radical innovation networks supported by an innovation community were significantly more effective, efficient and thus, more successful than radical projects without an innovation community (see Table 2.16). Incremental innovations networks did not show a significant difference in project performance regarding the presence of an innovation community. Therefore hypothesis 9 is fulfilled especially for radical innovation networks. The presence of an innovation community is a success factor particularly for radical innovation networks.

Table 2.15 Comparison of networks with and without an innovation community with respect to stage, geographical distribution of partners and communication channels

Stage of the network	Innovation networks with an innovation community (n = 23) (%)	Innovation networks without an innovation community (n = 77) (%)
Building a business case	9.5	9.2
Development stage	28.6	34.2
Testing & validating	42.9	38.2
Launch	19.0	18.4
Geographical distribution of the partners		
Local	12.5	6.7
Regional	25.0	23.3
National	41.7	50.0
International	20.8	20.0
Communication channels (multiple answers)		
Phone	100.0	93.6
E-mail	95.7	91.2
Telephone conferencing	47.8**	17.9
Internet forums	21.7**	1.3
Video conferencing	8.7	1.3
Blogs	0.0	1.3
Face-to-face meeting (two persons)	69.6	70.5
Task forces	60.9	53.8
Meetings with the whole group	69.6	50
Workshops, including external ones	52.3	34.6
Congress/conferences	56.5*	28.2
Others	13.0*	0.0

Source: Authors
**Significant with $p < 0.01$; *Significant with $p < 0.05$

Table 2.16 Presence of an innovation community (IC) on project performance

	All networks			Radical networks			Incremental networks		
	With IC	Without IC	F	With IC	Without IC	F	With IC	Without IC	F
	n = 23	n = 79		n = 17	n = 33		n = 6	n = 46	
Project performance[a]	5.90	5.38	5.97*	6.08	5.28	10.19**	5.38	5.31	0.03
Effectiveness[a]	5.86	5.24	7.38**	6.02	5.14	9.73**	5.40	5.58	0.23
Efficiency[a]	5.93	5.52	3.21+	6.13	5.42	7.23*	5.37	5.45	0.04

Source: Authors
**Significant with $p < 0.01$; *Significant with $p < 0.05$; +Significant with $p < 0.1$
[a]Scale: 1 = "strongly disagree" to 7 = "strongly agree"; mean value

2.5 Discussion

This research studied innovation networks of small and medium sized enterprises at the interpersonal level, focusing on roles, group interaction and group identity of networks of innovators. Regarding roles the promotor concept was transferred to the network context. Innovation communities were operationalized, taking role constellation, group identity and group interaction into account. Hypotheses with respect to innovation networks and innovation communities were developed based on the available literature. The role constellation in innovation networks and the impact of the respective promotor roles on project performance were hypothesized. In addition, group interaction and group identity within these networks were analyzed as success factors. A measurement concept was developed, partly based on existing studies. Finally, a quantitative definition of innovation communities was derived and tested. The hypothesized relationships were tested using a comprehensive sample of 107 German networks of SMEs and research institutes.

Our study shows that in contrast to German studies that had examined promotors within firms, network members to a large extent occupied multiple roles. This tendency even increases regarding network initiators. In contrast, separate promotor roles played a minor role and even had a negative influence on project performance when their presence increased in numbers. Our empirical results confirm the assumption that champions, but not promotors, play a fundamental part in innovation networks. As in other empirical studies (e.g. Bruce et al. 1995; Littler et al. 1995), the presence of one champion, here occupying all four promotor roles, had a positive influence on project performance. In addition, the numbers of champions was positively correlated with project performance. Champions were also more frequent in innovation communities. At the same time, we could show that group interaction and group identity factors are significantly higher in successful innovation networks. Testing the quantitative definition, the results indicate that the presence of an innovation community is a success factor especially in radical innovation networks. We could not confirm that innovation communities occurred more frequently in the early phases of the innovation process, needed partner organizations nearby, or relied (for the most part) on face-to-face communication.

2.5.1 *Limitations and Pathways for Future Research*

When looking at the role concept, it seems to be better suited to describe network initiators, because 34.7% of the network members, but only 14.0% of the network initiators did not occupy any role according to our definition. One reason for this rather high percentage of "no role owners" might be our strict role definition, only assigning a role if at least half of the items of a role were confirmed. Another reason might be that the questionnaire was too complicated tiring the respondents. Wenger et al. (2002) differentiate between a small core group participating intensively, active members and peripheral members. No role owners and their characteristics in innovation communities should be investigated in more detail. Other role concepts could be applied taking an interactional perspective into account, in with roles develop through an interaction process (e.g. Benne and Sheats 1948). In addition, it could further be investigated if and how roles, activity levels and even membership (fluctuation) can change over time in innovation processes (for communities of practice see Jahnke 2010). The stage model of development with respect to communities of practice (Wenger et al. 2002) could be transferred to the innovation community context. Another shortcoming is that the item scales of the role concept could not be tested confirmatorically (reliability, discriminant validity).

Regarding the two dimensions of group identity and group interaction, we are confronted with the usual methodological limitations of success factor analysis, for instance the fact that successful networks tend to overestimate the explanatory factors. A more detailed analysis has shown that the factors themselves are affected by multi-collinearity (for more detail see Rese et al. 2010a). For further investigations, it should be taken into consideration that there is a reciprocal relation between group identity and group interaction (e.g. cognitive similarity and communication, Rentsch et al. 2008). The measurement scales for group identity should be tested and further developed in other organizational group contexts. In addition, in order to confirm the subjective perception of the group members regarding group boundaries and transactions, an external perspective on the groups would be interesting.

Another research limitation is that one person was questioned for an entire network. The networks were rather small so that one person might in fact have complete insight; nevertheless, the networks are treated here as homogeneous and as favouring one single behaviour or strategy. Therefore, it would be interesting to include multiple perspectives from each network (e.g. Högl and Gemünden 2001). This was done in the case studies that are described in more detail in the following chapter.

2.5.2 Implications for Managerial Practice

The results of this study also provide some valuable insights for managers. They indicate that group interaction and group identity of key actors should not be neglected when developing new products in networks. In successful innovation projects, the facets for both concepts were (significantly) higher. The measurement scales could help to conduct an assessment of the status quo and show deficits regarding the network and its management. In addition, the results can help to identify the right partners. Partners should to some extent be generalists (champions). It is favourable if some of the innovation community members come from the same organization. The results also indicate that a "core" innovation community exists alongside peripheral actors that have yet to be described.

Acknowledgement The authors would like to thank Nicolai Sand for his support in developing the questionnaire and collecting the data.

Appendix

Item list (translated from German)

Promotor Roles

A key actor with a power promotor role. . .
- . . . supports the innovation project from a higher hierarchical level.
- . . . provides resources and financial support for the innovation project.
- . . . makes significant decisions within the innovation project.
- . . . motivates key actors to participate in inter-organizational exchange.
- . . . sanctions incorrect behaviour by key actors within the project.
- . . . ensures that opponents of the innovation project do not hamper its progress.

A key actor with an expert promotor role . . .
- . . . provides (technical) know-how.
- . . . pushes innovation-related technological developments forward.
- . . . addresses and solves innovation-related (technical) problems.
- . . . recognizes possibilities for further technological development of products and procedures.
- . . . assesses proposals for solutions and if necessary, develops alternatives.
- . . . questions opinions of the key actors, to ensure the best possible solution.
- . . . initiates new strategies and approaches to project tasks.

A key actor with a process promotor role . . .
- . . . is familiar with the partners, including knowledge of their modes of operation and corporate cultures.
- . . . assumes a central role in the innovation-related communication process.
- . . . distributes and coordinates the activities relevant to the innovation.
- . . . seeks, filters and saves relevant information and passes them on as appropriate.
- . . . translates and interprets information in the innovation-related communication process.
- . . . ensures that resources and information are used in a controlled and appropriate manner.
- . . . ensures that agreements between the key actors are kept.
- . . . coordinates meeting of key actors.
- . . . picks up on good ideas and makes sure that they match the strategy of the innovation project.

... ensures that deadlines and milestones are met.

A key actor with a relationship promotor role...

... has particularly good personal relationships with all important partners.
... has a particularly good inter-organizational and personal network.
... brings key actors together via his/her relationship network.
... can start cooperative efforts with partners, and support contract negotiations.
... moderates the cooperative effort of the key actors in difficult phases.
... works out compromises between key actors.
... promotes exchange of information between key actors to an especially high degree.
... ensures that all partners can contribute their ideas and opinions.
... ensures that personal conflicts do not negatively affect the project.
... initiates activities to improve the relationships between key actors.

Group Identity

Response scale: 1 = 'strongly disagree' to 7 = 'strongly agree'

Boundary clarity:

1. The group of key actors behaves like a team.
2. The key actors see themselves as "insiders".
3. The group of key actors is definitely perceived as an entity by outsiders.
4. All key actors currently involved fit into the group very well.
5. Every member of the group knows exactly who is member of the group and who is not.
6. The group of key actors is very stable, so there is little fluctuation.

External connectivity:

1. The group of key actors has a close exchange with other groups.
2. The group of key actors has many links to other partners or networks.
3. The group of key actors is open for new or additional partners or members.
4. Group members have access to contacts with relevant (technical) expertise concerning the radical innovation project.
5. Group members have very good access to decision-makers who are important for the radical innovation project.

Cohesion:

1. For the key actors, it is important to be part of the project.
2. The key actors are strongly tied to the project.
3. Every key actor is fully integrated into the group.

4. There are no personal conflicts within the group.
5. There is a personal affinity amongst the key actors.
6. The network is characterized by a strong personal cohesion amongst the key actors.
7. Each protagonist is proud to be a part of the group.

Common language and understanding:

1. The key actors agree upon the importance of the innovation.
2. The key actors use the same "language" (the same terminology, technical language etc.).
3. There are no conflicts due to conceptual misunderstandings.

Cooperative working atmosphere:

1. There is an atmosphere of confident amongst the key actors.
2. There is a cooperative working atmosphere in this group.
3. All members of the group can express their opinions openly.
4. The members of the group share ideas without being called upon to do so, they feel free to share their ideas.

Cognitive similarity:

1. The key actors are similar in terms of education and professional career.
2. The key actors are similar regarding occupational roles, functions or tasks.
3. The key actors have similar beliefs and opinions.
4. The key actors have a common "world view".

Group Interaction

Response scale: 1 = 'strongly disagree' to 7 = 'strongly agree'

Communication:

1. There is frequent communication between the key actors.
2. The key actors frequently communicate directly or personally.
3. The key actors frequently communicate in spontaneous meetings or telephone calls.
4. Project-relevant information is communicated and shared openly.
5. In given situations, important information is not withheld from others.
6. The key actors are satisfied with the information received from other members of the group.
7. Communication takes place even without the consent or knowledge of higher authorities.
8. Information circulates rapidly within the group of key actors.
9. It is easy to get answers to difficult questions within the group.

Coordination:

1. Clear goals that are understood by all are defined within the project.
2. These goals are generally accepted.

Balance of member contributions:

1. There is a balance in the contributions of the key actors.
2. The members of the group recognized the specific strengths and weaknesses of individual group members.
3. The key actors contribute to the achievement of the goals in accordance with their individual potentials.

Mutual support:

1. The key actors support each other as best as they can.
2. If conflicts arise, they are resolved easily and quickly.
3. Discussions and controversies are conducted constructively.
4. Suggestions and contributions from group members are dealt with and discussed respectfully.
5. The group is able to reach a consensus regarding important issues.

Effort:

1. Each of the key actors fully supports the innovation project.
2. The innovation project has the highest priority for each of the key actors.
3. The innovation project is important for the key actors as a group.

Project Performance

Response scale: 1 = 'strongly disagree' to 7 = 'strongly agree'

1. Until now, the radical innovation project has been very successful (effectiveness).
2. The project results so far were of high quality (effectiveness).
3. The key actors of the network are satisfied with the present result (effectiveness).
4. Until now, the product has required little rework (effectiveness).
5. For now, everything indicates a future success of the radical innovation (effectiveness).
6. Until now, all goals regarding the radical innovation project were achieved (efficiency).
7. From the network's perspective, one can be satisfied with how the project progressed (efficiency).
8. Until now, the project has been cost-efficient (efficiency).
9. The project is within schedule (efficiency).
10. The project is within budget (efficiency).

Project Innovativeness

Response scale: 1 = 'strongly disagree' to 7 = 'strongly agree'

1. The technology included in the innovation is/was very new to us and to our partners (technology).
2. The technology included in the innovation has replaced technologies used so far (technology).
3. The innovation is based on technological knowledge we have had very little experience with (technology).
4. The design of the innovation is based to a very low degree on technology already utilized (technology).
5. We have virtually no experience with the technical components used in the innovation (technology).
6. The innovation appeals to many customers whom we do not yet serve (market).
7. The innovation provides completely new customer benefits (market).
8. The innovation requires new cooperative efforts with market partners that we do not have yet (organization).
9. The innovation requires skills that the (network) management has not aquired so far (organization).

References

Agazarian YM (1989) Group-as-a-whole systems theory and practice. Behav Sci 13(3–4):131–154

Alderfer CP (1977) Group and intergroup relations. In: Hockman JR, Suttle JL (eds) Improving life at work. Goodyear, Santa Monica, pp 227–296

Alderfer CP, Smith KK (1982) Studying intergroup relations embedded in organizations. Adm Sci Q 27(1):35–65

Allen TJ (1964) The use of information channels in R & D proposal preparation (Working Paper No. 97-64). M.I.T. Sloan School of Management, Cambridge, MA

Allen TJ (1970) Communication networks in R & D laboratories. R&D Manage 1(1):14–21

Allen TJ, Cohen SI (1969) Information flow in research and development laboratories. Adm Sci Q 14(1):12–19

Allen J, James AD, Gamlen P (2007) Formal versus informal knowledge networks in R&D: A case study using social network analysis. R&D Manage 27(3):179–196

Anand BN, Khanna T (2000) Do firms learn to create value? The case of alliances. Strateg Manage J 21(3):295–315

Anderson NR, West MA (1998) Measuring climate for work group innovation: development and validation of the team climate inventory. J Organ Behav 19(3):235–258

Ashforth BE (1985) Climate formation: issues and extensions. Acad Manage Rev 10(4):837–847

Ashforth BE, Mael F (1989) Social identity theory and the organization. Acad Manage Rev 14(1):20–39

Ashforth BE, Mael FA (1996) Organizational identity and strategy as a context for the individual. Adv Strateg Manage 13:19–64

Bagozzi RP, Yi Y (1988) On the evaluation of structural equation models. J Acad Market Sci 16(1):79–94

Baier D, Queitsch M, Freund S (2006) Erfolgsfaktoren für das Innovationsmanagement in Netzwerken aus KMU und Forschungseinrichtungen: Eine empirische Untersuchung. In: Meyer J-A (Hrsg.) (ed) Kleine und mittlere Unternehmen in neuen Märkten – Aufbruch und Wachstum – Jahrbuch der KMU-Forschung und -Praxis 2006. Josef Eul, Lohmar, pp S197–S214

Bales RF (1950) Interaction process analysis: a method for the study of small groups. Addison-Wesley, Reading

Baum JAC, Calabrese T, Silverman BS (2000) Don't go it alone: alliance network composition and startups' performance in Canadian biotechnology. Strateg Manage J Special Issue: Strateg Networks 21(3):267–294

Benne KD, Sheats P (1948) Functional roles of group members. J Soc Issues 4(2):41–49

Blindenbach-Driessen F, van Dalen J, van den Ende J (2010) Subjective performance assessment of innovation projects. J Prod Innovat Manage 27(4):572–592

Bruce M, Leverick F, Littler D (1995) Complexities of collaborative product development. Technovation 15(9):535–552

Burgelman RA (1983) A process model of internal corporate venturing in the diversified major firm. Adm Sci Q 28(2):223–244

Calantone R, Garcia R, Dröge C (2003) The effects of environmental turbulence on new product development strategy planning. J Prod Innovat Manage 20(2):90–103

Cantner U, Graf H (2006) The network of innovators in Jena: an application of social network analysis. Res Policy 35(4):463–480

Casey-Campbell M, Martens ML (2009) Sticking it all together: a critical assessment of the group cohesion–performance literature. Int J Manage Rev 11(2):223–246

Chakrabarti AK, Hauschildt J (1989) The division of labour in innovation management. R&D Manage 19(2):161–171

Child J (1972) Organizational structure, environment and performance: the role of strategic choice. Sociology 6(1):1–22

Chiocchio F, Essiembre H (2009) Cohesion and performance: a meta-analytic review of disparities between project teams, production teams, and service teams. Small Group Res 40(4):382–420

Cohen SG, Bailey DE (1997) What makes teams work: group effectiveness research from the shop floor to the executive suite. J Manage 23(3):239–290

Conner DS (2003) Social comparison in virtual work environments: an examination of contemporary referent selection. J Occup Organ Psychol 76(1):133–147

Cook SDN, Brown JS (1999) Bridging epistemologies: the generative dance between organizational knowledge and organizational knowing. Organ Sci 10(4):381–400

Dawes J (1999) The relationship between subjective and objective company performance measures in market orientation research: further empirical evidence. Market Bull 10:65–75

Dhanaraj C, Parkhe A (2006) Orchestrating innovation networks. Acad Manage Rev 31(3):659–669

Eggers T, Kinkel S (2002) Die "virtuelle Fabrik" in weiter Ferne: Verbreitung und Nutzen von Produktionskooperationen und Produktionsnetzwerken im Verarbeitenden Gewerbe. In: Mitteilungen aus der Produktionswirtschaft, vol 25. Fraunhofer Institut für System- und Innovationsforschung, Karlsruhe

Ellemers N, van Knippenberg A, de Vries N, Wilke H (1988) Social identification and permeability of group boundaries. Eur J Soc Psychol 18(6):497–513

Ellemers N, Haslam SA, Platow MJ, van Knippenberg D (2003) Social identity at work: Developments, debates, directions. In: Haslam SA, van Knippenberg D, Platow MJ, Ellemers N (eds) Social identity at work: developing theory for organizational practice. Psychological Press, New York/Hove, pp 3–26

Ellemers N, de Gilder D, Haslam SA (2004) Motivating individuals and groups at work: a social identity perspective on leadership and group performance. Acad Manage Rev 29(3):459–478

European Commission (2006) The new SME definition. User guide and model declaration. Enterprise and Industry Publications, Luxembourg

Feldman DC (1984) The development and enforcement of group norms. Acad Manage Rev 9(1):47–53

Festinger L, Schachter S, Back KW (1950) Social pressures in informal groups: a study of human factors in housing. Harper, New York

Fichter K (2009) Innovation communities: the role of networks of promotors in open innovation. R&D Manage 39(4):357–371

Fichter K, Beucker S (2008) Innovation communities – Promotorennetzwerke als Erfolgsfaktor bei radikalen Innovationen (nova-net Werkstattreihe). Fraunhofer IRB, Stuttgart

Folkerts L (2001) Promotoren in Innovationsprozessen. Empirische Untersuchung zur personellen Dynamik. Deutscher Universitäts-Verlag, Wiesbaden

Folkerts L, Hauschildt J (2002) Personelle dynamik in innovationsprozessen. Neue fragen und befunde zum promotorenmodell. Die Betriebswirtschaft 62(1):7–23

Fornell C, Larcker DF (1981) Evaluating structural equation models with unobservable variables and measurement error. J Market Res 18(3):39–50

Friedlander F (1987) The ecology of work groups. In: Lorsch W (ed) Handbook of organizational behavior. Prentice-Hall, Jay Englewood Cliffs, pp 301–314

Frietsch R (2007) Patente in Europa und der Triade: Strukturen und deren Veränderung. Studien zum deutschen Innovationssystem (Nr. 9-2007). Fraunhofer Institut für System- und Innovationsforschung, Karlsruhe

Garcia R, Calantone RJ (2002) A critical look at technological innovation typology and innovativeness terminology: a literature review. J Prod Innovat Manage 19(2):110–132

Gefen D, Ridings CM (2003) IT acceptance: managing user – IT group boundaries. DATA BASE Adv Inform Syst 34(3):25–40

Gemünden H-G, Salomo S, Hölzle K (2007) Role models for radical innovations in times of open innovation. Creat Innovat Manage 16(4):408–421

Gerbing DW, Hamilton JG (1996) Viability of exploratory factor analysis as a precursor to confirmatory factor analysis. Struct Equ Model 3(1):62–72

Gerybadze A (2003) Gruppendynamik und Verstehen in Innovation Communities. In: Herstatt C, Verworn B (Hrsg.), (eds) Management der frühen innovationsphasen. Gabler, Wiesbaden, pp S145–S160

Gladstein DL (1984) Groups in context: a model of task group effectiveness. Adm Sci Q 29 (4):499–517

Hackman JR (1987) The design of work teams. In: Lorsch JW (ed) Handbook of organizational behavior. Prentice-Hall, Englewood Cliffs, pp 67–102

Hauschildt J (1999) Promotors and champions in innovations – development of a research paradigm. In: Brockhoff K, Chakrabarti AK, Hauschildt J (eds) The dynamics of innovation: strategic and managerial implications. Springer, Berlin, pp 167–182

Hauschildt J, Kirchmann E (2001) Teamwork for innovation – the 'troika' of promotors. R&D Manage 31(1):41–49

Hauschildt J, Schlaak TM (2001) Zur Messung des Innovationsgrades neuartiger Produkte. Zeitschrift für Betriebswirtschaft 71(2):161–182

Hirst G, Mann L (2004) A model of R&D leadership and team communication: the relationship with project performance. R&D Manage 34(2):147–160

Högl M (2005) Smaller teams-better teamwork: how to keep project teams small. Bus Horizons 48(3):209–214

Högl M, Gemünden H-G (2001) Teamwork quality and the success of innovative projects: a theoretical concept and empirical evidence. Organ Sci 12(4):435–449

Högl M, Weinkauf K, Gemünden H-G (2004) Interteam coordination, project commitment, and teamwork in multiteam R&D projects: a longitudinal study. Organ Sci 15(1):48–55

Högl M, Ernst H, Proserpio L (2007) How teamwork matters more as team member dispersion increases. J Prod Innovat Manage 24(2):156–165

Howard LW, Foster ST, Shannon P (2005) Leadership, perceived team climate and process improvement in municipal government. Int J Qual Reliab Manage 22(8):769–795

Howell JM, Higgins CA (1990) Champions of technological innovation. Admin Sci Q 35(2):317–341
Howell JM, Shea CM, Higgins CA (2005) Champions of product innovations: defining, developing, and validating a measure of champion behavior. J Bus Venturing 20(5):641–661
Human SE, Provan KG (1997) An emerging theory of structure and outcomes in small-firm strategic manufacturing networks. Acad Manage J 40(2):368–403
Jahnke I (2010) Dynamics of social roles in a knowledge management community. Comput Hum Behav 26(4):533–546
Katz D, Kahn RL (1966) The social psychology of organizations. Wiley, New York
Kelley D, Lee H (2010) Managing innovation champions. the impact of project characteristics on the direct manager role. J Prod Innovat Manage 27(6):1007–1019
Kim J, Wilemon D (2003) Sources and assessment of complexity in NPD projects. R&D Manage 33(1):16–30
Kirchmann EMW (1994) Innovationskooperation zwischen Herstellern und Anwendern. Deutscher Universitäts-Verlag, Wiesbaden
LaBahn DW, Ali A, Krapfel R (1996) New product development cycle time: the influence of project and process factors in small manufacturing companies. J Bus Res 36(2):179–188
Laursen K, Salter A (2006) Open for innovation: the role of openness in explaining innovation performance among U.K. manufacturing firms. Strateg Manage J 27(2):131–150
Lave J, Wenger EC (1991) Situated learning: legitimate peripheral participation. Cambridge University Press, Cambridge, MA
Leblebici H, Salancik G (1982) Stability in interorganizational exchanges: rulemaking processes of the Chicago board of trade. Admin Sci Q 27(2):227–242
Ledwith A, Coughlan P (2005) Splendid isolation: Does networking really increase new product success? Creat Innovat Manage 14(4):366–373
Lichtenthaler U, Ernst H (2009) The role of champions in the external commercialization of knowledge. J Prod Innovat Manage 26(4):371–387
Linton R (1936) The study of man. Appleton-Century, New York
Littler D, Leverick F, Bruce M (1995) Factors affecting the process of collaborative product development: a study of UK manufacturers of information and communications technology products. J Prod Innov Manage 12(1):16–32
Markham SK, Ward SJ, Aiman-Smith L, Kingon AI (2010) The valley of death as context for role theory in product innovation. J Prod Innovat Manage 27(3):402–417
McGarty C, Haslam SA, Hutchinson KJ, Grace DM (1995) Determinants of perceived consistency: the relationship between group entitativity and the meaningfulness of categories. Br J Soc Psychol 34(3):237–256
McGrath JE (1984) Group interaction and performance. Prentice-Hall, Englewood Cliffs
Mullen B, Copper C (1994) The relation between group cohesiveness and performance: an integration. Psychol Bull 115(2):210–227
Murphy KR, Davidshofer CO (2001) Psychological testing: principles and application, 5th edn. Prentice-Hall, Upper Saddle River
Nakamura M, Shaver JM, Yeung B (1996) An empirical investigation of joint venture dynamics: evidence from U.S.–Japan joint ventures. Int J Ind Organ 14(4):521–541
Nunnally JC (1978) Psychometric theory, 2nd edn. McGraw-Hill, New York
Papies S (2006) Phasenspezifische Erfolgsfaktoren von Innovationsprojekten: Eine projektbegleitende Längsschnittanalyse. Deutscher Universitäts-Verlag, Wiesbaden
Pinto MB, Pinto JK (1990) Project team communication and cross-functional cooperation in new program development. J Prod Innovat Manage 7(3):200–212
Podsakoff PM, MacKenzie SB, Paine JB, Bachrach DG (2000) Organizational citizenship behaviors: a critical review of the theoretical and empirical literature and suggestions for future research. J Manage 26(3):513–563
Postmes T, Spears R (2000) The formation of group norms in computer-mediated communication. Hum Commun Res 26(3):341–371

Powell WW, Koput KW, Smith-Doerr L (1996) Interorganizational collaboration and the locus of innovation: networks of learning in biotechnology. Admin Sci Q 41(1):116–145

Qurashi MM (1993) Dependence of publication-rate on size of some university groups and departments in UK and Greece in comparison with NCI, USA. Scientometrics 27(1):19–38

Reichers AE, Schneider B (1990) Climate and culture: an evolution of constructs. In: Schneider B (ed) Organizational climate and culture. Jossey-Bass, San Francisco, pp 5–39

Rentsch JR, Delise LA, Hutchison S (2008) Transferring meaning and developing cognitive similarity in decision making teams: collaboration and meaning analysis process. In: Letsky MP, Warner NW, Fiore SM, Smith CAP (eds) Macrocognition in teams. Ashgate Publishing, Burlington, pp 127–142

Rese A, Baier D (2011) Success factors for innovation management in networks of small and medium enterprises. R&D Manage 41(2):138–155

Rese A, Sand N, Baier D (2009) Relevance and contribution of promotor networks in cross-company innovation projects: an empirical investigation. 16th International product development management conference, Twente, 7–9 June 2009, p 20

Rese A, Sand N, Baier D (2010a) Group identity and group interaction in innovation networks. 17th International product development management conference, Murcia, 14–15 June 2010, p 20

Rese A, Sand N, Baier D (2010b) Innovation communities: investigating innovation networks at the interpersonal level. 13th QMOD conference, Cottbus, 30 Aug–1 Sept 2010, p 14

Ritter T (1999) The networking company: antecedents for coping with relationships and networks effectively. Ind Market Manage 28(5):467–479

Ritter T, Gemünden H-G (2003) Interorganizational relationships and networks: an overview. J Bus Res 56(9):691–697

Roberts EB (2007) Managing invention and innovation. Res Technol Manage 50(1):35–54

Roberts EB, Fusfeld AR (1981) Staffing the innovation technology-based organization. Sloan Manage Rev 22(3):19–33

Rosenfeld SA (1996) Does cooperation enhance competitiveness? assessing the impacts of inter-firm collaboration. Res Policy 25(2):247–263

Rost K, Hölzle K, Gemünden H-G (2006) Promotoren oder champions? Vor-und Nachteile der Arbeitsteilung in Innovationsprozessen. Arbeitspapier iou – Universität Zürich, Zürich

Rost K, Hölzle K, Gemünden H-G (2007) Pros and cons of role specialisation for economic process. Schmalenbach Bus Rev 59(4):340–363

Rothwell R, Freeman C, Horsley A, Jervis VTP, Robertson AB, Townsend J (1974) SAPPHO updated – project SAPPHO phase II. Res Policy 3(3):258–291

Sand N, Rese A (2011) Comparing promotor networks in radical and incremental innovation projects: roles and interaction as success factors. In: Baier D, Gaul W, Rese A, Würth R (eds) The entrepreneurship – innovation – marketing – interface, proceedings of the 3rd symposium, Cottbus. Swiridoff, Künzelsau, pp 233–250

Schon DA (1963) Champions for radical new inventions. Harv Bus Rev 41(2):77–86

Schreiner M, Kale P, Corsten D (2009) What really is alliance management capability and how does it impact alliance outcomes and success? Strateg Manage J 30(13):1395–1419

Stewart GL (2006) A meta-analytic review of relationships between team design features and team performance. J Manage 32(1):29–54

Stewart GL, Barrick MR (2000) Team structure and performance: assessing the mediating role of intrateam process and the moderating role of task type. Acad Manage J 43(2):135–148

Swaab R, Postmes T, van Beest I, Spears R (2007) Shared cognition as a product of, and precursor to, shared identity in negotiations. Pers Soc Psychol Bull 33(2):187–199

Tajfel H (1972) Social categorization (English translation of "La categorisation sociale"). In: Moscovici S (ed) Introduction à la psychologie sociale, vol 1. Larousse, Paris, pp 272–302

Tajfel H (ed) (1978) Differentiation between social groups: studies in the social psychology of intergroup relations. Academic, London

Tajfel H (1979) Individuals and groups in social psychology. Br J Soc Clin Psychol 18(2):183–190

Tajfel H, Turner JC (1979) An integrative theory of intergroup conflict. In: Austin WG, Worchel S (eds) The social psychology of intergroup relations. Brooks-Cole, Monterey, pp 33–47

Tajfel H, Billig MG, Bundy RP, Flament C (1971) Social categorization and intergroup behavior. Eur J Soc Psychol 1(2):148–178

Teusler N (2008) Strategische Stabilitätsfaktoren in Unternehmenskooperationen: Eine kausalanalytische Betrachtung. Gabler, Wiesbaden

Thoms U (2003) Langfristige Beziehungen zwischen Unternehmen. Zum Wert und zur Stabilität inter-organisationaler Partnerschaften. Deutscher Universitäts-Verlag, Wiesbaden

Thorgren S, Wincent J, Örtqvist D (2009) Designing interorganizational networks for innovation: an empirical examination of network configuration, formation and governance. J Eng Technol Manage 26(3):148–166

Turner JC (1978) Social categorization and social discrimination in the minimal group paradigm. In: Tajfel H (ed) Differentiation between social groups: studies in the social psychology of intergroup relations. Academic, London, pp 101–140

Turner JC (1985) Social categorization and the self-concept: a social cognitive theory of group behaviour. In: Lawler EJ (ed) Advances in group processes: theory and research, vol 2. JAI Press, Greenwich, pp 77–122

Turner RH (2002) Role theory. In: Turner JH (ed) Handbook of sociological theory. Kluwer Academic/Plenum, New York, pp 233–254

Un CA, Cuervo-Cazurra A, Asakawa K (2010) R&D collaborations and product innovation. J Prod Innovat Manage 27(5):673–689

van Dijk B, Hertog RD, Menkveld B, Thurik R (1997) Some new evidence on the determinants of large and small-firm innovation. Small Bus Econ 9(4):335–343

Wageman R, Hackman JR, Lehman E (2005) Team diagnostic survey. Development of an instrument. J Appl Behav Sci 41(4):373–398

Walter A (1998) Der Beziehungspromotor: ein personaler Gestaltungsansatz für erfolgreiches Relationship Marketing. Gabler, Wiesbaden

Walter A, Gemünden H-G (2000) Bridging the gap between suppliers and customers through relationship promoters: theoretical considerations and empirical results. J Bus Ind Market 15 (2/3):86–105

Wenger E, McDermott RA, Snyder W (2002) Cultivating communities of practice: a guide to managing knowledge. Harvard Business School Press, Boston

Whelan E, Teigland R, Donnellan B, Golden W (2010) How internet technologies impact information flows in R&D: reconsidering the technological gatekeeper. R&D Manage 40(4):400–413

Witte E (1973) Organisation für Innovationsentscheidungen: Das Promotoren-Modell. Otto Schwartz, Göttingen

Witte E (1977) Power and innovation: a two-center theory. Int Stud Manage Organ 7(1):47–70

Woodside AG (1994) Network anatomy of industrial marketing and purchasing of new manufacturing technologies. J Bus Ind Market 9(3):52–63

Workman M (2005) Virtual team culture and the amplification of team boundary permeability on performance. Hum Resour Dev Q 16(4):435–458

Zenger TR, Lawrence BS (1989) Organizational demography: the differential effects of age and tenure distributions on technical communication. Acad Manage J 32(2):353–376

Chapter 3
The Evolution of Innovation Communities: Case Studies

Abstract While the previous chapter clearly indicated that innovation communities have a significant positive effect on project performance and can therefore be considered a success factor of radical innovation, questions remain concerning the evolution (How does the size and structure of innovation communities change over time?), the performance (How do innovation communities exactly contribute to the success of innovation projects?), and the quality of innovation communities (Which features make them successful?). Against this background, this chapter presents findings from five case studies and answers the questions mentioned above.

First the contribution of innovation communities to innovation success will be conceptualized (Sect. 3.1). Then in a second step the construct of community performance is elaborated, consisting of the three sub-constructs of community composition, community identity and community interaction (Sect. 3.2). The research design and methodology for the case studies is described in Sect. 3.3.

The following Sects. 3.5–3.9 present the five cases studies that analyze the role of innovation communities in radical innovation processes. The cases were selected to cover various sizes and constellations of communities, and permit insights into different sectors and fields of technology, covering basic new technologies (like carbon nanotubes and dye-sensitized solar cells), new comprehensive usage systems (like the entire value chain of wood crops that are used for power generation), as well as specific products and applications of new technologies (like tilt sensors and smart home networks).

Finally, Sect. 3.10 draws conclusions from the specific cases and compares the role and the contributions that the innovation communities have on the evolution, the performance and the quality of innovative projects.

K. Fichter and S. Beucker (eds.), *Innovation Communities*,
DOI 10.1007/978-3-642-22128-6_3, © Springer-Verlag Berlin Heidelberg 2012

As presented in Chap. 2, radical innovation projects that have an innovation community are significantly more successful than radical innovation projects that do not. The results thus confirm the central assumption of the InnoCo project that innovation communities are a key factor for the success of radical innovation projects. While there is clear empirical evidence that innovation communities have a positive effect on project success, some questions remain concerning the exact course of action, the constitution and the evolution of innovation communities. To date, however, little is known about whether the contribution of innovation communities to project success varies in successive phases of the innovation process, for instance, and to which extent other contextual factors influence the relevance of the communities for project progress. Furthermore, knowledge of the nature and scope of methods and measures for community development and collaboration is so far limited.

Against this background, further in-depth case studies were conducted in the context of the InnoCo research project that can help to answer the following questions:

- Evolution: How does the size and structure of innovation communities change over time?
- Performance: How do innovation communities contribute to the success of innovation projects?
- Quality: Which features make innovation communities successful?
- Methods: Which measures and methods constitute and improve the performance of innovation communities?

Before answering these questions in the context of the aforementioned case studies, it is first of all necessary to create a conceptual framework and clarify key concepts like "radical innovation" and "project success" have to be defined, and the existing research on the ways in which innovation communities contribute to the success of innovation projects will be outlined.

3.1 Conceptualizing the Contribution of Innovation Communities to Innovation Success

Klaus Fichter

3.1.1 *Conceptualizing the Degree of Innovation*

The underlying definition of "radicality" draws on previous work by Salomo (2003), Salomo et al. (2007), and Gemünden et al. (2007), and describes radical innovation as determined by a high degree of change in various dimensions. Radical innovation can comprise completely new technological principles, materials, or architectures that enable a significant leap in performativity, meet previously non-existent or emerging needs, create new markets, or redefine and change entire industrial value chains. As the available literature suggests, it is useful to take market, technological and organizational aspects and changes in the environment into account, hence examining innovation as a multidimensional phenomenon (Gemünden et al. 2007, 410). Based on the prior work mentioned, a multi-dimensional construct for the degree of innovation was developed in the InnoCo project, developing questions and items along the way (Sand and Rese 2011). This construct is used as a foundation in the case studies presented in this work.

3.1.2 *Conceptualizing Project Success*

Within innovation research, there has been ongoing and passionate debate on how to conceptualize the success of innovation projects since the 1980s (Högl and Gemünden 2001, 438). Similarly to questions of conceptualization of innovation levels, it is assumed that a multidimensional approach reflects best the project's success. Based on extensive literature research, Högl and Gemünden (2001, 438) differentiate between "team performance" (task-related) and "personal success of team members" (person-related) in their concept. Since individual satisfaction or learning processes of team and community members are not the main focus of the work presented, the development of a theoretical construct here concentrates on "team performance".

Here, the issue of task performance of a team is based on the degree to which defined project goals are or have been achieved. It is useful to distinguish between the effectiveness and efficiency of team performance. Effectiveness relates to the extent in which project goals are reached, an immediate comparison between intended and achieved results. This for instance concerns meeting technical targets, specifications regarding product quality, or the client or user satisfaction desired. The efficiency of team performance, however, refers to a comparison of the intended inputs (money, time, resources, etc.) to factual inputs.

The items developed by Högl and Gemünden (2001, 447) for effectiveness and efficiency of team performance can essentially be adopted for this study. Only the items regarding user and customer impact need to be adjusted, as four of five cases examined are not in the market phase, so user satisfaction and image-related issues therefore cannot be applied. At this point, it is only possible to examine whether customer and user requirements have systematically been taken into account.

It is known from innovation and diffusion research that innovation projects' success can be judged very differently, within the innovating companies themselves (depending on the role, interests and involvement) as well as from outside the organization (clients, competitors, citizens, interest groups) (Fichter 2009a). It is therefore important to ascertain that the underlying concept of "project success" here refers to the views and evaluations from within the innovating organizations, including the teams and promotors involved. The views of clients or other possible stakeholders are not examined here independently as such.

3.1.3 Contributions to Project Success

The lead question of this book ("How and to what extent do innovation communities contribute to the success of radical innovation projects, and which community traits does their contribution depend on?") addresses the specific functions of promotor networks with a clear perspective on the success of a given project. As first case studies on innovation communities have shown, the existence, relevance and configuration of innovation communities depend on a wide range of circumstances (Beucker 2007; Fichter et al. 2007; Springer 2007).

Five main conditions can be articulated here (Fichter and Beucker 2008):

1. *Distribution and availability of relevant resources*: The greater the distribution of innovation-critical resources is and the less an individual player can control these resources, the greater the demand for cooperation and therefore, the probability for the emergence of an innovation community will be.
2. *Lack of formal structures*: The informal relations of innovation communities will gain importance especially if formal organizational innovation or project structures are lacking or weak. The more formal structures are missing, the higher the importance of innovation communities will be. This may be particularly true for early phases of innovation where there are no project structures or contracts yet.
3. *Motivation*: Informal relations of trust and understanding encourage community members in pursuing their innovative ideas, allow for mutual support and collaboration, especially with regard to overcoming innovation barriers. The stronger the motivational needs of individuals are, the more meaningful an innovation community will be.
4. *Conflict levels*: Promotors will meet a multitude of barriers. The importance of innovation communities increases depending on the extent in which more

formal, hierarchical control mechanisms are unable to resolve conflict. This also applies to conflicts within innovation teams and innovation communities. Greater levels of trust will facilitate conflict resolution within innovation processes.

5. *Situational configuration*: Innovation community membership can adapt according to the current stage of innovation or resource demands. They might be formed from existing, stable networks, or be created entirely from scratch.

With this foundation, a number of roles that innovation communities play in innovation processes and in successfully developing and implementing innovative projects can be identified (Fichter et al. 2007). Accordingly, innovation communities are especially important:

- In early stages of innovation, when there is no formal cooperation, network or structure yet established.
- In emerging fields of technology that make radically new applications possible or make the development of new value chains and markets necessary.
- In acquiring cooperation partners and power promotors that retain important resources for a given project.
- For "stalemates" or setbacks in innovation, to motivate promotors to follow their innovative ideas or projects.
- In situations in which the success of the innovation process mainly depends on the effectiveness and efficiency of promotor communication.
- In conflict-ridden settings, whenever allies are needed, tension needs to be reduced within formal contracts or collaborative teams, or when group climate needs to be improved.

Thereby we can assume that innovation communities, through their collaboration, contribute to a project's success by:

- Helping to promote an innovative idea from the informal stage into a formal organization.
- Motivating promotors and strengthening their role.
- Obtaining resources critical to performance, such as money, knowledge, reputation.
- Helping solve or mitigate project-threatening conflicts.

3.2 Performance: Teamwork Quality of Innovation Communities

Klaus Fichter and Severin Beucker

To further investigate the question of what innovation communities can actually accomplish for the success of a collaborative innovation project and what community traits this contribution depends on, the concept of an innovation community's "quality" needs to be introduced at this point. "Quality" here refers to those characteristics that determine the contribution of a promotor network to a given project's success. The more pronounced these features are, the higher the quality of the community and the greater its contribution to project success. The terms "quality" and "performance" of an innovation community are therefore used synonymously here.

We will assume here that the success of a collaborative innovation project of course does not only depend on the existence and performance of a specific innovation community, but rather on a number of factors. These include "classic" success factors from product development, such as product superiority, quality of implementation, or market potential (Sand and Rese 2011), success factors for innovation cooperation, such as the complementarity of know-how and resources (Gerybadze 2004, 195), as well as the "strategic fit" (Fichter 2009a). Even so, based on the results of a survey of innovation networks undertaken in the InnoCo project (see Chap. 2), it can be assumed that the existence of an innovation community can in fact significantly and positively influence the success of collaborative innovation projects, which therefore can be seen as one of the relevant factors. In the following, the focus will be exclusively on this factor of influence on cooperative innovation projects.

As Chap. 2 has shown, innovation communities can be described as informal networks of promotors with an average size of three to seven individual promotors. In this sense, innovation communities generally have an informal, inter-organizational character. In order to analyze which features determine the quality of innovation communities, there is a considerable amount of preparatory work and findings from team research to be drawn from.

Widely noted in team and innovation research, a specific construct of team quality or performance was proposed by Högl and Gemünden (2001). Building on previous work in team and network research, they developed the idea of "teamwork quality." This focuses on team interaction and cooperation comprises six points that concern both task-related and social interaction (Högl and Gemünden 2001, 436):

- Communication: Is there sufficient regular, informal, direct and open communication in the team?
- Coordination: Are individual contributions well structured and synchronized within the team?
- Balance of contributions: Can all team members fully weigh in their expertise?
- Mutual support: Do the team members help and support each other with their duties?

- Effort: Do team members dedicate all their efforts to the team's challenge?
- Cohesion: Are the members motivated to remain on the team? Is there a "team spirit"?

Högl and Gemünden tested this construct through an empirical survey of 145 software development teams. Their results show that the six aspects examined can in fact capture the quality of cooperation, and that there is a connection between teamwork quality and overall project success. They conclude that the concept of teamwork quality is useful as a "higher-order latent construct" to investigate and assess the quality of team cooperation, and can be measured by means of the aspects noted above as well as by indicators derived from those items (Högl and Gemünden 2001, 445).[1] The construct has since been used in numerous other teamwork studies (Easley et al. 2003; Högl and Proserpio 2004; Högl et al. 2007; Dayan and Benedetto 2009). Against this background, this study uses the concept of "teamwork quality" as its foundation.

The "teamwork quality" concept can be used almost identically to study interaction in innovation communities. A comprehensive concept for innovation communities, however, should not be limited to interaction-related issues, but rather be extended to other performance-related aspects as well. This is because innovation communities are made up of informal and generally inter-organizational teams, whereas team research so far has largely been based on the analysis of formal teams within organizations. While the members' roles in formal project teams within innovating firms are supposed to be clearly defined and the team identity is clarified through its formal objective, this may not be clear at all in the case of promotor networks that develop informally and usually also maintain their informal character. Moreover, due to the informal nature of the community, there usually is no formal obligation to contribute or even to remain a member at all (Gerybadze 2003). Besides the aspects of team interaction described in the "teamwork quality" construct aspects of community identity must also be considered in formulating a concept for innovation communities. This concerns questions like: Do the members share a common idea of who belongs to the community, and who is perceived to be an insider? Do the members agree on the goals and importance of the project? Are the members motivated to stay part of the team? Does a certain team spirit exist?

As previous case studies (Beucker 2007; Fichter 2006; Springer 2007) have shown, the possibility that innovation communities might develop in an intrinsically dynamic "self-assembling" process has to be taken into account, therefore it cannot be generally assumed that all the necessary technical skills or relationships to decision-makers needed to lead a collaborative innovation idea to success are available. In contrast, within formal project teams it can be taken for granted that the necessary skills are available and the roles are more or less clear (e.g., who is the

[1] The three items used by Högl and Gemünden (2001, 447) for the aspect "Balance of contributions" are only partly suited to reveal the main issue (Can the members bring forth their own expertise?). Therefore, for the study presented here, these items have been modified.

group leader, etc.), due to their organized and purposeful structure. With regard to the quality and capacity of an innovation community, therefore, issues of team composition and resourcing in innovation communities need to be incorporated in addition to interaction- and identity-related aspects.

Against this background, a concept of the quality of innovation communities that includes aspects of community composition, community identity and community interaction is to be developed in the following. This concept is based on the assumption that innovation communities' impact on the success of collaborative innovation projects depends on these three elements. As such a construct is primarily theoretical or hypothetical and can therefore not be immediately recognized empirically, it needs to be made accessible through measurable indicators (Bortz and Döring 2006). Any latent, unobservable variables have to be operationalized and transformed into observable and measurable ones (Backhaus et al. 2008, 519). These latent variables (aspects) of the construct "quality of an innovation community" are presented in the following.

The construct of "quality" proposed here is made up of three sub-constructs: (1) the element of community composition, (2) the element of community identity, and (3) the element of community interaction. The latter is similar to the "teamwork quality" construct as developed by Högl and Gemünden (2001) and contains six latent variables (aspects). The construct of Högl and Gemünden is, however, adapted in two separate ways. Firstly, the aspect of "cohesion" in the work presented here is moved in the sub-construct of community identity. Secondly, the aspect of trust and group climate has been added to the sub-construct of community interaction. These changes have been made due to the following considerations:

- In their explanation of the aspect "cohesion," Högl and Gemünden (2001, 438) point out that they refer to the extent in which a team member wishes to be or to remain a part of the team, and use items in their operationalization that primarily pertain to emotional aspects. The aspect "cohesion," in the broader context of the "quality of innovation communities" construct, is therefore actually better placed as an element of the sub-construct community identity, because it involves the identifiability of a community, meaning that it relates to perceptual and emotional aspects of the involved members and promotors in particular.
- The addition of the new aspect "mutual trust and group climate" can be explained by the relevance that trust has especially in informal and inter-organizational teams and group processes (Fichter 2008). Högl and Gemünden (2001) have referred to formal project teams within organizations for their construction of concepts. In that case, people are interacting within one and the same organization and due to a quasi-formalized mandate, practically ordered to exchange information and knowledge. Internal interaction within the team thus takes place in a "protected space," and in a clearly organized formal framework. This is not necessarily the case in the more informal interactions of innovation communities, so that the power of communication and the efficiency and exchange of relevant information to a much greater extent

Table 3.1 The concept of quality of innovation communities

Composition of the community
· *Completeness and complementarity*: Does the community have the necessary promotor roles available, and are all cooperation partners critical to a project's success represented?
· *Social capital*: Do members have access to decision-makers (funding), experts (knowledge) and agents of reputation?
· *Reconfiguration*: Can the community integrate new members, and let those go that are no longer needed if required?
Identity of the community
· *Clear community boundaries*: Do members have a common perception of who belongs to the community, and are they seen as a community from outside?
· *Functional identity*: Do members agree on goals and relevance of the project, and do they share the same "world view" in this regard?
· *Cohesion*: Are members motivated to stay a part of the team? Is there a team spirit?
Interaction of the community
· *Communication*: Does the team adequately employ regular, informal, direct and open communication?
· *Coordination*: Are individual contributions of the team well structured and synchronized?
· *Balance of contributions*: Can all team members fully place their individual expertise?
· *Mutual support*: Do team members support each other in completing their tasks?
· *Effort*: Do members sufficiently devote effort to the shared project?
· *Trust and group climate*: Do members trust one another, and is collaboration characterized by a positive group climate?

Source: Authors

depends on the degree of trust between promotors and on the general team climate in the community.

- So the sub-concept "community interaction" is composed of six quality variables, while the sub-constructs "community identity" and "community composition" include three variables each. The quality variables are derived from the basic definition of innovation communities as described in Chap. 1, from reviews of literature on team and network research, from the empirical findings on innovation communities to date (Beucker 2007; Fichter 2006; Springer 2007; Fichter 2009b; Rese et al. 2009), and from the results presented in Chap. 2. The comprehensive construct "quality of innovation communities" thus contains 12 aspects in total, presented in an overview in the chart above (Table 3.1).

These variables require further specification – the three individual sub-constructs will be developed in the following.

3.2.1 Features of the Quality of Community Composition

The sub-construct "composition" follows a "resource-based view" of network research (Duschek 1998; Sydow et al. 2003) and refers to the question of which skills, roles or resources a community requires for it to be able to successfully act within a joint innovation project framework. The construct hence includes characteristics that determine the situational and task-related resource demands of an innovation community. Based on promotor research findings (Hauschildt 1999; Hausschild and Gemünden 1999; Rost et al. 2007), recent empirical findings on innovation communities (Beucker 2007; Fichter 2006; Springer 2007; Fichter 2009b; Rese et al. 2009), on findings regarding the importance of "social capital" in networks (Aulinger 2005; Fichter 2008) and findings of team research (Högl and Gemünden 2001; Easley et al. 2003; Högl and Proserpio 2004; Högl et al. 2007; Dayan and Benedetto 2009), three quality variables can be outlined further with regard to the composition and resource demands of the innovation community; these are presented below.

Completeness and Complementarity of Skills

As a survey of innovation networks in the InnoCo project has shown, successful innovation projects have more "champions" within their community than less successful projects, i.e. more people holding two or more promotor roles simultaneously (see Chap. 2). Apparently, the presence and number of champions positively influences innovation communities' success. It has further been established that the success rate is increased if all promotor roles (power, expert, process, relationship promotor) are represented in the community. In contrast, it does not seem to benefit performance if the same promotor role (e.g., expert promotor) is represented more than once. This suggests that promotor roles in the innovation community should be as comprehensive and complementary as possible, and that it enhances success rates if community members are able to fulfill multiple roles at the same time (champions).

This aspect of completeness and complementarity of skills is also supported through research on innovation cooperation, according to which it is one of the key factors for success (Gerybadze 2004, 195). Even if the idea of complementarity therein is based on organizational cooperation, the same can in principle be said to apply to inter-organizational human networks. Another aspect can also be added to the basic idea of completeness and complementarity. This concerns the question whether all the relevant companies or organizations in the network need to be represented by promotors. This form of "organizational scattering" therefore appears to be critical to success precisely because it constitutes the possibility to approach decision makers and "resource controllers" in these organizations.

That this kind of access to decision-makers can be critical to success in certain phases of innovation is shown in the case studies of Beucker (2007, 27), Springer (2007, 24) and Fichter (2009a, 368). It can be assumed, therefore, that the success rate of an innovation community will increase if promotors from all organizations relevant for the development and implementation of the innovation are represented in the community. This last point also relates to another key aspect of innovation communities' potential, namely to the issue of what social capital the members of the promotor networks themselves can make available.

Social Capital

For analyzing and explaining the success of entrepreneurs and business networks, the literature on entrepreneurship and network research has for some time been reverting to the concept of "social capital" (Aulinger 2005). This can equally be applied productively to promotor networks. While the aspects of completeness and complementarity of skills presented above relate to the ability of community members to interact appropriately to situation and overall goals, and to overcome obstacles using their specific power potential, the concept of "social capital" focuses on the assets of contacts, trust, reputation and social obligations of the people involved, i.e. on resources inherent to social relationships. The concept of "social capital" has a long tradition in social and political sciences (Putnam 2000). Building on the theoretical approaches and conceptualizations presently available (Bourdieu 1983; Coleman 1990; Nahapiet and Goshal 1998; Aulinger 2005), "social capital" can be defined as follows:

> Social capital is the total sum of resources (contacts, trust, reputation, obligations) that an individual or social unit can make available, due to social relations, and can generate income from. (Fichter 2008, 107)

Regarding innovation communities' impact on success rates, the social capital of community members plays a decisive role especially in the community's external relations to decision makers and "resource controllers" in collaborating companies and other performance-critical organizations (Van de Ven et al. 1999, 59). The level of social capital of individual community members is at the same time an expression of their central position in intra- and inter-organizational network relationships. The concept of social capital can therefore also be connected to concepts of "centrality," as is done in the structural analysis of social networks (Trappmann et al. 2005, 25). The "centrality value" of promotors therefore results from the quality and versatility of their social relations to project-relevant individuals. A key promotor allows for speedier action in innovation projects, because, for instance, information or necessary resources (e.g., budget or project approvals) can be collected or attained much faster.

Reconfiguration Capacity

Previous case studies (Beucker 2007; Fichter 2006; Springer 2007) have shown that innovation communities can change their composition over time. The results suggest that there are usually "key promotors" functioning as initiators of the innovation project and innovation community that remain members of the community during various innovation phases. But there are usually other members of the promotor network too, who change according to the phase of innovation. This seems plausible in the sense that in radical innovation the time lapse between a first innovative idea and its actual implementation is generally a longer period of 5 years or more, so that the project passes through phases with different challenges and obstacles along the way. So a certain degree of fluctuation is more or less natural in this case (Van de Ven et al. 1999, 40), and to a certain extent also appears to be functionally necessary, because different promotor skills are required depending on the tasks and obstacles at hand.

Against this background it can be assumed that the contribution of an innovation community also depends on its ability to reconfigure, in order to meet the needs of specific contexts, i.e. staying open towards new members whose specific expertise may be helpful in a new innovation phase. This kind of "transparency" of community boundaries does not only concern accommodating new promotors, but also the ability to part with community members who have come to be unhelpful or counterproductive or who no longer hold functional roles, again due to changing conditions (e.g., entering a new innovation phase). This latter point also touches upon possible conflicts between functional roles and personal, even friendly, relations that can easily arise in promotor networks.

3.2.2 Features of the Quality of Community Identity

The sub-construct of "community identity" defines the boundaries of a community. It includes factors determining internal and external visibility, and the "binding forces" within. These aspects are important to the performance of an innovation community in the sense that there is usually an inter-organizational or inter-departmental, informal network of people holding several professional roles. First of all, they are the employees or owners of an organization participating in a given innovation project. These persons' primary occupational or functional identity is in this sense determined by their formal membership or role in this organization or department. In addition to this primary identity or organizational "attachment," the individuals in question are also usually part of secondary organizational structures, such as temporary, intermediate innovation teams, in which they take on the role of promotors. That they will then even take part in further inter-organizational promotor networks, and work on advancing projects with promotors from other departments or organizations almost constitutes a third functional identity or role in itself. Therefore, innovation communities can also be described as a tertiary organization form of innovation management. This multiple functional "entanglement" of promotors, i.e. the numerous links between formal organization or project structures on the one hand, and informal processes on the other hand, has been identified in the existing case studies on innovation communities (Fichter 2006). Hence, the question of community identity is not trivial. It should rather be assumed that an innovation community only contributes to the success of a project as long as it has a certain mutual understanding (of who is in, who isn't), a specific functional identity and a sense of integration. In this respect, findings from team and inter-group research can be usefully applied (Alderfer 1987; Friedlander 1987).

While community composition focuses on quality traits of situation- and task-related community assets, i.e. more "objective" functional issues, community identity focuses on the more sensitive or emotional aspects of the individuals involved. The sub-construct of community identity has three aspects, to be detailed below.

Clear Group Boundaries

As an organizational or inter-departmental kind of informal promotor network, innovation communities can be described as a tertiary organization form of innovation management. Because of this "third-rate" status regarding the different organizational identities and roles, the identity and cohesion of innovation communities is fragile and cannot generally be considered a given constant. So it can be assumed that the performance of these networks relies on the members of the community having a mutual understanding of who belongs and who does not, who fits in well and who does not, and of the question whether the promotors are acting as a team. The outward image gives some indication of whether an informal group identity is being developed, for instance if the promotor network is perceived as a "community" by non-members.

Functional Identity

The importance of different characteristics of collaborating teams for individual and institutional performance has already been stressed since the 1980s in team and cross-team research (Alderfer 1987; Friedlander 1987). Friedlander studied the limits and boundaries of working groups within one company and describes them as teams that have a common "higher" goal and can only be conceived as part of an overall organizational system (Friedlander 1987, 304). He specifically included informal or ad-hoc groups in his examination, making it possible to transfer the concept to innovation communities. Besides the issues of clarity and permeability of group boundaries (see above, under the aspect of reconfiguration), he emphasizes the impact of group cohesion (see below) and functional identity: "Work groups need a strong sense of their own functional identity. A group knows what it is through its task purpose and task competence. [...] Functional identity contributes to the sense of a 'home base' and the security it provides" (Friedlander 1987, 305). The aspect of functional identity therefore concerns a general understanding of community members regarding the objectives and relevance of the innovation project, and relates to aspects that can further affect the project-based perspective and outlook of promotors. A common understanding of community members helps solve project- and cooperation-related problems and tasks effectively, thereby contributing to the performance of the promotor network in total.

Cohesion

The cohesion and solidarity of a group also has significant impact on its performance (Högl and Gemünden 2001, 438). Character and importance of cohesion can be described as follows: "Group cohesion refers to the degree to which work group members form a strong collective unit and are drawn inward to the centrality of the group. Cohesion is related to the willingness of members to commit energy and resources to the group" (Friedlander 1987, 305). Mullen and Cooper (1994) mention three types of cohesion: (1) interpersonal attraction of individual team members, (2) commitment to the common task, and (3) team spirit. Without a sense of "belonging," without the members' motivation to become or to remain part of the community, a promotor network will hardly succeed in making a substantial contribution to the success of the project. It can therefore be assumed that a strong inter-personal attraction of the community members, a high degree of individual commitment to team goals, and a strongly developed team spirit all strengthen community cohesion and positively influence the success of a collaborative innovation project.

3.2.3 Features of the Quality of Community Interaction

The sub-construct of community interaction quality focuses on traits that point to an innovation community's internal cooperation level. This encompasses the quality of interpersonal exchanges and cooperation processes. It is assumed that the quality of interaction within communities dramatically affects the success of their collaborative innovation projects. The main traits describing community interaction quality are:

- Communication
- Coordination
- Balance of contributions
- Mutual support
- Effort
- Confidence and group climate

This sub-construct is partly based on the construct of "teamwork quality" as developed by Högl and Gemünden (2001). As explained above, the construct of Högl and Gemünden has been adapted in two points, however. First, the aspect of "cohesion" in the present work has been "relocated" to the "community identity" domain (see above). Second, the aspect "trust and group climate" has been added to the "community interaction" domain.

Communication

The factor communication, adapted from Högl and Gemünden (2001), relies on the terms of communication and information. These are described as elementary components of working together as a team or network. Information and opinions are exchanged within a team through communication. To describe communication, the frequency, formalization, structure and openness are identified as its main features. While frequency describes the ratio of sharing information, formalization determines how spontaneous the communication is or whether it obeys formal rules.

Structure on the other hand, relates to the way team members communicate with each other and whether they do so directly or via an intermediary. The openness of communication is a measure of their ability to bring in their full knowledge into such collaborations without reservations. Whenever this openness is affected, cooperation and the processing of information within the team can be fundamentally disturbed. This particular aspect of communication was highlighted by Gerybadze (2003, 153) who introduced the term "social understanding level." In this sense, social communication in innovation communities encompasses much more than a mere exchange of information about the project status. A mutual "interpretation framework" establishing common grounds for exchange between individuals and of common understanding of sometimes ambiguous information, therefore, is just as conditional to functioning communication.

Coordination

Another central factor of team or community cooperation is the coordination of tasks and individual members' contributions. According to Högl and Gemünden (2005, 99), the management of these contributions is a relevant feature of teamwork, as this can be regarded as a mechanism to integrate various skills and capacities, necessary to successfully manage complex or indeterminate challenges (Högl and Gemünden 2001, 440). The nature and degree of coordination again depend on the particular goals of that community and its resulting tasks. Sydow (2006) further identifies the distribution of tasks, resources and responsibility as a key element of coordination within the network. Functioning coordination is seen as the result of a fair and collective process of negotiation, distributing tasks and contributions between network members based in particular on their skills and capabilities. It can therefore be assumed that it augments the capacity of an innovation community if the individual contributions of team members are well structured and synchronized.

Balance of Contributions

Another indication of the quality of team interaction is the balance of team members' contributions. Here, two of the features of this balance are of interest. The first concerns individual members' emotions, as well as their capacity to participate and present their own competencies sufficiently in the context of community cooperation: “It is essential to the success of innovative projects that all team members feel free to bring in their task-related expertise. If discussions and decision-making are dominated by some team members, and others in the team are unable to contribute their views and ideas, this will have negative consequences on the team’s performance (...)” (Högl and Gemünden 2001, 440). The second feature is connected to the evaluation of contributions by other members, i.e. the question if they are perceived to be balanced throughout the team or whether the specific potential of others in the team is acknowledged. The quality trait “balance of contributions” therefore describes the importance of a well-balanced network of relationships between team members, as well as their ideas and experiences, which as a whole enables a wider range of knowledge crucial to task performance and problem-solving (Högl and Gemünden 2005, 99).

Mutual Support and Conflict Resolution

Mutual support of team players, and their ability to resolve conflict within the innovation community are further defining features that determine the quality of community interaction. Accordingly, group collaboration should not so much be characterized by competition among the participants, but rather by their willingness to cooperate. This is considered necessary in particular in the context of complexity and uncertainty innovation teams are often faced with. While cooperating team members exchange information and support each other, more individualistic notions of competition oftentimes lead to mistrust, frustration and conflict (Högl and Gemünden 2005, 100). It is therefore plausible to assume that the degree of mutual support significantly influences team performance altogether (Högl and Gemünden 2001, 440). The same applies to a team’s ability to constructively allow debate and controversy, to reach consensus on key topics, and resolve conflicts. This ability has a significant impact on group climate and on the performance of the community.

Effort

The individual commitment or “effort” that community members display in collaborative innovation projects and in cooperation within the innovation community also impacts the success of promotor networks. Here, the focus is on the expectations that team members have of others regarding their performance. “Workload sharing and prioritizing of the team’s task over other obligations are indicators for the effort team members exert on the common task” (Högl and Gemünden 2001, 438). Labour, commitment and prioritization of shared tasks are therefore issues that should be treated as consensus in team collaboration, and should accordingly be accepted by all (Högl and Gemünden 2005, 100). This equally applies to sanctions by the team when individuals' conduct strays from the group's operative standards and norms. For cooperation in an innovative community to function well, it is therefore important that members know the expectations regarding their work performance, and develop some common understanding in this regard. A consistently high level of effort of all the community members can enhance the quality of interaction and, hence, the community's contribution to the success of the project.

Trust and Group Climate

Finally, the trust between the members of a promotor network, and the group climate are further important factors in terms of interaction quality. Trust has since the 1990s become an important explanatory element within network theory. Several authors take trust to be the key "constitutional" requirement for networks (Weyer 2000, 7; Aulinger 2005, 244). As Eberl (2003, 42) points out, "(without) trust among network members, this is the basic message, networks cannot work." Weyer, as a key representative of network research in social science, regards trust as the central "means of coordination" for networks (Weyer 2000, 7). For innovation communities that constitute a personal network rather than an organizational one, personal trust between members is most important. Other kinds of trust (self-confidence, organizational trust, etc.) will therefore not be considered at this point (see Fichter 2008, 31). With view to the informal nature of promotor networks and the quality of cooperation, the "action-oriented" definition of Ripperger (1998, 7) aptly conceptualizes trust as follows: "Trust is the deliberate provision of risky inputs while at the same time surrendering explicit contractual security or control measures against opportunistic behaviour, in expectation that the trustee also be motivated to voluntarily desist from opportunism." In the absence of formal security and sanctioning mechanisms in informal innovation communities (Gerybadze 2003) and with high levels of complexity in these collaborations, trust is an essential constituent for functional cooperation in promotor networks.

Notably, trust and group climate play together closely. Group climate can comprise widely different issues such as power, hierarchy, or trust. Friedlander (1987, 306) points out that participatory structures in fact promote group work. Exaggerated hierarchy in innovation communities and teams may negatively impact the quality of interaction, as they favour the execution of power. Exercising power is most often justified with overcoming obstacles or securing potential for action, but in fact this is only useful in cases when this potential is actually threatened. If this is not the case, the display of power more likely damages the cooperation within the team (Scholl 2005, 35). Trust on the other hand is crucial to all factors within the construct of interaction quality.

3.2.4 Summary: Constituting Factors of Performance

The three sub-constructs presented are designed in such a way as to represent all of the factors that determine an innovation community's performance, and thus their role regarding the success of an innovation project. Based on the 12 quality criteria clarified above, it can for the following work be assumed that the contribution of a promotor network to the success of a cooperative innovation project depends on these three key factors:

- Situational or task-based composition of the community regarding their skills, roles and resources
- Existence and extent of community identity
- The level of interaction quality

This leads to the following overall construct of the performance of innovation communities (Fig. 3.1):

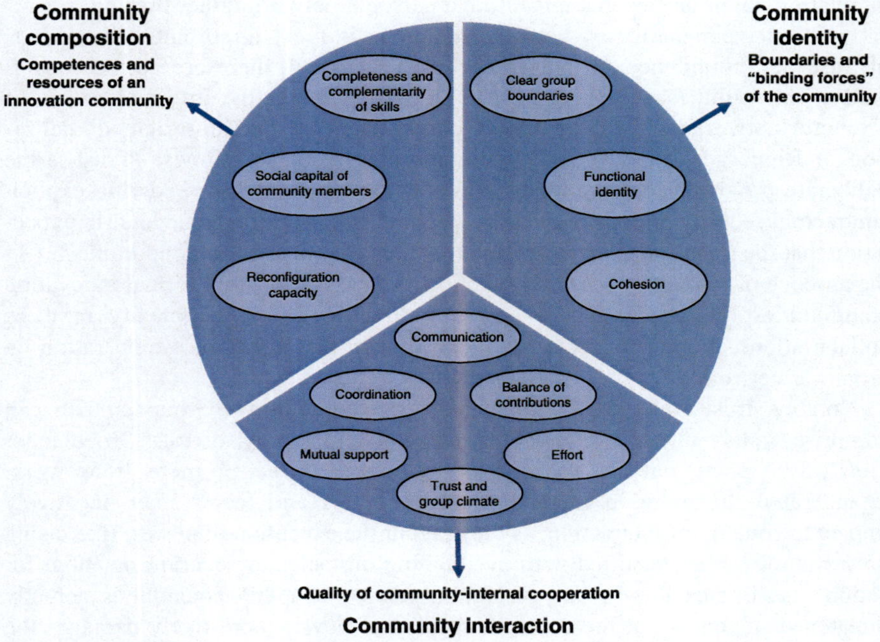

Fig. 3.1 Influencing factors on the performance of innovation communities (Source: Authors)

3.3 Research Design and Methodology of Case Studies

Severin Beucker

Based on the conceptual framework developed in Sects. 3.1 and 3.2, in the following, the research design and methods for a more detailed analysis of innovation communities' evolution and of their contribution to innovation success is compiled. The research design draws on findings from the empirical survey presented in Chap. 2, and takes the key questions formulated at the beginning of this chapter into consideration.

- Q1: Evolution: How does the size and structure of innovation communities change over time?
- Q2: Performance: How do innovation communities contribute to the success of innovation projects?
- Q3: Quality: Which features make innovation communities successful?
- Q4: Methods: Which measures and methods constitute and improve the performance of innovation communities?

The cognitive interest of this in-depth investigation focuses on questions of "how" and "what" – on the form and manner in which innovation communities evolve, function, and contribute to innovation projects' success. The investigation concentrates on studying a current phenomenon of innovation development (innovation communities) and exploring its role in the context of cooperative innovation projects. Monitoring participants' behaviour, as might be necessary in a laboratory setting, is essential here. Against this background, the case study method recommends itself for this kind of examination (Yin 2003, 5f).

The significance of innovation communities for collaborative innovation projects stipulates that the main research emphasis be placed on the innovation community. Hence, the central unit of analysis should be the community itself, examined in its context of a concrete innovation project or project scheme. The innovation project then constitutes a second level of analysis. Here, the focus is on the relation between the innovation community and the cooperative innovation project. The first three key questions therefore also reference the innovation community's influence of project success. This influence can be established based on the constructs *project success* and *performance* of the innovation community.

Another question guiding the case study design was which methods and measures can be employed that contribute to the development and potential performance of an innovation community. Of specific interest here is the question whether the use of specific methods and measures not only helps to improve the performance of the innovation community, but also the management of the innovation project concerned. The methods employed by a given community therefore represent a third unit of analysis within the case studies.

This results in the following research design, presented in Fig. 3.2.

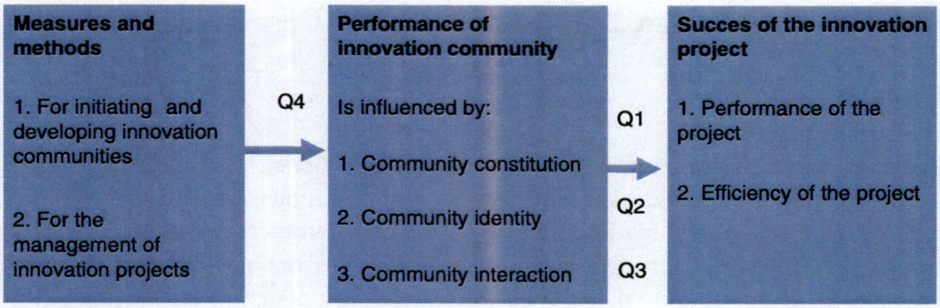

Fig. 3.2 Research design for the case studies (Source: Author)

For analyzing and evaluating the success rate and performance of innovation communities in radical innovation projects, a method is required that can explain current events the researcher has no influence on in an exploratory, descriptive manner. Furthermore, a qualitative and quantitative line of reasoning is desirable with regard to attaining scientifically reproducible results and a certain level of abstraction, as well as to developing and testing the theory and generalizing its scope and validity in further investigations (Eisenhardt 2007).

Against this background, we have chosen a multiple case study approach (Yin 2003, 46) that permits the analysis of several similar cases, thereby increasing the validity and relevance of the results and facilitating the construction and verification of a pertinent theory, due to the greater number of cases available for evaluation. This approach is used to strengthen the findings on "group identity and group interaction" by examining various examples, based on the empirical study (Chap. 2) developed in the InnoCo project and the construct of "quality of innovation" of innovation communities as described in the previous chapter. These select cases of radical innovation networks are analyzed using semi-structured interviews as well as closed-question questionnaires.

3.3.1 *Criteria for the Selection of Cases*

For the selection of cases, 107 networks covered in the empirical study (see Chap. 2) were considered as a first step. Since the aim of selecting these case studies is to gain new insights into how innovation communities contribute to radical innovation projects and their performance, further criteria for the selection of case studies were developed, besides the radical nature of the innovation project itself:

- Several phases of innovation covered: All cases should allow the investigation of permanence or change of the innovation community in the course of different phases of innovation. Therefore, projects that have already reached an advanced stage, e.g. development, implementation or market phase, should be selected.

- Consideration of different promotor roles within the innovation community: Innovation communities chosen for case studies should include different promotor roles (power, technical, process, or network promotors), as well as champions. As the underlying assumption here is that the proportion of champions has a great influence on identity, interaction and innovation levels within the community in question (see Sect. 2.4), both communities in which (nearly) all members are champions and those in which less than half the members are champions should be considered.
- Community size: Since we can assume that the size of the community also has an influence on identity, interaction and performance, it is helpful to distinguish three sizes and take this dimension into account for the selection of the six cases to be examined – Innovation communities with three to four promotors or champions, with five to six promotors or champions, and with seven to eight promotors or champions.
- Geographical distribution of community members: Previous theoretical work (Gerybadze 2003; Fichter 2009b) as well as first empirical studies of innovation communities (Fichter 2009b) point to the fact that regular face-to-face communication and personal interaction notably impact the identity and interaction quality of the group. In what way the geographical distribution of the community influences the team should therefore also be investigated in the case studies. We can distinguish two types of communities: those whose members are in close proximity (regional ICs), and those whose members are more widely distributed (supraregional ICs).
- Different techniques and methods employed: One main focus of the InnoCo research project is the knowledge that can be gained from the nature and scope of the applied methods. What effects do operational principles have on the identity, interaction and success of innovation communities? Hence, the cases should be selected in such a way that the range of methods available to the communities varies widely, in order to facilitate a comparison between innovation communities specifying low, medium and high numbers of methods for innovation management and community development.
- Further distinguishing features: Additional characteristics to be considered are, for instance, the industry sector, size, and regional distribution of the organizations involved.

Finally, the availability of network spokespersons for qualitative interviews permitting further in-depth analysis is an important criterion for case selection in itself. This willingness was already queried prior to the empirical study (see Chap. 2), and then confirmed in a personal conversation by TU Cottbus representatives with the main contact persons of the respective innovation network. Subsequently, this contact was passed on to the Borderstep Institute and the University of Stuttgart's IAT which then conducted the case analyses themselves. In another personal meeting, the availability of the contact person was reconfirmed and also determined whether there is indeed an innovation community with further contacts attached.

3.3.2 Case Selection Procedure

Based on a cluster analysis, the innovation networks identified in Chap. 2 for the empirical study were divided into 50 radical and 57 incremental networks. Further analysis revealed that 17 of those 50 radical innovation projects feature an innovation community and are therefore in principle eligible for case analysis (see Sect. 2.2.3 and 2.4.4). In a second step, of those 17 cases identified as having an innovation community attached, six were selected for the multiple-case study. This number was set to cover the widest possible range of cases, in accordance with the criteria listed above, and at the same time limit the labor-intensive evaluation of case studies. As mentioned above, the willingness of the communities' promotors and champions to participate in interviews was also crucial to the selection of cases. This is obviously a key requirement, as the preparatory dialogues and interviews associated with a case study are time-consuming and require the participation of several key figures from the network. All selected innovation networks demonstrated their availability for interviews with at least three of their promotors. The six cases finally selected are summarized the following Table 3.2.

3.3.3 Execution of Case Studies

For the case studies, an interview guideline for the detailed survey of promotors and champions and, additionally, an advanced questionnaire have been developed, the latter based on the structure of the survey from the empirical study and its items (see Sect. 2.6). This parallel, qualitative analysis using the questionnaire and the interview manual, has enabled us to obtain additional data regarding the traits of the innovation project and the actors within the innovation community, which could then be used for further quantitative analysis.

Developing the Interview Guideline

The purpose of the interview guideline is to carefully analyze the cases identified in the previous chapter and to determine in greater detail the success and performance rates of innovation communities in radical innovation projects. Also, the change in community composition over time and the measures and methods used to improve the efficiency of innovation communities are interesting (see also Chap. 5).

Table 3.2 Overview on selected case studies

1	Radical innovation project	Innovation community (number of promotors ≥ 3, identity and interaction ≥ 5)[a]	In advanced innovation phase in early 2010	Proportion of champions in the innovation community	Size of innovation community (number of members)	Geographical distribution of members	Scope of methods employed	Case	Industry/sector	Innovation phase
1	Yes	Yes	Yes	(Almost) only champions	4	Regional	Medium	Connected Energy (Berlin)	ICT, research & development (R&D), ICT engineering	Concept
2	Yes	Yes	Yes	(Almost) only champions	5	Interregional	High	Dye-sensitized solar cells (Germany, Switzerland)	Chemistry, engineering, glass, R&D	Development
3	Yes	Yes	Yes	(Almost) only champions	4	Regional	Low	Tilt sensor manufacturing (Stuttgart)	Engineering, vehicle construction, measurement and control engineering, R&D	Realization
4	Yes	Yes	Yes	Proportion of champions < 50%	5–6	Interregional	High	Geothermal energy kit (3 Federal States, e.g. Saxony-Anhalt)	Plant engineering, geotechnical services, R&D	Concept
5	Yes	Yes	Yes	Proportion of champions < 50%	6	Regional	Medium	Wood Energy (Lusatia/ Eastern Germany)	Vehicle construction, engineering, agriculture	Realization
6	Yes	Yes	Yes	Proportion of champions < 50%	6	Interregional	Low	Carbon nano-tubes (Germany)	Chemistry, engineering, electrical industry	Development

Source: Author
[a]For items and thresholds see Chap. 2

The structured interview guideline therefore contains the following topics:

- Introduction and project presentation
- Classification of the interview partner
- Description of the cooperative project
- Classification of the business or industry environment
- Composition of the innovation community
- History of and collaboration in the innovation community
- Identity of the innovation community
- Quality of the innovation community
- Performance analysis of the innovation community
- Measures and methods used to improve efficiency of the innovation community
- Performance analysis of the innovation community for the project in question

Items and Scales for the Additional Questionnaire

In addition to the interview guideline described above and drawing from the experiences made in the quantitative survey (see Chap. 2), the questionnaire used there was further developed and amended with additional questions for the following review and further analysis of the construct of "quality" of innovation communities. In particular, questions that relate to the following topics were added:

- Group composition: This includes questions about the completeness and complementarity of skills, the social capital, as well as the "reconfiguration" abilities within the community.
- Group characteristics: Items included here were functional identity, cohesion and the quality of community identity.
- Internal relations: Several points were completed here with regard to the assessment of the quality of innovation communities, e.g. items on coordination, mutual support, or balanced contributions.

The scales of the items had already been adjusted in the original questionnaire (from 1, for "fully disagree" to 7, "fully agree"). A comparison of the questionnaire adapted for the case studies with the one used in the empirical study (see Chap. 2) is laid out in Table 3.3.

Access to the Field and Conducting of Interviews

The access to the innovation networks for in-depth case studies was originally arranged through the research partners at TU Cottbus. The six networks identified were first interviewed by TU Cottbus staff in a telephone conversation to ascertain their willingness to participate in further interviews as part of a more in-depth

Table 3.3 Comparison of questionnaires from the empirical study and in-depth case analysis

Original questionnaire from the empirical study (see Chap. 2)	Adapted questionnaire for case analysis
A: Introduction	Preliminary questions, e.g. regarding the assignment of the project to a specific industry sector etc., were dropped, as these had already been established in the first quantitative survey
B: Radical nature of innovation	A: Degree of novelty in innovation: Questions and items were transferred to check whether self-assessment in this point has remained the same
C: Characteristics of partners participating in the radical innovation in question	B: Characteristics of partners participating in the innovation project in question: Further assignment of partners to their respective industries was omitted, as this had already been established in the first quantitative survey. The naming and classification of the most important project partners however was kept to establish possible discrepancies or changes regarding key figures
D: Characteristics of innovation	The classification of the time frame and phases of the innovation project was dropped, as this was already covered by the interview guideline
E: Identification of key figures	C: Identification of key figures: The pattern of questioning was adopted (traits of promotors involved), and merely amended with one additional item on the distribution of resources in the innovation project. In the first and seconds sub-constructs (power and technical promotors) respectively, one item each was removed
	D: Group composition: This construct with six items altogether (completeness, complementarity, and reconfiguration capacity of the community) was newly developed and added to the original questionnaire to comprise the quality traits of community composition, as defined in Sect. 3.2.1
F: Characteristics of the community members (key figures identified above)	E: Identity of the key figures mentioned above: This section was considerably revised to comprise items regarding the constructs of "functional identity" and "cohesion" (see Sect. 3.2.2). The construct of "permeability of group boundaries" was completely omitted, while the six items within the construct of "clear group boundaries" were newly added
G: General characteristics of community interaction	Omitted (not applicable)
H: Forms of communication in group collaboration	Omitted (not applicable)
I: Specific characteristics of group collaboration	Omitted (not applicable)

(continued)

Table 3.3 (continued)

Original questionnaire from the empirical study (see Chap. 2)	Adapted questionnaire for case analysis
J: Characteristics of internal relations of key figures identified above	F: Interaction of key figures: With regard to the newly developed construct of "quality of innovation communities" and its sub-constructs (see Sect. 3.2), this section was also considerably revised (compare Chap. 2 for the according elements in the original survey). The sub-construct "communication" and its items were adopted, except for the last item, as was the sub-construct "trust and group climate." The original sub-constructs "coordination," "mutual support" (now: "mutual support and conflict resolution," one item modified), and "effort" (one item added) were partly adapted, and the sub-construct "balance of contributions" with three items was included. These modifications allow for a better analysis of the "community interaction" construct (see Sect. 3.2.3)
K: Methods to enable the collaboration of key figures	Omitted, as the question of methods employed is already covered in the interview guideline
L: Success of innovation and collaboration	G: Success of project collaboration: One item was added in the sub-construct "effectivity"

Source: Author

analysis. Subsequently, the Borderstep Institute and the University of Stuttgart's IAT conducted further dialogues with the networks' key players to clarify the following points:

- Short introduction, explaining why the contact has been approached, and presentation of the procedure planned.
- Determining whether the project in fact qualifies as a radical innovation project and whether the information already provided in the quantitative survey is still correct and up-to-date.
- Determining whether at least three promotors or key figures exist within the community.
- Determining whether at least three of these key figures are available for in-depth interviews.

In all six cases, this tentative audit revealed positive results for all of the above points, allowing the case analysis to start. In the case of the geothermal energy kit, the case analysis had to be terminated after two of the interviews had been conducted, because the network was in a phase of transition at the time and no extra interviewee could be won over, contrary to the original assessment of the community. Due to the advanced level of the project at this time, it was no longer possible to find a replacement for this case. Therefore, the following comments to only five of the cases.

The main objective of the case studies was to conduct interviews with at least three promotors and key figures of the innovation community in question. Altogether, in each of the five cases examined four interviews with community members were conducted, which gave us 20 interviews in total. These were conducted between August and December 2010, either in person ("face-to-face") or in the form of a telephone conversation. The course of the conversation followed the question structure outlined in the interview guideline, while small divergences from the given order were accepted in favor of a more fluent conversation. Most of the interviews (16 out of 20) were recorded on tape with the approval of the interviewee.

3.3.4 Evaluation of Interviews and Questionnaires

The analysis of the interviews and additional questionnaires was carried out independently for each of the five cases investigated (see Sects. 3.5–3.9), and in a cross-case examination as well (see Sect. 3.9). The transcribed interviews were coded and evaluated using the data and text analysis software MAXqda. The underlying codebook was compiled in compliance with the central questions and hypotheses formulated.

The evaluation was aimed to make the interview partners' statements serviceable for interpretation and qualitative analysis. To that purpose, the key questions (see Sect. 3.3) were assigned to corresponding codes, and these in turn assigned to specific passages of the interviews in order to interpret the contents of statements within.

In addition, the results from the questionnaires were subjected to a quantitative analysis that included both an individual and an inter-case examination. Results from this analysis are compiled in Sect. 3.10.

3.4 Delimitation of Key People, Promotors and Innovation Communities

Klaus Fichter and Severin Beucker

An analysis of cases and contributions to the success of innovation communities requires that it is determined how innovation communities, and promotors as their central actors, can be identified and delimited with respect to other actors and networks. The preselection of the case studies made clear that for this purpose, a further differentiation of the "key actor", as defined in Sect. 1.2.1, is needed.

Basically, key persones can influence the course of an innovation process in two different ways: first, by promoting the development and implementing an innovative solution to a problem; and second, by impeding or preventing one. In the context of the present case studies, the focus is on the promoting people. The promoting key people break down into two different personality types:

(1) promotors, and
(2) supporters.

3.4.1 Promotors

If a key person promotes a concrete innovation project, and meets at least half of the criteria of a promotor's role (cf. Appendix Chap. 2), that person will be categorized as a "promotor". A promotor is always a key person, but a key person will not always automatically be a promotor.

In the pre-selection of the cases to be analyzed on the basis of the results of the network survey (cf. Sect. 3.3.2), it has been shown that in the selected cases, all promotor roles – power, expert, process and relationship promotors – in fact occur (cf. Sect. 3.3.2). The ensuing case analyses thus have the task of determining whether the assignment of the roles of the promotors is seen in the same way by the other actors in the innovation community, or whether there are differences here. Moreover, it should be ascertained which promotor roles are crucial for the success of the innovation community, and particularly of the innovation process, and how many "champions" are to be found in the communities, i.e., actors who fulfil several promotor roles simultaneously.

The first analysis also showed that not all members of an innovation community can be clearly assigned to a specific promotor role (power, expert, process or relationship promotor), based on the answers in the questionnaire (see Sect. 3.3.3), even though they may be considered and described as promotors in the interviews with the other members of the community. For such cases, the evaluation of the present case studies will establish the following convention: In cases where a person is described as a promotor by several interviewees, but where fewer than half of the criteria assigned for qualification as a promotor in each of the four roles are attributed to him or her in the questionnaire, that person will be categorized as a promotor in that category in which he or she has received the most attributed criteria.

3.4.2 Supporters

As the interviews in the case studies will show, there are, apart from the promotors, other key people who have a significant effect on the initiation, progress or even the success of the innovation project, by supporting the innovation process through their activities, competence or decision-making power. While they, unlike the promotors, do not drive the innovation process actively and in many ways, and hence do not meet the definition of a promotor, they at the same time provide sporadic support of great importance for the success of the innovation project. Such actors should therefore also be included in the analysis, when it comes to key people of the focal innovation process. They are herein called "supporters" of the innovation process.

3.4.3 Innovation Communities

While the "supporters" belong to the broader network of the focal innovation process, the concept of the "innovation community" is limited to the promotors' group. Therefore, we will at this point define the following requirements with respect to promotors and innovation communities which must be fulfilled for the selection of case studies:

1. *Delimitation of a focal innovation process*: A concrete innovation idea or a concrete innovation project can be reliably delimited.
2. *Existence of at least three cooperating promotors*: In the context of the focal innovation process, at least three people can be identified who fulfil one or more of the promotor roles, and who cooperate closely.
3. *Fulfilment of the promotor's function*: A person is considered a promotor if at least half of the criteria (items on the questionnaire) of any promotor role are affirmatively answered or met.
4. *Fulfilment of community identity and interaction quality*: A community can be considered to exist if both constructs reach an average score of at least 5 on a seven-point scale (cf. Appendix in Chap. 2 and Sect. 3.3.3).

3.5 Connected Energy: Developing Smart Home Solutions for Energy Saving

Linda Bergset

Fact Sheet

Case Study: Connected Energy

Innovation object:	Smart home automation technology, connecting to the Smart Grid
Industry sectors:	Energy suppliers, automation technology, building and property management
Period analyzed:	2009–2011
Initial situation:	Identifying power-saving potentials in private homes is becoming increasingly important, not only to tenants, but to energy suppliers and policymakers as well.
Current phase:	Development phase
Research highlights:	Radical innovation project in which the existence of an innovation community strongly influences overall project success, high degree of interaction and communication between various industry sectors and function roles needed

Key persons involved:	8
Size of innovation community:	4
Group composition:	Almost only champions
Regional distribution:	Regional (Northern Germany)

Community performance:	5.6*
Contribution to project success:	6*
Project success:	5.3*

*On a scale from 1 (very low/poor) to 7 (very high/good).

3.5.1 Context and Object of Innovation

Decentralized Energy Management and Intelligent Home Networking

In view of climate change and a growing scarcity of energy resources, we not only need an expansion of renewable energy sources, but also improved energy efficiency through the optimization and reduction of energy consumption. Germany's households account for around a quarter of the total energy consumption in Germany, a share of great energy and climate policy significance. Especially the rental flat sector, accounting for 58% of the German housing stock (Federal Statistical Office 2008), still has a large untapped energy savings potential. In addition to insulation of older buildings and energy-efficient construction methods for new buildings, the challenge in mobilizing this potential is to make current energy consumption in the form of heating, ventilation and electricity in homes, and especially in rental housing, transparent to the residents.

What is more, incentives and opportunities are needed that influence and control energy consumption. Not only the absolute reduction in energy consumption in households is desirable; postponing consumption can also contribute to energy efficiency. One approach is that energy suppliers and network operators incorporate households into a system of levelling energy demand (load management) by means of linking flexible electric power rates and so-called "smart meters" with decentralized energy management, further increasing the available electric power reserves. The required total capacity of power stations in Germany could thus be limited without endangering supply security.

Decentralized energy management will require taking into account the integration of a growing number of decentralized power generation and storage units, such as mini CHP (Combined heat and power) plants, solar thermal systems, photovoltaic systems and electric vehicles. In this way, electric power could be delivered cheaper or more expensively, depending on the fluctuations in renewable energy supplies.

A prerequisite for such a decentralized energy management is smart home networking, including the installation of smart meters. An intelligent home network consist of three components: "connection of the house to the Internet via broadband data lines, in-house data wiring for computer communications between different rooms, [and] networking and control of various operations within the house to create 'smart homes'" (BITKOM 2008, 8). Rental flats are of great interest with regard to energy consumption, since significant efficiency potentials and economies of scale can be achieved for properties using intelligent controls of networked energy and storage. Using intelligent control algorithms and remote maintenance functions, an energy saving potential of up to 35% is possible (BITKOM 2008). Smart meters that permit electricity consumption to be recorded precisely in 15 min units help utilities make more accurate forecasts for future load management, affording customers greater transparency regarding their electric power consumption (BITKOM 2008). The result of such a system is not only energy savings; it will in future also allow customers to save costs if the system is linked to time-variable or discounted rates within a demand-response approach of the energy suppliers.

Some European countries, such as Italy, Sweden and Denmark, have been pursuing a policy of nationwide smart meter instigation for several years (BITKOM 2008). In Germany, policies for similar developments have been formulated,[2] and different approaches for disseminating decentralized energy management using intelligent networks, intelligent metering and home networks are currently being tested. The largest current project is being funded by the Federal Ministry of Economics (BMWi) and the Federal Environment Ministry (BMU) through the E-Energy Programme, an umbrella under which a number of smart grid-related projects are being carried out in various so-called model regions in Germany.[3] This case study examines the project "Connected Energy – SHAPE",[4] which is also being sponsored by the BMWi as part of the "Connected Living" association. Connected Living was founded in 2009[5] and aims at the development and standardization of intelligent home networking technology. To this end, companies from the entertainment, communications, safety, health, financial and energy sectors are to be involved, in order to develop inter-industry solutions and business models for intelligent home networks.

Object of Innovation

The goal of the innovation project Connected Energy-SHAPE is to determine the energy-savings potentials in private homes by using smart home networking and home automation solutions. A decentralized energy management system for households is being developed, built around existing approaches and based on the complex interaction of separate components, and will be implemented in prototype form. An intelligent, adaptive learning-based automation technology is to control the heating, ventilation and equipment of a flat via a home manager, and derive the most energy-efficient control option from the behaviour and needs of the inhabitants. This technology is supported, where available, by controllable user

[2] Cf. "Eckpunkte für ein integriertes Energie- und Klimaprogramm", http://www.bmwi.de/BMWi/Redaktion/PDF/E/eckpunkt-fuer-ein-integriertes-energie-und-klimaprogramm, p. 15 (Retrieved: March 17, 2011)

[3] Cf. http://www.e-energy.de/documents/BMWI_Brosch_E_Energy_d_16_6_web.pdf (Retrieved: March 21, 2011)

[4] SHAPE: Service-Oriented Home-Automated Platform for Energy Efficiency Increase.

[5] Cf. http://www.connected-living.org/news/singleview/article/der-verein-connected-living-wurde-gegruendet/ (Retrieved: March 21, 2011).

devices that the power consumption of specific appliances is measured by, and this information is then passed on to the central control unit via a "room manager". For whole buildings, a building manager is additionally provided, which can, for example, optimize the central heating and ventilation systems and other energy-consuming units in conjunction with decentralized energy generation (mini-CHPs, solar thermal systems, photovoltaics, etc.), storage facilities (air conditioners, fridges, electric cars etc.), and flexible power rates. Although the direct integration of decentralized electricity generation is not provided in the focal innovation project, the possibility will be considered so that it can be implemented in follow-up projects. The consumption data is conveyed to a by-the-minute billing system. Here, it is possible to make use of experience from mobile phone accounting, since a large number of mobile phone rates have been managed by the second for decades.

The decentralized energy management system is to be kept open for potential external provider services, such as inter-household or inter-neighbourhood optimization of locally produced and consumed energy. The intent therefore is to apply standardized hardware and software interfaces yet to be established. The services will consist of a Home Service Platform (HSP) and a central Service Provider Platform (SPP) located in the house, connected via a "home gateway" or processing platform. The exchange of information between the connected, but often autonomous units of the home network, e.g. control units like the flat and building manager, context sensors, decentralized power generators, and multi-utility meters, is conducted entirely on this platform. This system draws from the already existing SerCHo architecture.[6] The HSP will take on the configuration of power management, and will be able to retrieve various data and services, e.g. accessing energy consumption data, or querying flexible electricity rates from the SPP. Part of the SPP is an online portal from which an "application store" can be accessed, displaying the current prices and energy costs. Invoice previews and booking of additional services can also be carried out via the SPP. The energy management system will also calculate information relevant to decision-making regarding cost, convenience and energy efficiency, and transfer it to a consumer terminal, such as a TV, smart phone, or PC.

Specifically, the technology is to be implemented in two different areas of the project. An exhibition space will present a range of equipment and services as comprehensively as possible, as well as a vision for the future of intelligent home networking technology. Moreover, the currently implementation possibilities are to be explored in a field test with a fairly large number of flats. Taking the existing supply of equipment and variable rates into account (Fig. 3.3).

[6] Cf. www.sercho.de for the approach of the SerCHo research project.

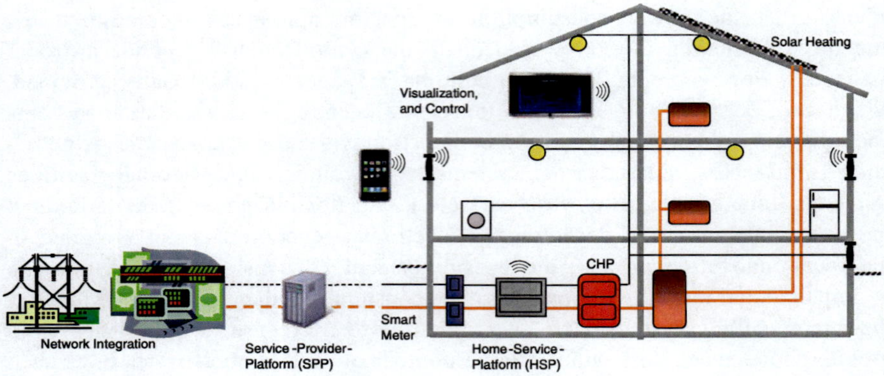

Fig. 3.3 Overview of the components of a connected home (Source: http://www.connected-living.org/)

Degree of Innovation

This innovative project can be described as radical because it both promotes new major inter-industry solutions "which are to open up completely new products and markets" (Project Manager, Research Institute, 51), and at the same time requires "systemic innovation" (32). The idea of a "new in-house operating system" (Project Manager, Clearing and Settlement Service Provider, 60) requires conditions that are not yet at hand, including important changes in flexible electric power rates, an adjusted calibration law, and in the standardization of interfaces:

> What's actually novel is not so much the components themselves, but rather the interaction of those components in decentralized power management, with this very complex environment of variable rates. (Project Manager, Research Institute, 35)

Beyond the load and time-variable rates, the innovation on the energy side is that one is dealing with "very small amounts of energy" (Innovation Manager, Energy Supply Company, 38) which, firstly, need to be recorded by the minute, using new metering systems that have until now mainly been used in telecommunications, and which, secondly, need be aggregated between households so as to develop their full potential. While industrial customers have been involved in the load management of power supply companies for some time, small consumers have not (BITKOM 2008).

> The really interesting thing behind this is that we are certainly not talking about megawatts here, in some cases not even about kilowatts, but about watts and watt-hours and their aggregation, so that something is actually occurring that could be relevant to us. That is certainly one aspect, and connected with it, the significance of clearing and settlement systems, for example, . . . for units that are much smaller than those that we normally sell or trade in. And that means we need a tool kit to manage and aggregate miniscule units of energy and money, to get them up to sensible sizes. (Innovation Manager, Energy Supply Company, 40)

3.5.2 The Innovation Process

History and Environment of the Innovation Process

The association Connected Living can be considered to be the group that, in 2009, initiated and provided the framework for the innovation project Connected Energy – SHAPE and its early activities.

> The association Connected Living is intended to be the platform for the organization and public presentation of our joint initiative at trade fairs and in public relations, but also, within its working groups, meant to define innovative topics that will then lead to a project proposal. The Federal Economics Ministry is keen to push this whole issue forward, because they see the area of home networking as one of the key areas, among many others, where Germany has the skills required but where the individual participants are relatively isolated from one another and therefore should be brought together. (Managing Director, Technology Company, 10)

Although the energy issue, partly due to the abundance of possible issues, was "not so high on the agenda" at the outset (Project Manager, Clearing and Settlement Services, 55), some of the actors, during their discussions with the promotors, recognized the potential for a project which could successfully relate to a growing overall interest in energy efficiency. In July 2009, the association members came together in a workshop and developed scenarios for their respective working groups:

> And these scenarios were very influential because it quickly became clear in the discussion what the common idea of such a project could be, and who could assume which role in such a project. (Project Manager, Research Institute, 92)

Building on this common understanding, the members of the Energy Working Group mainly worked on developing an outline and an application for the project Connected Energy – SHAPE, based on a "document of understanding." About a year later, the project was granted approval, and could formally begin in July 2010. In the early phase of the approved project, preparatory work was carried out. Specifically, the goal was to achieve a more detailed assessment of the legal and economic environment, and to examine the basic framework conditions. On this basis, a more concrete scope of the envisioned system could be established.

The focal innovation process in this case is easy to pinpoint, since it is tangibly connected to the innovation project Connected Energy – SHAPE, which has funding from 2010 to 2012. Thus, the process is clearly divided into three phases:

- Initiation phase: early 2009– mid 2009 (between the founding of the association and its Energy Working Group, and the time of application)
- Conceptual phase: mid 2009– mid 2010 (application phase)
- Development phase: mid 2010–2011 (start of the approved project)

All organizations involved in the project are listed in the cooperation agreement:

- Five companies (a technology company, a clearing and settlement service provider, two energy supply companies and a telecommunications company)

- Two research facilities (one university institute and one independent institute)
- One association

Central Factors of Influence of the Focal Innovation Process

The most immediate influence at the beginning of the focal innovation process was the association Connected Living. However, a number of other driving forces also had a crucial effect on the innovation process, as described below.

Founding of the Connected Living Association: The great majority of the partners of the innovation project (five out of seven) were involved in the founding of Connected Living. These organizations also constituted the majority of the co-founders (five out of nine). The managing director of a technology company describes the co-founders as the "main drivers" (146) of the home networking concept during the initial period. In addition, another organization involved in the project advised the founders of the association regarding the structure of the working groups and the scientific advisory board, thus assisting the foundation process. The association for the first time allowed the project partners to come together based on their common interests. So the project was essentially made possible by the formation of the association.

Funding by the Federal Ministry of Economics (BMWi): The funding of the project was directly connected to the establishment of the association. The Federal Ministry of Economics and Technology made earmarked subsidy funding available to a selection of projects planned by the association to enable them to contribute directly to the goals of the association. That lowered the barriers, and reduced the risks for individual companies, while also saving time "in the overall development process" (Managing Director, Technology Company, 48). Although not all partners were funded equally, the project manager of one of the research institutions (44) stresses the critical nature of this source of funding: "Otherwise, there would have been no project at all."

Policy objectives in the energy area: The funding of the project in turn depends on the policy goals of the German Federal Government, which have been mentioned repeatedly by the key people in this case study. The Economics Ministry has judged the issue of home networking to be an important field of activity for ICT strategy in Germany, with special emphasis on the issue's link to the question of energy efficiency.[7] Furthermore, the importance of energy efficiency and integration of renewable energy sources into German energy policy has repeatedly been stressed (e.g., the German energy concept).

Insufficient framework conditions: Although policymaking is a driving force, the existing legal framework for the development of home networking to a certain

[7] Cf. The Stuttgart declaration of the fourth National IT Summit, Stuttgart, December 8, 2009, http://www.bmwi.de/BMWi/Redaktion/PDF/Publikationen/Technologie-und-Innovation/it-gipfel-2009-stuttgarter-erklaerung,property = pdf,bereich = bmwi,sprache = de,rwb = true.pdf (last accessed: March 28, 2011).

extent constitutes an obstacle. Here, the project partners have come to see their role – beyond the issue of technological progress – as one of policy consultants as well, providing legislation with recommendations for necessary adjustments:

> In the last few work sessions, we did identify a couple of barriers and are also considering, at the appropriate time, when we've consolidated a bit, telling the funders once more that, especially in the areas of smart metering and remote meter read-out, there are still a few regulatory hurdles that are impeding mass marketability. And we would like to have that documented, and tell the funders once again: Hey, do your homework, Economics Ministry, why don't you try to get some legal stipulations through to make sure that this stuff can really enter the mass market! (Project Manager, Clearing and Settlement Service, 140)

In the area of variable rates there are legal ambiguities, too, which to some extent explain the reluctance of the energy sector regarding more flexible rates and other services:

> There are indeed some reservations, especially among energy suppliers who were not united in promoting this whole development. In fact, within the companies, there are different tendencies. (Managing Director, Technology Company, 101)

In this context, the project partners are hoping to demonstrate the future potential through practical experience gained in the project:

> We see ourselves as a partner who can provide inspiration for the design of such a flexible payment rate system, because we are proving what potentials result from which price – in other words, what cost savings and energy savings potentials can be realized. (Project Manager, Re-search Institute, 32)

Convergence of different industries: One of the main reasons for the formation of networks in this focal innovation process is the combination of elements from different industries.

> It is, like I said, a complex blend of technical components, software, the services behind them, potential markets – all that is more than one actor could handle alone. Or, of course, one person could cover it, theoretically, if they had a huge company providing all these aspects under one roof, but the charming and innovative part is precisely that completely new business areas are coming together here. We have partners in the project who come from clearing and settlement services in the telecommunications sector and who can calculate short-term rate changes very well. We have clean technology developers. We have ... a big telecommunications provider who will also be able to implement similar services. So the interesting thing is precisely the interplay of these different partners. (Project Manager, Research Institute, 41)

For the managing director of a technology company, the project is feasible only through a intercompany project, because "here, a relatively large number of different problems have to be solved" (39) that "exceed the real goals of individual companies" (41). Some of the project partners see the topic area as providing new areas of business for their own technologies. In addition, other parts of the value-added chain provide certain industry companies in the chain with new areas of business:

> We're going to the other side of the meter, to the customer side of the meter, in very small areas where we (a) don't have the experience, and (b) don't know the necessary technologies. In other words, we don't have this as a business application, and we don't know how to turn it into one. (Innovation Manager, Energy Supply Company, 42)

Fig. 3.4 Applications in a smart home (Source: www.connected-living.org)

Contacts and social capital: Both in terms of the contacts of the Connected Living association, and of the various sectors represented by the project partners (telecommunications, energy, automation), the social network is important to the innovation process. Experience from the E-Energy projects mentioned above is available to the partners through other work relationships. The results of previous projects in which the partners have participated, together with intelligence from their contacts, constitute the project's knowledge base (including the SerCHo project mentioned above). Through many years of experience with similar projects, contacts with the project developers and with the BMWi have also been well established (Fig. 3.4).

3.5.3 Development of the Innovation Community

The following section will examine who the key people energetically promoting and advancing the innovation project Connected Energy – SHAPE are. In this context, the present study mainly focuses on the composition, identity and quality of interaction at the time of the survey (autumn of 2010). In addition, we will also examine the history and different phases of the focal innovation project (2009–2011) to provide a complete overview of the innovation community.

Background: History of the Community to Date

The participants in the innovation project did not know each other before the foundation of the home network association Connected Living. Two of the key figures identified have been heavily involved from the beginning, and got to know each other during the initial phase. One key person is the managing director of one of the companies that founded the association. The second is the project manager of a research institute. As a representative of his institution, he advised the founders and accompanied the creation process due to his experience in the area of start-ups and innovation projects.

Key People and Promotors

Eight key people who significantly promoted the project can be identified within the Connected Energy – SHAPE project. Of these eight, five have been classified as promotors in the questionnaires. The three other key individuals can be identified as supporters who back the innovation process in certain aspects, but not consistently, and therefore cannot be classified as promotors. Most of the people who participated in project meetings regularly were identified as key people by the interviewees; in addition, one other person was included who played a role within the association at the beginning for the initiation of the innovation process. A more detailed assessment of the supporters is presented in the section "Evolution: Steps and Phases of Community Development."

As is shown in the overview in Table 3.4, the innovation community consists of four promotors, two of which have assumed a single promotor role, while the other two have assumed several promotor roles ("champions"). The promotors are mainly active in business practice (three out of the four), bringing in expertise from various sectors, and are all considered expert promotors. In addition, another promotor comes from the scientific community and assumes several promotor roles, so that he too can be classified as champion. The four people in the community have a relatively consistent perspective regarding the key people who belong to the core of the project. The first champion of this innovative community, a project manager of a research institute, is the coordinator of the project and fulfils three promotor roles: as power, process, and relationship promotor. All members of the community share this view. The second champion is the general manager of a technology company, and is referred to by all members of the innovation community as both an expert and a power promotor. Finally, the project manager for a clearing and settlement service provider can clearly be classified as an expert promotor (he is seen as such by three of the four members).

The fourth person is the innovation manager for an energy supply company, who has been assigned several important characteristics of various promotor roles in all the questionnaires, but does not meet the formal requirements for any one specific role. However, if both sources, the questionnaires and the interviews, are used for the analysis, it is apparent that the other members of this innovation community view this key person as part of the inner circle, and appreciate his contributions as highly significant. Since this key person is, both in the interviews and in the questionnaires, mainly assigned the role of an expert promotor, according to

Table 3.4 Key actors in the innovation process Connected Energy-SHAPE (2009–2011)

	Geographical location	Key person	Promotor	Champion	Member of the innovation community
Business practice					
Managing director, Technology company	Berlin	Yes	Power and expert promotor	Yes	Yes
Project manager, Clearing and settlement service provider	NRW	Yes	Expert promotor	No	Yes
Innovation manager, Energy supply company	Hamburg	Yes	Expert promotor	No	Yes
Innovation manager/ Researcher, Telecommunications company	Berlin	Yes	No	No	No
Scientific community					
Project manager, Research institute	Berlin	Yes	Power, process and relationship promotor	Yes	Yes
Researcher, University	Berlin	Yes	Expert promotor	No	No
Association					
Project coordinator, Home networking association	Berlin	Yes	No	No	No
Chief scientist, Home networking association	Berlin	Yes	No	No	No
Number		**8**	**5**	**2**	**4**

Source: Author

sheer number of characteristics, he is classified as such here. Overall, all four promotor roles are represented in the innovation community.

One other expert promotor, a researcher from a university institute, is considered an important contributor in the innovation projects by some innovation community members (mainly in the interviews, but to some degree also in one questionnaire), but is not described as part of the inner core. One could therefore say that a slightly larger network exists which contains within it the smaller core of the innovation community.

Evolution: Steps and Phases of Community Development

The innovation process of Connected Energy – SHAPE had, at the time of the case study in autumn 2010, been in progress for about one and a half years.

Table 3.5 The evolution of the innovation community Connected Energy – SHAPE

	Initiation (early 2009–mid 2009)	Conceptual phase (mid 2009–mid 2010)	Development phase (mid 2010–2011)
Participating organizations	Technology company	Technology company	Technology company
	Clearing and settlement service provider	Clearing and settlement service provider	Clearing and settlement service provider
	Energy supply company 1	Energy supply company 1	Energy supply company 1
		Telecommunications company	Energy supply company 2
	Research institute, university	Research institute, university	Telecommunications company
		Home networking association	Research institute, university
			Home networking association
Innovation community	Managing director, Technology company	Managing director, Technology company	Managing director, Technology company
	Project manager, Clearing and settlement service provider	Project manager, Clearing and settlement service provider	Project manager, Clearing and settlement service provider
		Innovation manager, Energy supply company 1	Innovation manager, Energy supply company 1
	Project manager, Research institute	Project manager, Research institute	Project manager, Research institute
Other promotors participating in the preliminary phase of innovation		University researcher	University researcher

Source: Author

During that short period, the promotors remained more or less the same throughout three phases (see Table 3.5). Three of the four promotors in the innovation community who accompanied the entire process can therefore be described as core promotors. Two of these core promotors are champions; the third is the project manager for a clearing and settlement service provider, who, while he was not present at the time of the founding of the association, has helped to design the application for the focal project in the Energy Working Group from the beginning. The initiation phase in the case study is the relatively brief period from the founding of the Energy Working Group in the association to the start of the work on the draft application.

The reconfiguration capacity of the group is considered a significant factor in the composition of an innovation community. In the Connected Energy – SHAPE innovation project, too, a certain, albeit small, expansion of the core group can be ascertained. The conceptual phase during which the application was designed and prepared continued for about 1 year, from mid 2009 to mid 2010. During this phase, the fourth promotor was accepted into the inner circle of the innovation community. This key person, the innovation manager for an energy supplier, can be designated as an expert promotor. During the phase examined in greater detail in this case study (the early part of the development phase), the constellation of the innovation community looked exactly as it had during the conceptual phase. However, one additional person who entered the innovation process later, a project coordinator from the home network association, was described as a key person by one member of the innovation community. This can be explained by the fact that he filled out his questionnaire later, and was thus commenting on a slightly later phase of the innovation process in comparison to the interview. This expanded view shows the dynamics of community boundaries over time.

The number of organizations involved in the innovation project is to varying degrees slightly larger than the number of organizations represented in the innovation community. This slightly larger network that accompanies the innovation process makes the existing differences of opinions regarding the group's boundaries reasonable. Figure 3.5 illustrates the different information provided by respondents on the key persons within and outside the innovation community, to the exclusion of the formal contacts that exist between all the organizations involved. The innovation community (the four red dots) is easier to distinguish from the other

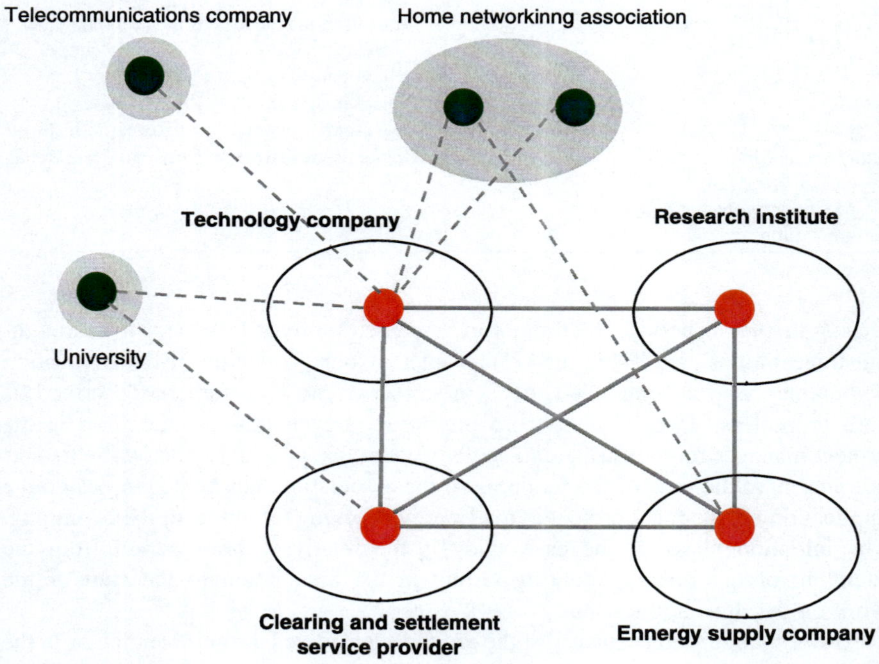

Fig. 3.5 Promotors (*red dots*) and supporters (*green dots*) in the innovation project Connected Energy (Source: Author)

key persons (green dots) based on the information given in both the interviews and the questionnaires; the non-members are generally mentioned only occasionally.

The researcher from a university institute identified as the expert promotor supports the innovation project selectively in some development tasks. One key member can be called a supporter at the relationship level, mainly during the conceptual phase. Moreover, an association coordinator is brought in during the development phase, and is soon viewed as a relationship promotor by one innovation community member. A third supporter, the project manager for a telecommunications company, has contributed to the technical debate since early on in the development phase.

The Composition of the Promotor Network

Three of the four promotors of the innovation community advance the project as expert promotors with their technical and industry-specific knowledge; one of them is a champion since he also assumes the role of power promotor. Although all the promotors in the innovation community bring with them contacts and relationships from their respective industries and areas, and thus complement each other well, only one key person from the entire group is a relationship promotor. The project coordinator, one of the two champions, assumes three promotor roles (power, process, and relationship promotor) and is in the special position of holding two roles (process and relationship promotor) that none of the other promotors fulfil. His networking involves accompanying research in related research projects, and keeping good relations with promotors and clients. The skills of the project coordinator as an "application poet" are generally credited as having secured funding approval. He is on a whole seen as a driving force behind the innovation project.

One crucial factor for the right composition of the innovation community is said to have been a workshop following the foundation of the home networking association in 2009:

> In my view, it was that scenario workshop last summer which clarified the different roles and the potential contributions. Without that, it would probably not have been that easy to form a coherent project consortium. We in fact very quickly realized that there are partners that complement each other rather than compete with each other. (Project Manager, Research Institute, 136)

The interviewees repeatedly emphasize the fact that the actors and their skills complement each other very well in the innovation project. The innovation community in the Connected Energy – SHAPE project is relatively small; nonetheless, half the key persons are to be considered champions. The size seems to be the one required:

> I have worked in other consortia, including at the EU level, and if you have consortia there that are too large, and too many people who want or need to have a say on an issue, then it's often hard to get moving. (Project Manager, Clearing and Settlement Services, 112)

Interestingly, in the interviews, only one partner's contribution to the project is seen as somewhat unclear or doubtful – an organization not represented in the innovation community by any promotor:

> One difficult situation was that we then took on an additional partner . . . in the consortium, even though many failed to see what the real value of his contribution was. (Project Manager, Clearing and Settlement Service Provider, 132)

Table 3.6 lists the information provided by the promotors interviewed regarding the composition of the innovation community. The mean value of all items is 5.3, which thus meets the requirements defined in Sect. 1.1 for the construct of an innovation community. Nonetheless, there are still some deviations that require explanation. The first differences are of a fundamental nature and can be found in almost all parts of the questionnaires returned by the respondents. Champion 1, the project coordinator, consistently deviates upwards in his questionnaire. This is presumably due to the fact that, as the moving force behind the innovation project,

Table 3.6 Responses of community members regarding the composition of the community

Composition of the group of the above persons	Champion 1	Champion 2	Expert promotor 1	Expert promotor 2	Total
Completeness and complementarities of skills					
The group of people named above has all companies and organizations important to the success of the innovation project represented within	7	7	5	3	
Social capital					
The group of people named above has very good relations with decision makers in partner organizations	7	5	5	6	
The group of people named above has excellent contacts with decision makers who are important to the joint project, even beyond the partner organizations involved	6	6	5	6	
Group members have access to persons with relevant (technical) expertise with respect to the joint innovation projects	7	6	5	6	
Reconfiguration					
The group of people named above is open to new partners and members	5	4	5	4	
The group of people named above was or is in a position to let go of members who are unhelpful or obstructive to the success of the project	7	4	4	2	
Arithmetic mean for community composition	**6.5**	**5.3**	**4.8**	**4.5**	**5.3**

Source: Author. Scale: 1 (disagree completely) to 7 (strongly agree)

he has a tendency to see his project in a positive light, and to present it accordingly. On the other hand, expert promotor 1 grades most characteristics throughout the questionnaire in the upper middle range (4 or 5), although in the interview, he describes these topics very positively. This kind of response behaviour is consistent with a more cautious evaluation process, which might be due to his personality. A second type of deviation can be observed for the item "reconfiguration capacity." Here, the overall assessments are relatively low. This can be explained by the fact that the innovation community is developing within the relatively fixed and formal framework of a subsidized project that does not permit staff "restructuring" at will. This was also emphasized in some of the interviews.

The Identity of the Innovation Community

Clarity of group boundaries, functional identity and the existence of a common spirit are all relevant to a strong identity of the innovation community. In the interviews, the promotors of the innovation project Connected Energy – SHAPE confirmed their sense of unity and common understanding of project objectives. However, it can be argued that each partner tends to convey targets rooted in his or her own corporate background, which can impede the development of a common purpose.

Table 3.7 shows that the mean value of the respondents is greater than five, even for the identity construct, which result meets the defined requirements of an innovation community (see Sect. 2.2.3). For the construct cooperation, the estimates are relatively high (mean value: 5.8). For the construct identity, the deviation of Expert Promotor 2 is particularly notable. He is the last member to join the community, and thus was not so involved as the other three promotors in the community who were its main initiators. Particularly noteworthy is his assessment of the "common language" item that he only awarded a score of 2 to. He also addresses this issue in the interview several times:

> For a long time, I had the sense of having come late. It was only a year, but . . . for me . . . I perceived it as being well-established for quite some time, and had the feeling that there was really something heavy going on here. . . and they were using a lot of cryptic abbreviations, with everyone always bringing in the worst ones from his own field – OSGI, UPNP and so on, which really put me off at first. . . . For me, the subjective feeling was: There's an existing group, and now I show up, and don't even know the words . . . That probably happens quite often, I've seen it in other processes, you go into a process or a project or whatever, into an evolving something, and first of all you see something's happening here, and the vocabulary has already been set up. (Innovation Manager, Energy Supply Company, 90–92)

Another promotor confirmed that a common vocabulary was developed very early on, and has been important for the evolution of a common understanding of the project:

Table 3.7 Responses of community members regarding aspects of community identity

Identity of the group	Champion 1	Champion 2	Expert promotor 1	Expert promotor 2	Total
Clarity of group boundaries					
The group of people named above behaves like a team	7	6	5	5	
The people named above see themselves as "insiders"	6	6	4	5	
The group of people named above is clearly perceived as a unit from the outside	7	6	4	5	
All people named above fit into the group very well	7	6	4	3	
Each member of the group knows exactly who is a member of the group and who is not	7	7	4	5	
The group of actors is very stable, turnover is minor	7	6	5	7	
Arithmetic mean for clarity of group boundaries	**6.8**	**6.2**	**4.3**	**5.0**	**5.6**
Functional identity					
The members agree on the importance of the innovation	7	6	5	5	
The members agree on the objectives of the joint innovation project	6	6	5	5	
The members use the same "language" (terminology, technical language, etc.)	7	6	4	2	
There are no conflicts due to conceptual misunderstandings	7	7	3	4	
Arithmetic mean for identity	**6.8**	**6.3**	**4.3**	**4.0**	**5.3**
Cohesion					
For the people named above, it is important to be part of the project	7	6	5	6	
The members are strongly connected to the project	7	6	5	6	
All of the people named above are fully integrated into the group	7	6	5	6	
There are no personal conflicts within the group	7	7	5	7	
There is personal affinity or sympathy between the actors	6	6	5	6	
The network is characterized by a strong solidarity between the individual members	7	6	4	5	
Each member is proud to participate	6	6	4	5	
The people named above feel responsible for the group and defend it	7	6	4	6	
Arithmetic mean for cohesion	**6.8**	**6.1**	**4.6**	**5.9**	**5.8**
Arithmetic mean for community identity	**6.8**	**6.2**	**4.4**	**5.2**	**5.6**

Source: Author. Scale: 1 (disagree completely) to 7 (strongly agree)

... In these first working group meetings, we simply found a common language, so as to say: What is the problem? How can we address it? (Project Manager, Clearing and Settlement Service Provider, 153)

Interaction in the Innovation Community

Behind the construct of interaction quality of the innovation community is the idea that the success of the innovation project is more probable with a high quality of interaction. The construct "interaction" includes the following six aspects: communication, coordination, balance of contributions, mutual support and conflict resolution, effort, trust and group climate.

Table 3.8 shows that, apart from some fluctuations, the level of interaction is relatively high. The mean value for the construct is 5.7. In the interviews, both coordination and exchange within the group were described as informal. The sincerity of communication as well as the regularity and frequency of meetings and exchanges were repeatedly emphasized. The fact that not all promotors rated the "frequency of communication" as high in the questionnaires may be due to the fact that a large part of communication took place bilaterally and in informal meetings outside the project.

References were made to an "open, fair climate" and "positive atmosphere." Good cooperation in a group takes time. This is evident from the statements of the innovation manager of an energy supply company who speaks of "growing mutual trust" (84) and a climate "more open than at the beginning" (86). The questionnaires also confirm that mutual trust and group climate within the promotor network is very good (mean value 6.1). The fact that trust itself is connected to cohesion and to the ability to resolve conflicts is evident in this statement from one of the interviews:

In the last *6* months of the initial phase, we also had the problem that there were attempts ... of course, there were partners who were trying to get in here with huge budget demands, and we pushed them out accordingly, so I think that all of that had a very confidence-building effect, since we ascertained that you could rely on an agreement once it was made. (Project Manager, Research Institute, 133)

Also, the construct "effort" is consistently classified as very good by the respondents (mean value: 5.9).

Measures and Methods of Community Development

Table 3.9 lists the measures and methods applied in the innovation process of the project Connected Energy – SHAPE. Although the distribution is relatively uniform, an emphasis can be ascertained, partially on information procurement (11 measures) and dissemination (10 measures), and partially on interaction (12 measures). Since the project is being carried out in an inter-industry environment with new value chains, involves a system innovation, and calls for a number of legal

Table 3.8 Responses of community members regarding aspects of community interaction

Interaction among community members	Champion 1	Champion 2	Expert promotor 1	Expert promotor 2	Total
Communication					
There is frequent communication between the actors named above	7	6	4	5	
The people named above communicate frequently in spontaneous meetings or phone calls	7	6	4		
They communicate frequently in person	7	5	5	3	
Project-related information is communicated openly and shared	7	6	5	5	
Important information is not withheld from others	7	6	5	5	
The members are satisfied with the information received from other group members	6	6	5	5	
Communication takes place without the consent or knowledge of higher entities	7	6	5	7	
Information circulates quickly within the group of people named above	7	6	5	5	
Arithmetic mean for communication	**6.9**	**5.9**	**4.8**	**5.0**	**5.6**
Coordination					
The sub-tasks that each of the people named above in the innovation project performs are clearly defined	7	6	5	4	
The sub-tasks that each of the people named above performs in the innovation project are well coordinated	7	6	4	4	
The different tasks and duties of individual members are well coordinated	7	7	4	4	
Arithmetic mean for coordination	**7.0**	**6.3**	**4.3**	**4.0**	**5.4**
Balance of contributions					
All of the people named above can contribute their skills to the common project	6	6	5	5	
Everyone in the group contributes to the achievement of project objectives with their skills and capacities	7	6	5	5	
The contributions of the people named above to the innovation project seem well balanced	7	6	4	5	
Arithmetic mean for balance of contributions	**6.7**	**6.0**	**4.7**	**5.0**	**5.6**

(continued)

Table 3.8 (continued)

Interaction among community members	Champion 1	Champion 2	Expert promotor 1	Expert promotor 2	Total
Mutual support and conflict resolution					
The members support each other as much as possible	7	6	4	6	
Debates and controversies are carried out constructively	7	6	4	6	
Suggestions and contributions of group members are taken seriously and discussed	7	6	4	7	
When conflicts arise, they are resolved quickly and without problems	7	6	3	4	
The group is capable of reaching a consensus on central issues	7	6	5	5	
Arithmetic mean for mutual support and conflict resolution	**7.0**	**6.0**	**4.0**	**5.6**	**5.7**
Effort					
All of the people named above fully support the innovation project	7	6	5	6	
The innovation project has a high priority for all of the people named above	7	6	5	6	
The group is making substantial efforts to move the project ahead	7		5	6	
The group is completely satisfied with the individual efforts of each person	6	6	5	6	
Arithmetic mean for effort	**6.8**	**6.0**	**5.0**	**6.0**	**5.9**
Trust and group climate					
There is a climate of trust between members	7	6	5	6	
There is a climate of cooperation in the group	7	6	5	6	
All members can express their views openly	7	6	5	7	
The group members feel free to share ideas unquestioned	7	6	5	6	
Arithmetic mean for group climate	**7.0**	**6.0**	**5.0**	**6.3**	**6.1**
Arithmetic mean for community interaction	**6.9**	**6.0**	**4.6**	**5.3**	**5.7**
Arithmetic mean for community performance	**6.8**	**6.0**	**4.6**	**5.2**	**5.6**

Source: Author. Possible answers from 1 (disagree completely) to 7 (strongly agree)

changes, it is clear why information procurement from various areas and the use of analytical instruments feature so prominently. This fact becomes even clearer due to the fact that almost all measures require full or partial expertise. Although the promotors emphasized the frequent, informal and bilateral communication in interviews, it is evident that a number of highly structured methods were used to advance the innovation process.

Table 3.9 Measures and methods in the innovation process Connected Energy – SHAPE

	Functional focus				Contribution to network and community development		Preconditions
	Information dissemination & training	Info. gathering and analysis	Organization	Personal inter-action	Contribution to network and community development	Contribution to which quality aspect of the community?	Methodological expertise required?
Strategy workshop of the association[a]	X	X	X	X	High	Composition, identity, interaction	Partially
Scenario development[b]		X		X	High	Identity, interaction	Yes
Trend analysis		X			Medium	Composition	Yes
Roadmap	X	X	X		Medium	Identity, interaction	Yes
Project meetings	X	X	X	X	High	Identity, interaction	No
Project outline (based on first "Living Document")	X	X	X	X	High	Identity, interaction	Yes
Development of the software & hardware architecture	X	X		X	High	Interaction	Yes
Meeting and reporting to funding institutions	X		X	X	Medium	Identity, interaction	Partially
Formal kick-off meeting			X	X	Medium	Interaction	Partially
Trade fair presentations	X	X		X	High	Interaction	Partially
Workshops with external partners	X	X		X	Medium	Composition, interaction	Yes
Bilateral, personal communications	X	X	X	X	High	Interaction	No
Industry presentations in the group	X	X		X	Medium	Composition, interaction	No
Project-related excursions	X	X		X	High	Identity, interaction	No
Total	**10**	**11**	**8**	**12**			

Source: Author

[a] In a 2-week workshop of the Connected Living Association, the participants defined the terms and objectives of the project, partitioned them into several working groups, and established a project architecture und roadmap (cf. Section "Additional Methods")

[b] The early scenarios established which partners would be required for the project, thereby helping to establish the right composition of the group

In the interviews, the promotors were largely satisfied with the use of different methods, but the structured development of a common language (vocabulary, technical terms, etc.) was lacking, particularly for the expert promotor who had come late to the innovation community.

3.5.4 Effects of the Innovation Community on the Project's Success

Past Success of the Innovation Project

The success of the innovation project Connected Energy – SHAPE is classified as high (mean value 5.3) by the promotors. Efficiency is assessed as relatively high, with a mean value of 5.6, which is confirmed in the interviews (Table 3.10):

Table 3.10 Evaluation of the success of the innovation project Connected Energy – SHAPE

Success of the innovation project	Champion 1	Champion 2	Expert promotor 1	Expert promotor 2	Total
Effectiveness					
The joint innovation project has been very successful to date	7	6	5	5	
The requirements of clients or users have systematically been considered	4	5	5	5	
Recent results are of high quality	6	4	4	5	
The project partners are satisfied with the result	6	4	4	4	
There has been little need for follow-up work to date	7	4	3	6	
At present, everything points to the future success of the joint innovation project	7	5	5	5	
Arithmetic mean for effectiveness of the innovation project	**6.2**	**4.7**	**4.3**	**5.0**	**5.0**
Efficiency					
The objectives of the joint innovation project have been achieved	7	5	5	5	
From a network perspective, one can be satisfied with the progress of the innovation project	6	5	4	5	
To date, the project has been carried out cost-effectively	7	6	4	7	
The project is progressing as scheduled	7	6	4	5	
The project is within the budget	7	6	4	7	
Arithmetic mean for efficiency of the innovation project	**6.8**	**5.6**	**4.2**	**5.8**	**5.6**
Arithmetic mean for project success	**6.5**	**5.1**	**4.3**	**5.4**	**5.3**

Source: Author. Scale: 1 (disagree completely) to 7 (strongly agree)

> In a small group, you can make decisions faster, and do it either this way or that. During the application process, we had to make certain decisions on which direction we were heading, and then it really made a difference that it was a small circle and everyone had a very clear area of responsibility, without much overlapping, which would have meant longer discussions or rounds of voting. (Project Manager, Clearing and Settlement Service Provider, 114)

The effectiveness of the project is classified somewhat lower, with a value of 5.0. This could be due to the fact that the project was still in a relatively early stage at the time of the case study, and that there were few concrete results visible at that point. One champion saw the funding received as a proof of success in itself, since no other project in the home networking association had received any in the first round. The future of the project was also assessed as positive:

> I'm very optimistic that it's going to be a positive project, so that all partners will be able to achieve their results as planned. Although the sponsor has told us again that we have a very tight schedule, of course, we basically only have 2 years. ... But I think if the work continues as it has in the initiation phase – I mean we have met several times, have had very good sessions and also achieved clearly structured results – so if things continue like this, I think it will be very, very positive. (Project Manager, Clearing and Settlement Service Provider, 163–165)

Conspicuous in the response behaviour of respondents was the success rating of Champion 1. He evaluates project success far more positively than the other respondents do. This could be due to the fact that he is the project coordinator, and the project's success is more important to him since its initiation was to a large extent due to his initiative and effort.

Contribution of the Innovation Community to the Success of the Project

In Table 3.11, the question of the contribution of the innovation community to the success of the innovation project is presented as addressed by the four core promotors. The estimates are consistently high to very high. The expert promotors awarded a value of 5, and the two champions a value of 7, so that the average was 6. The research institute project manager (Champion 1) sees this contribution as crucial: "Without this cooperation, there would have been no project at all" (112).

The leverage created by the members of the innovation community and their according contribution to project success has several facets. Firstly, the home

Table 3.11 Evaluation of the success of the innovation project Connected Energy – SHAPE

	Champion 1	Champion 2	Expert promotor 1	Expert promotor 2	Total
What is your estimate of the contribution of the group of people named above to the success of the innovation project?	7	7	5	5	6

Source: Author. Scale: 1 (very low) to 7 (very high)

networking association provided a suitable framework for the key persons – in fact it was where they met in the first place –, one which very good use was made of. The promotors also made good use of their contacts in acquiring funding and support within the association, determining precisely which characteristics were particularly desirable for project funding, and then clearly presenting these aspects to the sponsors:

> So we persistently refined and revised the objective within the context of the outline and of the application, so that now really everybody was backing it, and then we presented it to the sponsors and developers in the core consortium. There were always these key people, and I think that ultimately helped us to acquire funding relatively quickly. . . . One highlight for me was at the trade fair CeBIT where we had a session in which we gave several successive presentations, where I believe some representatives of the sponsors were present, as well, who could see, yes, this SHAPE project has really moved forward. That people got along with each other so well, that it was this well coordinated. . . all that really looked very good to the outside – at least compared to the other initiatives that were at that time still more or less in the discovery phase. (Project Manager, Clearing and Settlement Service Provider, 100, 117)

Secondly, in terms of group size and composition, the innovation community was a powerful constellation. The small size of the group and their clear and distinct responsibilities were beneficial to fast decision-making. The clear division of responsibilities in turn was due to the specific composition of the community: business-oriented promotors from various industries on the one hand, and a more "open process," or science-oriented promotor on the other (Champion 1).

> We soon noticed that here we had partners complementing each other and not just facing each other as competitors. (Project Manager, Research Institute, 136)

Thirdly, a combination of conflict management and cohesion of the group has helped to reduce external pressure:

> Yes, initially, there were a few conflicts with other consortia in the association, so of course it is good to be able to pick up the phone quickly and support other partners. For instance, I supported one partner when there was a conflict with someone from the association management, where there was basically a tendency of getting rid of him as group leader at the time. By the same token, when some unjustified things came up against other members, people supported each other very well, in order to maintain the line that we had decided on for the project. (Project Manager, Clearing and Settlement Service Provider, 130)

3.5.5 Conclusions

This case study has examined the innovation project Connected Energy – SHAPE which emerged from the home networking association Connected Living as an innovation community. The project was designed to develop a home automation technology with integrated appliances to determine the energy savings potential of households, at the same time contributing to standardization in that field. It is a radical system innovation, considering the changes required in the structural and legal framework.

The in-depth examination identified an innovation community that is small, but effective in action. The possible promotor roles – power, expert, process, and relationship promotor – are all represented, and two of the four promotors are champions. The two champions and one expert promotor are also core promotors, having initiated the project and been involved in it from the start. The second expert promotor was soon included. It can therefore be assumed that the composition of the innovation community has significantly contributed to the overall stability of the project. The identity of the group is characterized by a strong sense of solidarity, and the interaction is considered to be trusting and supportive.

The innovation community has contributed to the success of the project by its competent interaction with the home network association, the complementary composition of the group, and its capacity for resolving conflicts.

3.6 Dye-Sensitized Solar Cells: A Research Network for a New Photovoltaic Technology

Severin Beucker

Fact Sheet

Case Study: Dye-Sensitized Solar Cells

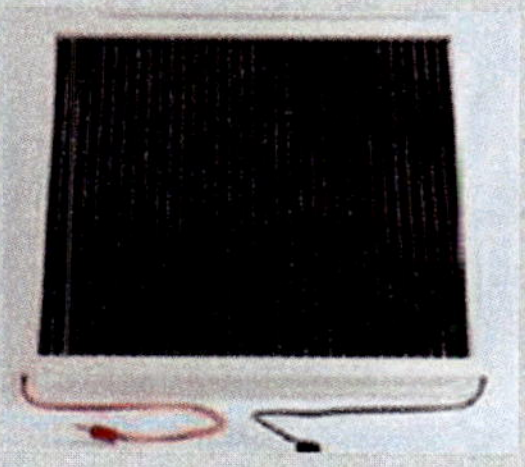

Innovation object:	Glass-based, monolithic solar cells
Industry sectors:	Solar industry, chemical industry, various manufacturers
Period analyzed:	2008–2011
Initial situation:	The photovoltaics industry has become a profitable and fast-growing market, with numerous areas of application. However, new materials and manufacturing processes are needed for higher efficiency and new applications.
Current phase:	Development phase
Research highlights:	Radical innovation project, changing photovoltaics markets and energy policies, high level of cooperation, large scope of methods and technologies employed

Key persons involved:	8
Size of innovation community:	5
Group composition:	Almost only champions
Regional distribution:	International (Germany and Switzerland)

Community performance:	5.2*
Contribution to project success:	5.8*
Project success:	5.3*

*On a scale from 1 (very low/poor) to 7 (very high/good).

3.6.1 Context and Object of Innovation

The photovoltaics industry has in recent years become a profitable, fast growing market, reinforced by powerful new photovoltaic technologies that do not use the traditional, but expensive and scarce polysilicon. Thin-film technology, new semiconductor systems, organic solar cells and dye-sensitized solar cells are increasingly emerging as alternative approaches in photovoltaic power generation. However, they require not only new raw materials, but also different manufacturing structures and value added chains.

For the past 15 years, research into dye-sensitized solar cells (DSSCs) has been progressing worldwide in a number of research networks. Here, unlike in conventional solar cells, a metal-organic dye embedded in thin mixed layers transforms light into electrical energy. Compared with existing photovoltaic technologies, DSSCs are characterized by particularly good behaviour with regard to indirect radiation, high temperature resistance, the possibility of semi-transparent design, and their colour, making them ideal for use in façade, indoor and design applications.

The technology is available in either of two approaches: First, glass-based dye-sensitized solar cells that, as partially translucent solar cells, can open up new areas of application in building-integrated photovoltaics; and second, the polymer-based option. Especially the polymer-based technology is a potentially cost-effective, lightweight and flexible solar cell that could in future enable a new variety of applications, e.g. the power supply for charging mobile devices such as mobile phones or notebooks, or also for integration into packaging materials and textiles, or in roof foils. However, the polymer-based option currently provides lower output performance, but on the other hand, at the corresponding scale, has lower production costs.

The basic manufacturing processes for DSSCs are based on simple screen-printing and sintered glass processes, or, in the case of flexible polymer-based cells, on a "roll-to-roll" process. These steps do not require clean room technology, as is necessary for the production of silicon-based photovoltaics. This makes DSSCs a cost-effective approach for mass production. Moreover, due to the materials used and their low energy payback time, they have a high potential for resource conservation. Glass-based DSSCs can be manufactured by simple screen-printing and sintered glass technology, or in the case of flexible polymer-based cells, in a film-based roll-to-roll process.

The development of the technology to market maturity will mean meeting such key conditions, as the development of new value chains and markets. A German-speaking DSSC R&D group consisting of two medium-sized companies, a major chemical corporation and other research partners, is engaged in a race with Japanese, Australian and U.S. consortia. In a joint venture, the companies of this association intend to initiate the joint marketing of DSSCs. One crucial step is the development of a strong and stable innovation community of manufacturers and suppliers reflecting the future DSSC value added chain, and recruiting trend-setting users and decision makers to develop test markets.

Object of Innovation

The object of innovation of this case study consists of the development of glass-based, monolithic DSSCs, and the associated innovation network.

The monolithic DSSC approach differs in important respects from the development pursued in previous projects, in that it is based on a crystal-based structure with sintered sealing layers. In the monolithic DSSC, the simpler layer structure permits the entire cell to be manufactured via the screen-printing process, which holds considerable potential for rationalization and cost reductions in future production (Fig. 3.6).

The development of the monolithic DSSC is a challenging task, since technologies from different scientific disciplines, such as physics, chemistry, material sciences and engineering, must be combined.

In order to develop the technology with its high degree of complexity for the market, the R&D network described will pursue two key objectives, which are to be supported by the creation of an innovation community:

1. The development of an efficient and stable network of manufacturers and suppliers, to build a value-added chain for DSSCs. In addition to the further development of cell technology, this network should primarily coordinate all the key steps and processes for future production.
2. The inclusion of future trend-setting service and application users of DSSCs into the innovation process, so as to create at an early date the conditions for test markets (e.g., for façade integration or in consumer electronics), in which products can be tested.

Fig. 3.6 Prototype of a monolithic DSSC with the dimensions 30 x 30 cm (Source: Company II)

The novel and complex nature of the innovation project is highlighted by the following assessment of one of the interviewees:

> . . . It is both: it is of course, novel, complex, which means complex in terms of . . . technical aspects from various fields of knowledge [which have to] be brought together and combined in a manner that fits. (Innovation Coach, Research Institute I, 66)

In particular, the development of an efficient supply chain is a major challenge that requires recruiting further selected partners for the existing network, such as in the form of suppliers or pilot customers, and bringing them together. As one of the partners in the innovation network puts it:

> . . . Of course we don't have the know-how to develop such modules, and for this reason, it was important to work . . . with . . . a company that has expertise in the area of the modules. And then,. . . as we looked around, we came upon the company [Company I] ....especially due to the fact that they had just that expertise. (Managing Director A, Company I, 37)

The development of monolithic DSSC and the associated innovation networks can therefore be characterized as a novel and radical innovation project: Although each partner in the innovation project itself has extensive expertise regarding the separate technical and chemical components of the DSSC, all surveyed innovators indicate that innovation requires a new, not previously existing cooperative effort with partners, and that through them, a completely new customer network will emerge.

3.6.2 The Innovation Process

History and Factors Influencing the Innovation Process

Research into DSSCs has been going on since the late 1980s. Their discoverer was Prof. Michael Grätzel, who taught at the École Polytéchnique Fédéral in Lausanne, Switzerland. He transferred the principle of photosynthesis to metallo-organic dyes, and in years of research used them to develop a functional and reproducible photovoltaic cell. Since then, the research landscape has been diversified and internationalized considerably, and several cooperating, and also competing research networks have emerged. The intermediate state of the research community is documented by Beucker and Fichter (2007).

The innovation process analyzed in the present case study originated in the ColorSol research project,[8] which was completed in 2008. This project had three objectives: (1) The cooperative technology development of glass-based DSSCs in selected fields and their applications; (2) Support for the technology and market development by means of an innovation and technology analysis, in which the

[8] The ColorSol Combined Research Project for Sustainable Product Innovation through Dye Sensitized Solar Cells was subsidized from April 2006 to October 2008 by the German Federal Ministry of Education and Research (BMBF).

technology and product roadmaps would be designed, life cycle assessments carried out, and the success factors of product innovation in the solar industry identified; and (3) Measures for the transfer of results in the solar industry were designed.

This research project and the partners cooperating in it have great influence on the innovation network analyzed below, which is to be explained by means of the following effect factors.

The existing network: The consortium of the project ColorSol consisted of application-oriented research institutes, companies from the glass and chemical industries, and a planning office for photovoltaics. The project thus established a high-performance network of research and development institutions on the one hand and potential users of the technology on the other. Thus for example, close contact was maintained with organizations, networks and projects from the research environment, in order to get a grasp of DSSC technology and the emerging technological, economic and social trends in their environment, and also to exchange technological experience and discuss user requirements. Moreover, market potentials and target costs were identified, and discussions carried out with potential alliance partners for production and marketing. In this process, investment requirements and marketing structures were ascertained. Towards the end of the project, talks on establishing a joint company were conducted, which would produce and market glass-based DSSCs, in which not only the project partners, but also representatives of venture capital companies were involved. The innovation process of the present example is based on this longer-term, high-performance network.

The need for reorientation: With the completion of the ColorSol project, it became clear that a subsequent development and differentiation of DSSC technology would be necessary, in order to position them successfully in specific application fields and in the market. The fact that a systematic development of the innovation and partner network would be needed to achieve these goals was also recognized. In this context, the practical project InnoCo-DSSC was launched, with the intention to proceed with a realignment of the content of the research, and the creation of such innovation communities. At the beginning of the DSSC InnoCo practice project, several of the ColorSol research partners began to analyze and evaluate technological variants of the cell, resulting fields of application, and partner constellations. One of the interviewed new partners describes it in this way:

> So then we presented our DSSC technologies to a panel of various people who might have something to do with this matter, independent of the rest of the ColorSol project, so to speak, and then we were suddenly not only a supplier, a potential supplier of chemicals, but providers of this entire technology. (Managing Director B, Company II, 56)

The involvement of new partners thus led not only to the development of the focal innovation process described in the next section, but also to a shift of roles within the innovation network.

The benefits of networking in the pre-competitive market and technology environment: Although the DSSC market and technology environment is largely pre-competitive, there are still many competing technological concepts and research associations. In addition, there are various technical concepts and research approaches, such as the thin-film modules, already on the market, or approaches completely based on photoactive polymers, which enhance the innovation and cost pressure on DSSCs. The market and technology environment for traditional photovoltaic technology, such as in buildings or mobile applications, results in narrow corridors for the prices to be achieved by future products on the market. It is therefore important to identify the specific advantages of the DSSC, and to tap additional product benefits and possible applications for new market segments. This may for example be found in architecture and building integration (e.g. façades, windows, glass partitions, etc.), since these applications have hitherto been realizable only partially or not at all on the basis of existing technologies. The formation of a network that promotes the development and positioning of such new fields of application and market segments could provide an advantage for the organizations involved, since they would not have had sufficient knowledge and capacity for such a step on their own. This is illustrated in a passage from an interview:

> ... Basically, that is research, too... because it also involves mustering the capable people, ... so for us, it's also an element of success ... if you recruit the good people, the chances are greater that something will come out of it, than if everyone researches in his own corner, or doesn't even hear about someone else, who may already have the solution in his drawer, but it will never come out, that would be a real shame. The interaction of people, that is, basically, the true measure of success. (Managing Director B, Company II, 220)

Especially during the pre-competitive phase, networking can have great benefits for the partners in terms of allocating necessary resources and skills to survive in competition with other consortia. This also motivated the partners in the innovation projects analyzed.

Research funding for renewable energies: Renewable energy, specifically photovoltaics, are a major focus of German and European research funding and energy policy. Despite declining subsidies for solar electricity in Germany, photovoltaics is considered a great hope for the achievement of renewable energy targets. Both government and industry recognize that production costs for electricity from photovoltaics must drop in order to provide competitive electricity. These are important conditions for the development of DSSCs, since this technology is associated with great potentials for cost reduction and also opens up many new application areas, such as in architecture, due to its specific characteristics. The research funding plays an important role in the innovation process, by reducing investment risk:

> ... Indeed the R&D risk is not negligible in this whole story, and we'll also have to try a lot of different things, so at the stage we are now ... it is true that subsidies really make sense. I mean, the BMBF [Federal Ministry of Education and Research] has also recognized this, otherwise we wouldn't even been able to launch this OPV [Organic Photovoltaics] tender *2* or *3* years ago... so that Germany could play in this league accordingly. (Managing Director A, Company I, 57)

Research funding also supports the cooperation partners in their positioning in the scientific and market environment, and influences the emergence of pre-market structures and powerful partnerships. The specific enabling environment for renewable energies and the associated initial advantage that the partners of the innovation network will secure are therefore an important factor in the case at hand.

The Focal Innovation Process

The focal innovation process examined in the context of this case study includes the development of a monolithic DSSC and the associated production technology, and the creation of a powerful innovation community to support this process. The focal innovation process is part of the practice of the InnoCo DSSC project, and was carried out between 2006 and the spring of 2011.

The DSSC approach pursued in the practice project differs from the developments of previous projects (e.g. ColorSol), which were to some extent based on a glass-based structure with sintered sealing layers. In the case of the monolithic DSSC process, it is possible with a simpler layer structure to produce the entire cell using the screen-printing process, which makes use of a considerable potential for rationalization and cost reduction. Since the technology described was, at the beginning of the project, still in its pilot stage, one of the key objectives was to develop the cell and the future production technology further. Here, the necessary skills were to be concentrated, and specific sub-projects defined amongst the partners.

If we include the history of the innovation process described above, four distinct phases can be defined, which have had an influence on the focal innovation process:

- The maturation phase from the late 1980s until around 2005
- The initial phase: 2006–2007
- The conceptual phase: 2007 to about 2008
- The development phase: spring 2008 to spring 2011

The *maturation phase* of the focal innovation process can be said to include the entire history of DSSC research described in the previous chapter. Some of the actors who participated in the focal innovation process had worked since that time with professor Grätzel at the École Polytéchnique Fédéral in Lausanne. During this long maturation phase, many foundations for the understanding of the DSSC were laid, and an extensive and highly specialized network of research organizations and companies emerge that addressed the basic and application research in the field.

With the passage of the Renewable Energy Sources Act (EEG) in Germany in 2000, a massive demand for photovoltaic technology was created. Parallel to that, research activities in Germany were intensified so as to reduce the high cost of photovoltaic systems and to give German companies and research institutions competitive advantages over foreign companies. It quickly became clear that in addition to the established approaches of crystalline silicon cells (first generation), other, more cost-effective approaches would be needed to realize the desired expanded capacities of photovoltaics. In this context, research in the fields of thin film cells (second generation) and organic photovoltaics (third generation), which also includes the DSSC, was intensified. There were several research projects in the

German language area dealing with the study of cell and production technology as well as economies of scale through mass production; one was the above-mentioned ColorSol project. The maturation phase will not be subject of the further research. The complex and diverse networks resulting from this phase would require an additional analysis.

The *initial phase* begins with the start of the ColorSol project, in which a number of organizations were brought together that were to play an important role in the later focal innovation process. Progress was achieved in the technology of glass-based DSSCs and in knowledge of the market, and a network of potential value-added partners and subsequent users of the technology emerged.

Towards the end of the ColorSol project however, it became clear that on the one hand, targeted further development and differentiation of the technology of glass-based DSSC would be necessary, and on the other, more partners would be needed to enable realization of future production, and to cover specific areas of application of this new photovoltaic technology. A former project partner from ColorSol describes the need to deepen and develop the ColorSol results:

> We in the innovative community ... actually sought a connection with ColorSol, ... to continue to operate ColorSol, that was actually the original intention. (Innovation Coach, Research Institute I, 28)

That was also the beginning of the *conceptual phase* of the focal innovation process, in which some of the ColorSol partners pursued the continuation of the innovation and partner network in a concerted manner. A high level of dynamics characterized this phase, since there were different ideas about the objectives to be pursued, and the actors were at the same time involved in various networks. In parallel, we had the opportunity to accompany the development of an innovation network via the InnoCo research project, and thus to establish a new, independent project with partners from the completed ColorSol project, supported by accompanying research. This new project goal first had to be explained to the partners, since it involved a break with the previously agreed-upon research process, which was focused more on technology development. One of the key actors from a research institute, who was at the same time one of the initiators of the project, describes his role as follows:

> ... The other companies don't have to pay our way. So we weren't in the role of a traditional consultant, who works on a contract basis. We were actually the founders of this practical area. (Innovation Coach, Research Institute I, 111)

Through the initiative of the participating research institute, the practical field and the research network for DSSCs under examination here was initiated, and ultimately carried over into a formal partnership with a work plan, a division of labour and a cooperation agreement in the form of a research proposal.

This point in turn marks the transition to the *development phase*, which began in the spring of 2008 and lasted until the spring of 2011. During this phase, technological variants of the cell and the resulting application fields and potential partner constellations were first analyzed and evaluated in a cooperative effort of three partners. This began with the definition of one's own role and one's positioning within the research network. The Managing Director of a specialty chemicals company describes this as follows:

> . . . I first had to try to find innovation partners, . . . but also to penetrate the entire innovation chain. Furthermore, I had to worry . . . about where we actually were; with this DSSC, we are at the very beginning of the value-added chain. (Managing Director A, Company I, 21)

This initial analysis was followed by the realization that the partner network had to be extended, so as to create a powerful innovation community. In the context of a strategic partner analysis, further actors were identified to complement the innovation community, thus closing the gaps in the supply chain. The important thing at the time was especially to present the integration of a partner with systems expertise in the field of DSSC as well as with knowledge in the up-scaling and the large-scale production of specialty chemicals, the precursors of the DSSC. Contacts with potential partners already existed to a degree via the existing network, so that they could be addressed directly.

In addition to recruiting other partners, another crucial step in the development phase was to define specific projects between partners and to initiate technological developments (e.g., the optimization of chemicals and the development of production technology) with them, and to encourage the extension of the network of future service and application users.

This phase temporarily came to an end in early 2011, with a final multipliers workshop and a resulting network of research institutes and companies, which was formed in order to realize a 300 sqm demonstration façade for DSSCs in 2013. This step is of great importance for DSSC technology, as it would be the first large-scale installation of this cell type in Europe, for the implementation of which a high-performance added value chain would be necessary.

Organizations Involved in the Innovation Process

A large number of organizations are involved in the advanced DSSC research network (see Fig. 3.7). The organizations participating in the focal process of innovation include:

- A research institute in the area of innovation research
- A medium-sized company in the area of specialty chemicals involved in the field of ionic liquids, one of the key components of the DSSC
- A medium-sized Swiss company in the area of specialty chemicals and apparatus construction with systemic knowledge of DSSCs
- A large company in the chemical industry active in the research and development of dyes

The key player is the research institute from the area of innovation research. It coordinated all the essential activities of the focal innovation process.

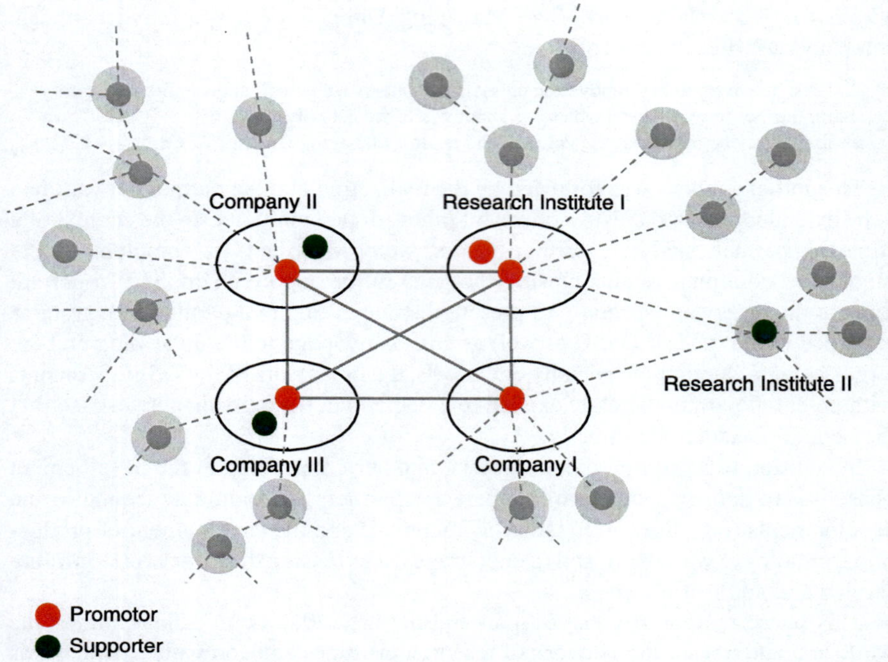

Fig. 3.7 The DSSC advanced research and innovation network (Source: Author)

3.6.3 Development of the Innovation Community

The focus of this case study is the question of how and to what extent innovation communities contribute to the success of radical innovation projects. This is to clarify what key actors and promotors were involved in the focal innovation process of DSSCs from the spring of 2008 to the spring of 2011.

Key People and Promotors

In the DSSC case, a total of eight key people can be identified (see Table 3.12). On the one hand, they come from research institutions and on the other from chemical industry companies. Of these eight, two key people each are from an innovation research institute, a large corporation in the chemical industry, and a Swiss-based medium-sized specialty chemicals and equipment company. The other two key actors come from a medium-sized chemical company and another research institute.

Six key people meet the definition of promotor. The expert promotors are most strongly represented in the network (four), followed by power promotors (three) and relationship and process promotors (two each). Of the six promotors, two fulfil the criteria of a champion. None of the key people assumed all promotor roles simultaneously. The case also shows that not all key people must be promotors or

Table 3.12 Key actors in the DSSC innovation process (2008–2011)

	Geographic location	Key person	Promotor	Champion	Member of the innovation community
Scientific community					
Innovation coach, Research institute I, Inno. research	Berlin	Yes	Power, process, relationship promotor	Yes	Yes
Scientist A, Research institute I, Inno. research	Berlin	Yes	Process promotor	No	Yes
Scientist B, Research institute II, Photovoltaics	Baden- Württemberg	Yes	No	No	No
Business					
Managing director A, Medium-sized company I, Specialty chemicals	Baden- Württemberg	Yes	Power and expert promotor	No	Yes
Managing director B, Managing director, Medium-sized company II, Specialty chemicals/ equipment	Switzerland (Vaud[a])	Yes	Power, expert and relationship promotor	Yes	Yes
Managing C, CTO, Medium-sized company II Specialty chemicals/ equipment	Switzerland (Vaud)	Yes	Expert promotor	No	No
Innovation manager, Large corporation III Chem. industry	Switzerland (Basel)	Yes	Expert promotor	No	Yes
Product managing director, Large corporation III Chem. industry	Hesse	Yes	No	No	No
Number	**8**	**8**	**6**	**2**	**5**

Source: Author
[a]Lausanne is the capital of the Canton of Vaud

members of the innovation community. Thus, two key people are not promotors. They can be classified as supporters and belong partially to the wider network of the innovation community. Finally one key person meets the definition of a promotor but is not neither a promotor nor a member of the innovation community. For that reason he was also classified as a supporter.

The case of the DSSC thus comprises an extended network of key actors, which consists of eight persons. Of these key actors, five in turn formed an innovation community, which can be clearly distinguished from the network by its intensive internal cooperation and its high sense of commonality.

Evolution: Steps and Phases of Community Development

The emergence and development of the DSSC innovation community can be traced according to the three phases of the innovation process described in Sect. 3.6.2. The following factors have therefore been particularly influential for the formation and development of the innovation community:

- The development of an existing network consisting of partners of the concluded ColorSol project, and new actors.
- The need for a new direction caused by the differentiation of the technology of glass-based DSSCs and their field of application, which demanded a specific reorientation of the research content.
- The advantages of network formation in a pre-competitive market and technology environment, due to the fact that the partners in the community can gain competitive advantages through cooperation with other consortia.
- The subsidies for research in the field of renewable energy, which made funds available, and which helped the community to a position in the scientific and market environment, and supported the development of pre-market supply structures and powerful partnerships.

Table 3.13 has made it clear that the number of participating organizations and key people in the innovation process is not consistent. We will dispense here with any detailed analysis of the maturation phase, since several major innovation networks were operating in parallel.

In the initial phase in 2006, the research project ColorSol was created as an independent network with the participation of Research Institute I, Research Institute II and Medium-Sized Company I, from the chemical industry. In addition, other partners were instrumental in this project (see also Sect. 3.6.2), but later in the focal innovation process and the innovation community, they no longer played any significant role. The innovation community analyzed here grew out of this stage.

During the conceptual phase, Research Institute I and Medium-Sized Company I began, with the knowledge of, but independently of the other project partners, to

Table 3.13 The evolution of the DSSC innovation community

	Initial phase (2006–2007)	Conceptual phase (2007–2008))	Development phase (2008–2011)
Participating organizations	Research institute I, Research institute II and Medium-sized company I	Research institute I, Medium-sized company I	Medium-sized company I, Medium-sized company II, Research institute I
Innovation community	Innovation coach, Scientist A, Managing director A, Scientist B	Innovation coach, Scientist A, Managing director A	Innovation coach, Scientist A, Managing director A, Managing director B, Innovation manager
Other promotors participating in the preliminary phase or the innovation process	Another research institute and other companies	–	Managing director C, Product manager

Source: Author

define a separate project for the further development of DSSCs and the associated production technology, with a new goal orientation. They began to selectively promote and seek partners for this project. During this phase, the link to Managing Director B of Medium-Sized Company II in Switzerland emerged. The innovation community was thus expanded by a major player and champion, and starting in 2008, moved into the development phase, at which stage the other key actors, Managing Director C of Company II and the Innovation and Project Manager of Major Corporation III in the chemical industry, joined.

The following findings can be derived from the evolution of this innovative community:

- The community is characterized by the continuity of a few promotors (the Innovation Coach, Scientist A and Managing Director A), who have been working together since 2005 on the research topic. These promotors played a role in all three phases, and ensured the stability of the organization and the communications of the community.
- One of the promotors who is also a champion (Innovation Coach, Research Institute I) played a crucial role in the initial and conceptual phase (see also Sect. 3.6.2), through his systematic search for suitable partners and his definition of development projects. Only through the integration of new resources did the network become a powerful innovation community.
- The integration of another champion (Managing Director B) extended the existing community by a vital key person. Through his network and his experience, he contributed to the performance of the innovation community. He also provided the interface to an extensive network of innovation in the area of DSSCs, which has since the 1990s conducted intensive research in this area.
- The funded and systematic structure of the DSSC community in the InnoCo research project was a key success factor. This support in the form of accompanying research provided important personnel and organizational resources to support the construction of the community through analytical and methodological skills (see Sect. 3.5.4).

The Composition of the Innovation Community

The DSSC community has six promotors, of which two can be classified as champions, i.e., they fulfil multiple roles (see Table 3.12). These champions are, first, the Innovation Coach from Research institute I (power, process and relationship promotor) and, second, Managing Director B from the Medium-Sized Swiss Company (power, technical and relationship promotor). Both play key roles for the innovation community because of their expertise and their networks, as the following interview passages show. One stated, regarding the Innovation Coach from Research Institute I:

> ... The [Research Institute I] has actually assumed the key role as an interface;... they really drove the whole thing forward and also actively maintain the contacts, and sent

> e-mails out accordingly... and provided the information in advance... and that set off a flurry of internal traffic between the partners ... (Managing Director B, Company II, 105)

Especially, the role of the process and relationship promotor is emphasized. Also highlighted is the Managing Director of Company II, who, in addition to that of a power promotor, also assumed the role of relationship promotor:

> For example, he has contacts with battery manufacturers etc. ... But there are contacts which are sort of latent ... and [Managing Director B] has that, they deliver their dyes and other things to [X] R & D facilities, he has an incredibly rich network which is of course very important for the whole project, and also for later, I think. (Innovation Coach, Research Institute I, 159)

The two champions both have a central role for the focal innovation process. They organize the process, make decisions, provide resources, and have an extensive network of contacts in research and science, as well as social capital. Altogether, the innovation community, if one includes the other promotors in the evaluation, has all relevant characteristics represented within. In addition to the above-mentioned power, process and relationship promotors, especially the expert promotor is often present, which ensures a high level of expertise.

According to Table 3.14, the four promotors questioned hold the view that the innovation community has all the people important for the success of the innovation project represented within it, so that the community is thus completely and complementarily occupied. All interviewed promotors rated this point with at least a 5, on a scale of 1 (disagree completely) to 7 (strongly agree).

The promotors surveyed give the social capital of the community comparable scores. All are of the opinion that the group has very good relations with actors in partner organizations who make decisions regarding the joint innovation project, and that the group also has very good contacts with decision-makers outside the participating partner organizations who are important for the joint project; all give this a score of at least 5. They also all view very positively the access of the group to people with relevant experience for the innovation project.

The promotors interviewed also see the reconfiguration capacity of the group as largely positive. The community demonstrated this ability particularly in the transition from the conceptual to the development phase (see Sect. 3.6.2). All participants evaluate the openness of the group with at least 5. The question as to the ability of the group to let go of members who are not helpful or necessary for the success of the project, or to effect such a separation, is also mostly rated positively. Only Managing Director A does not share this view to the same extent (a score of 4). This could be because he perceived the necessary focus of the community differently than the other members surveyed, and therefore considered separation or reconfiguration long overdue.

The Identity of the Innovation Community

Crucial to the identity of the community is the clarity of group boundaries, the functional identity, and the cohesion of the group members (see Section Identity of

Table 3.14 Responses of community members surveyed on the composition of the community

Composition of the group of the above-named people	Innovation coach (Champion)	Managing director A	Managing director B (Champion)	Innovation manager	All
Completeness and complementarity of skills					
The group of above-named people has all companies and organizations represented within that are important to the success of the innovation project	6	7	5	5	
Social capital					
The group of above-named people has very good relations with actors in partner organizations that make decisions regarding the joint innovation project	6	7	5	5	
The group of above-named people has excellent contacts with decision makers who are important to the joint project, even outside the partner organizations involved	6	6	5	5	
Members of the group have access to persons who have relevant (technical) expertise with respect to the joint innovation projects	6	7	6	6	
Reconfiguration					
The group of above-named people is open to new/additional partners/ members	6	5	6	6	
The group of above-named people was or is in a position to let go of members who are not helpful or necessary for the success of the project, or to ensure a separation from them	6	4	6	7	
Arithmetic mean for composition of the community	**6.0**	**6.0**	**5.5**	**5.7**	**5.8**

Source: Author. Possible answers from 1 (disagree completely) to 7 (strongly agree)

the Innovation Community). Two of the four promotors see the group boundaries as very clear, while two others would not agree as strongly (values of 4.7 and 4.4). Especially in the case of Managing Director B, this can be due to his integration in a very extensive research network on DSSCs. This probably makes a clear demarcation of the community more difficult for him.

All respondent promotors expressed strong agreement regarding questions concerning the functional identity of the group (see Table 3.15). The promotors are largely in agreement about the importance of innovation and the goals of the joint innovation project. They also all use the same "language," and there are no major conflicts due to conceptual misunderstandings. The Innovation Coach

Table 3.15 Responses of community members surveyed on aspects of community identity

Identity of the community	Innovation coach (Champion)	Managing director A	Managing director B (Champion)	Innovation manager	All
Clarity of group boundaries					
The group of above-named people behaves as if it were a team	6	3	5	6	
The above-named people see themselves as "insiders"	6	6	2	6	
The group of above-named people is clearly seen from the outside as a unit	4	2	3	6	
All above-named people fit very well into the group	6	3	6	6	
Each member of the group knows exactly who is a member of the group and who is not	7	7	6	7	
The group of actors is very stable, little membership turnover	7	7		6	
Arithmetic mean for clarity of group boundaries	**6.0**	**4.7**	**4.4**	**6.2**	**5.2**
Functional identity					
The actors agree on the importance of innovation	6	5	6	6	
The actors agree on the objectives of the joint innovation project	6	5	6	6	
The actors use the same "language" (use the same terms, technical language, etc.)	6	7	6	6	
There are no conflicts due to conceptual misunderstandings	6	7		7	
Arithmetic mean for identity	**6.0**	**6.0**	**6.0**	**6.3**	**6.1**
Cohesion					
For the above-named Individuals, it is important to be part of the project	6	1	4	6	
The actors are strongly connected with the project.	6	2	5	6	
All the above-named people are fully integrated into the group	6	6		6	
There are no personal conflicts within the group	6	4		7	
There is personal affinity/sympathy between the actors	6	5		6	
The network is characterized by a strong cohesion of its individual actors	6	3		6	
Each actor is proud to be part of the group	6	2		6	

(continued)

Table 3.15 (continued)

Identity of the community	Innovation coach (Champion)	Managing director A	Managing director B (Champion)	Innovation manager	All
All above-named people feel responsible for the group, and protect it	6	2		6	
Arithmetic mean for cohesion	**6.0**	**3.1**	**4.5**	**6.1**	**4.9**
Arithmetic mean for identity of the community	**6.0**	**4.3**	**4.9**	**6.2**	**5.3**

Source: Author. Possible answers from 1 (disagree completely) to 7 (strongly agree)

describes how the identity of the community has been encouraged, and an open communication style emerged:

> You're at the first meeting, then a second meeting with [Company II], you've had an exchange of views, and expressed some ideas, and developed them, and then later, even at this stage, discussed convergent and divergent interests ... I think those were really the crucial steps. (Innovation Coach, Research Institute I, 209)

With regard to functional identity, the community can thus be regarded as a relatively homogeneous group with a clear understanding of the goal and significance of the common innovation process.

Somewhat less consistent is the view of the cohesion of the community. While two of the promotors (the Innovation Coach and the Innovation Manager) expressed strong agreement (values of 6.0 and 6.1), Managing Director A evaluated the situation somewhat less optimistically (value 3.1). In his view, the cohesion is not so strong because, due to the emerging competition with the Company II, he sees his own company's business model as endangered. This is shown in the following comment:

> When you deal with mutual trust, with the material, without fear. Yes, but that trust, I see that as the biggest hurdle, because [Company I] to some extent has a similar business model as [Company II]. Yes, that would be more ideal at that point, if we had someone who only produced modules. But [Company II] is a hybrid there, and so that's a problem. (Managing Director A, Company I, 163)

This shows that the relationship between the two Managing Directors is difficult, despite the open communication style at the beginning of the project. This may be due to a lack of confidence and the threat of competition – an impression that is confirmed in the next section. Managing Director B commented on these issues with regard to his assessment as to the inadequate cohesion of the community.

Interaction in the Innovation Community

The construct of the performance of innovation communities assumes that the contribution of a network of promotors to the success of an innovation project increases with the quality of interaction. With reference to the DSSC community, what needs to be clarified is how its communications, coordination, contribution of members, mutual support and trust, and group climate are assessed.

Overall, the quality of community interaction is considered good, with a mean value of 5.1 (see Table 3.16). However, with respect to the individual promotors and questions, there are significant deviations that require a more detailed analysis. In particular, the significant deviation in the assessment of Managing Director A of Company I should be examined. Managing Director B of Company II only partially answered the questionnaire in this area, since in his opinion, he was not capable of satisfactorily answering the questions he left out. The Innovation Coach and the Innovation Manager on the other hand showed very similar response behaviour which requires no comparative analysis.

The area of communications is assessed positively by all respondents (mean value: 5.0). Managing Director A on the other hand provided a significantly different response to the question regarding satisfaction with the information obtained from other members of the group (value: 2). This response points to a fundamental problem of this innovation community, which was discussed in the previous section, and is reflected in several answers regarding the interaction between community members. Managing Director A felt less integrated into the community as the innovation process progressed. This is particularly shown by his strongly divergent responses to the following questions: He sees the sub-tasks which others perceive as existing in the innovation project as not clearly defined (value: 2), their contributions to innovation activities as unbalanced (value: 1), and feels that there is no atmosphere of trust between people (value: 1). This is also confirmed by statements from the interview, in which he differentiated between the partners and their tasks:

> The management was actually okay. ... The important thing is that the partners trust each other, which I really have to stress, so for me, that's the key thing, and then the rest will build itself up on its own... (Managing Director A, Company I, 238)

However, that trust seems to be severely disturbed, as a further passage demonstrates:

> Confidence grows, too, if you can generally see that someone else isn't going to screw you .... Yes, and you can also see that if ... As long as you don't have much exchange, that trust can't grow, because you ..., Yes, mistrust is something like fear, and fear often has something to do with the fact that you have incomplete knowledge about something. (Managing Director A, Company I, 249)

He sees the reason for that mistrust largely with respect to Managing Director B from Company II, which has gradually deteriorated due to experience over the course of the cooperative effort:

> ... I tried it once with [Managing Director B]..., then we would have had no interest in marketing it ourselves, for a start. And then of course, it would have been nice if he had said yes, that's great, then why don't you do it with me. And we also offered that to them, yes, but even that confidence-building measure didn't work. (Managing Director A, Company I, 244)

Table 3.16 Responses of community members surveyed on aspects of community interaction

Interaction among community members	Innovation coach (Champion)	Managing director A	Managing director B (Champion)	Innovation manager	All
Communications					
There is frequent communication between the above-mentioned people (actors)	6	4	4	6	
The actors communicate frequently via spontaneous meetings/phone calls	5	4	5	5	
The actors communicate frequently directly/personally	6	4	4	6	
Project-related information is communicated openly, and shared	6	3	6	6	
Important information will not be withheld from others in certain situations	6	3		6	
The actors are satisfied with the information received from other members of the group	6	2		6	
Communication takes place without the consent/knowledge of higher authorities.	6	3		7	
Information circulates very quickly within the group of the above-mentioned people	6	4	5	6	
Arithmetic mean for communication	**5.9**	**3.4**	**4.8**	**6.0**	**5.0**
Coordination					
The sub-tasks that each of the above named people in the innovation project performs are clearly defined	6	2		6	
The sub-tasks that each of the above named people in the innovation project performs are well coordinated	6	3		6	
The performance of the tasks and work of different people is well coordinated	6	3		6	
Arithmetic mean for coordination	**6.0**	**2.7**		**6.0**	**4.9**
Balance of contributions					
Each of the above-named people can contribute their skills fully in the common project	6	2		7	
Everyone in the group contributes to the achievement of the goals with their skills and capacities	6	2		7	
The contributions of the above-named people to the innovation projects are perceived as balanced	6	1		6	

(continued)

Table 3.16 (continued)

Interaction among community members	Innovation coach (Champion)	Managing director A	Managing director B (Champion)	Innovation manager	All
Arithmetic average for balance of contributions	**6**	**1.7**		**6.7**	**4.8**
Mutual support and conflict resolution					
The actors support each other as much as possible	6	5		6	
Discussions and controversies are carried out constructively	6	2		6	
The suggestions and contributions of group members are treated with respect and discussed	6	1		7	
When conflicts arise, they are resolved quickly and without problems	6	5		6	
The group of actors is capable of reaching a consensus regarding key issues	6	2		6	
Arithmetic mean for mutual support and conflict resolution	**6.0**	**3.0**		**6.2**	**5.1**
Effort					
Each of above-named people fully supports the innovation project	6	4		6	
The innovation project has a high priority for each of the people	6	4		6	
The group of above-named people make substantial efforts to move the project ahead		3			
The group is completely satisfied with the individual efforts of each person	6	3		6	
Arithmetic mean for effort	**6.0**	**3.5**		**6.0**	**5.2**
Trust and group climate					
A climate of trust exists between the people	6	1	6	6	
A cooperative working climate exists in the group	6	4	6	6	
All members of the group can express their views openly	6	5	6	6	
The group members share ideas without being asked, or feel free to share ideas	6	6	6	7	
Arithmetic mean for group climate	**6.0**	**4.0**	**6.0**	**6.3**	**5.6**
Arithmetic mean for community interaction	**6.0**	**3.1**	**5.3**	**6.2**	**5.1**

Source: Author. Possible answers from 1 (disagree completely) to 7 (strongly agree)

He attributes this to the existing business model of Company II, and the slow transformation from a pre-competitive to a competitive market environment. Managing Director B of Company II assesses the situation differently. He sees the relationship between the two as largely normal, and says that Managing Director A played an important role within the community:

> That's [Managing Director A], who is basically the cheerleader, and the one with whom we maintain contacts, we also have a relationship with [Company II] as a supplier, we buy some materials from them, or we sometimes give them free samples or test models . . . and they send us material, and we test them . . ., that's certainly helpful for the project. . . (Managing Director B, Company II, 117)

Overall, the quality of interaction within the innovation community can be rated as good; however, the answers of Managing Director A show significant deviations (mean value: 3.1).

Performance of the Promotor Network

Based on the interviews conducted, the questionnaires answered, and additional available information, the overall performance of the DSSC innovation community can be assessed as high (mean value 5.2, see Table 3.17). This is shown by the compilation of the arithmetic mean values of the composition, the identity and the interaction of the community. A special case is provided by the divergent assessment of Managing Director A with respect to identity and interaction. This has been addressed in the above sections.

Measures and Methods of Community Development

The four interviewed promotors of the DSSC community identified measures and methods that were used during the focal innovation process to develop and support the community. A summary of the measures and methods used is given in Table 3.18.

Table 3.17 Performance of the dye solar community

Performance of the community	Innovation coach (Champion)	Managing director A	Managing director B (Champion)	Innovation manager	All
Arithmetic mean for composition of the community	6.0	6.0	5.5	5.7	5.8
Arithmetic mean for identity of the community	6.0	4.3	4.9	6.2	5.3
Arithmetic mean for community interaction	6.0	3.1	5.3	6.2	5.1
Arithmetic mean for performance of the community	**6.0**	**3.9**	**5.0**	**6.0**	**5.2**

Source: Author. Possible answers from 1 (disagree completely) to 7 (strongly agree)

Table 3.18 Measures and methods in the DSSC innovation process

	Functional focus				Contribution to network and community development		Prerequisite
	Info. dissemination & training	Info. gathering & analysis	Organi-zation	Inter-action & encounter	Contribution to network/ community development	Contribution to which quality aspects of the community?	Methodological expertise required?
Project meetings, communication	X	X	X	X	High	Identity, interaction	No
Personal meeting				X	High	Interaction	No
Memorandum of Understanding/Code of Conduct			X		Medium	Identity	Partially
Strategy workshop for partner selection		X	X	X	High	Composition, Identity	Yes
Innovation workshop for users	X	X		X	High	Composition, interaction	Yes
Multipliers' workshop with added-value partners	X	X		X	High	Composition, interaction	Yes
Technology roadmap	X	X	X		Medium	Identity, interaction	Yes
Process consulting/ innovation coaching			X	X	Medium	Interaction	Yes
Exchange of material/ mock-ups	X	X		X	Medium	Interaction	Partially
Meetings/conferences	X	X		X	Low	Interaction	No
Trend analysis		X			Medium	Composition, interaction	Yes
Definition of joint development projects			X	X	Medium	Identity, interaction	No
Number of functional foci	**6**	**8**	**6**	**8**			

Source: Author

A total of 12 measures and methods were identified, spread relatively evenly among the functional areas. Measures and methods with a focus on the acquisition of information and interaction were somewhat more frequently used (with eight mentions). The three listed workshops – the strategy, innovation and multipliers workshops – played a special role in innovation. They were seen by all promotors questioned as important instruments that made a major contribution to development of the innovation community through interaction and identity formation. Moreover, they made an impact on the composition of the community, since they enabled partners for further projects to be recruited.

3.6.4 Effect of the Innovation Community on the Project's Success

In the following section, the contribution of the DSSC innovation community to the success of the innovation project will be evaluated. For that purpose, it is necessary first to ascertain whether and to what extent the DSSC focal innovation process has been successful to date.

Success to Date of the Innovation Project

In the opinion of the promotors interviewed, the innovation project has been successful to date. The arithmetic mean of all success aspects obtained and all community members interviewed yields a success assessment value of 5.3, on a scale of 1–7 (see Table 3.19). It is notable here that Managing Director A of Company I awards a lower rating than the other promotors (mean value: 4.4). Possible reasons for this have been discussed in the above sections. The champion Managing Director B provided no assessment in this area, on the grounds that on these issues, he was not sufficiently knowledgeable to provide information. In the interview however, he expressed himself very positively regarding the success of the innovation project:

> ... Even if the product is not yet there and is not yet complete, the intermediate segments of the product ... are also a marketable thing ... And we have been well served by the fact that various providers or general partners were all there in one place, and we are certainly in a good starting position for a future project or a further project, or a successor project. (Managing Director B, Company II, 177)

In his assessment, Managing Director A from Company I distinguished between the success of cooperative effort with various partners, and particularly with regard to the cooperation with the Major Corporation III, arrived at a positive evaluation:

> In the bilateral project with [Major Corporation III], we have now really got some momentum. We have exchanges of opinion every month, and uh, are trying – at least we, from the dye side, we from the IL [ionic liquid] side, to address the whole issue really

Table 3.19 Evaluation of the success of the DSSC innovation project

Success of the innovation project	Innovation coach (Champion)	Managing director A	Managing director B (Champion)	Innovation manager	All
Effectiveness					
To date, the joint innovation project has been very successful	6	4		6	
The requirements of customers or users have so far been considered systematically	4	5		6	
The recent results are of high quality	6	5		6	
The partners in the joint innovation projects are satisfied with the result.	6	4		6	
To date, there has been little need for follow-up work		2		6	
At present, everything points to the future success of the joint innovation project	6	5		5	
Arithmetic mean for effectiveness of the innovation project	**5.6**	**4.2**		**5.8**	**5.2**
Efficiency					
To date, the stated objectives within the joint innovation project have been achieved	6	4		6	
From the perspective of the network, one can be satisfied with the progress of the innovation	6	5		7	
To date, the project has been carried out cost effectively.	6	6		5	
The project is progressing on schedule	6	2		6	
The project is on budget	6	6		5	
Arithmetic mean for efficiency of the innovation project	**6.0**	**4.6**		**5.8**	**5.5**
Arithmetic mean for success of the innovation project	**5.8**	**4.4**		**5.8**	**5.3**

Source: Author. Possible answers from 1 (low) to 7 (very high)

> consistently. ..., it now really has some substance, you would have to say. (Managing Director A, Company I, 280)

Despite these differentiating critical statements, Managing Directors A, too, evaluates the course to date of the innovation project as successful.

Contribution to Success by the Innovation Community

The promotors interviewed see the contribution of the innovation community to the success of the project as great (see Table 3.20). Three promotors award it a score of six on a scale from 1 (very minor contribution) to 7 (very great contribution); one

Table 3.20 Evaluation of the success of the DSSC innovation project

	Innovation coach (Champion)	Managing director A	Managing director B (Champion)	Innovation manager	All
What is your estimate of the contribution of the above-named group of people to the success of the joint innovation project?	**6**	**6**	**5**	**6**	**5.8**

Source: Author. Possible answers from 1 (low) to 7 (very high)

respondent gives it a 5. These responses and the information from the interviews permit the unequivocal conclusion to be drawn that the success achieved would not have been possible without the close cooperation of the promotors.

That raises the question as to what the reason is for the specific contribution of the innovation community to the success to date of the innovation project. Based on the interviews, the following key contributions of the promotor network are identified:

1. *Defining goals and organizing the exchange of knowledge*: A basic function of the innovation community is to combine the expertise and knowledge of different actors, and to provide an organizational framework and timeframe in which research and development tasks can be processed in an atmosphere of trust and according to agreed-upon rules. Through its activities in the context of accompanying research, Research Institute I helped establish this organizational framework and timeframe by organizing workshops, defining project goals and project plans, and setting the "rules of the game" (the "Memorandum of Understanding"). The innovation coach describes the value of his own work for the innovation community as follows:

   > And they profited from [Research Institute I], I think they would certainly say that. And that also has the advantage – but now we're getting into the details – of making some progress within the community in efficiency, or with colour design, by the possibility of experimenting with the ionic liquids or combinations of dyes... (Innovation Coach, Research Institute I, 97)

   The function of Research Institute I in setting an organizational framework and timeframe is especially evident in light of its neutral position in the innovation community – as the only institution with no commercial interest in the subsequent marketing of the technology.
2. *Positioning and definition of further activities*: Although no market-ready products could be developed within the time frame of the accompanying research, the innovation community nonetheless contributed to bringing the achievement of that goal closer, and providing clarity regarding the tasks and steps to be taken in future. This is clear from a statement by Managing Director B:

   > ...And we have been well served by the fact that various providers or general partners were all there in one place, and we are certainly in a good starting position for a future project or a further project, or a successor project, we are definitely in a good starting position. (Managing Director B, Company II, 177)

The community thus also accepts a competition-related function with respect to the research environment and the larger network for research on DSSCs, since it is to define follow-up activities in new partnership contexts. The Innovation Manager from Major Corporation III also sees it that way:

> But it still makes sense as [Major Corporation III] to be active there, because of course we develop technology that could possibly become important.... in other projects. After all, we are dealing with nanomaterials for the cell. (Innovation Manager, Major Corporation III, 102)

3. *Bringing together actors and contacts critical for success*: Another important function of the innovation community was to recruit actors critical to the success of the community, and bring them together (see also Sect. 3.6.2). In the conceptual phase, the strategic partner search and recruitment was one of the essential tasks. But in the context of the workshops conducted, other important contacts were made additionally, which may be crucial for the future marketing of the products. Managing Director B describes this as follows:

   > ... We now have the cells and [Research Institute I] can accomplish something in the marketing channel, ... they have the necessary skills to disseminate it further or [hold] a workshop. So I have to say from our point of view, it is quite a success. Because we thus have the contacts, because we also exchanged samples, ... there [was] an interaction that wasn't there before, ergo for us, it's a success. (Managing Director B, Company II, 212)

   The Innovation Manager of Company III highlights a further point, and stresses that the value of the innovation community was to have built a suitable network during the early, pre-competitive phase, as this provided a decisive advantage over competitors:

   > To be there right from the outset, not quite from the outset, but still relatively early, and to understand the technology, to make the materials available, ... to be active in a community, to build up customers and so on, so as to be a player at all afterwards. (Innovation Manager, Major Corporation III, 96)

   An important function of the community was therefore precisely to create this early, pre-competitive network.
4. *Confidence building and conflict management*: It was stressed several times in the interviews that trust is an important factor for cooperation between actors. The innovation community can promote trust, since it is a secure, manageable space with selected actors, with whom it is possible to speak about potential conflicts and competition. The Innovation Coach describes this with respect to the promotors in the innovation community as follows:

   > And I think there is sort of an atmosphere between Companies I and II, if one were to go onto the market now, they would first talk to the other partner. ... We actually raised that at an early stage, so as to make just such pitfalls ... visible, and ... to reach agreements on how to handle them. That also has to do with trust. (Innovation Coach, Research Institute I, 260)

The innovation community fulfils a key role in bring possible conflict issues into the open, and can, by reaching certain agreements for the possibility of conflict (e.g., in the form of a "Memorandum of Understanding" or "Code of Conduct"), help deal with them.

3.6.5 Conclusion

The focal innovation process analysis in this case study involves the further development of glass-based, monolithic DSSCs, and the associated innovation networks. The development of glass-based, monolithic DSSCs to the production stage can be classified as a radical innovation.

The case analysis shows that the innovation process was largely the work of four promotors who worked together in an innovation community. This innovation community was not constant throughout the course of the focal innovation process in terms of its composition. It grew out of a research project in which two of the promotors knew each other (the initial and conceptual phase), and was then, during the development phase, expanded specifically to include actors from another network. This process clearly shows the phase and target-specific development and reconfiguration of innovation communities.

The community demonstrates a competent composition, a high level of identity, and good quality of interaction among its members. The case study also shows that the success of the innovation process and the community is strongly influenced by the promotors and their activities and cooperation.

The case study confirms the assumption that the success of a radical innovation project clearly depends on the existence and performance of an innovation community. The contribution of the community to the success of the DSSC project involved largely the definition of goals and the organization of the exchange of knowledge via cooperation of the promotors, enabling them to position themselves for further R&D work, to define it, and to bring together actors and contacts critical to its success, while supporting confidence building and conflict management.

3.7 Tilt Sensors: Development and Market Introduction of a New Generation of Sensors

Stefanie Bunzel

Fact Sheet

Case Study: Tilt Sensors

Innovation object:	Development of novel, high-precision tilt sensors
Industry sectors:	Automotive, engineering, production technology, medical technology, aerospace, consumer electronics
Period analyzed:	2006–2010
Initial situation:	New high precision tilt sensors are components that are becoming more interesting for a wide range of industries and product applications – even mobile phones.
Current phase:	Realization
Research highlights:	Radical innovation project, wide range of applications and industry contact points, successful contribution of the innovation community to the overall innovation process

Key persons involved:	6
Size of innovation community:	4
Group composition:	Almost only champions
Regional distribution:	Regional (Baden-Württemberg, Germany)
Community performance:	6.1*
Contribution to project success:	6.5*
Project success:	6.2*

*On a scale from 1 (very low/poor) to 7 (very high/good).

3.7.1 Context and Object of Innovation

The accurate detection of the tilt of machines and objects in relation to the horizontal (i.e., with respect to the earth's field of gravitation) has always been an important issue in many industries. There are a variety of technical operating

Fig. 3.8 Typical applications for tilt sensors (Source: http://www.2e-mechatronic.de)

principles that allow the measurement of a tilt angle. Often, so-called accelerometers are used, which can calculate the inclination in static systems on the basis of the acceleration of the earth. But there are also measurement systems based on conductometric, optical, inductive or piezoresistive principles, which provide data of various types on the existing tilt.

Tilt sensors have for a number of years penetrated many different markets and areas of application (Fig. 3.8). In addition to such classic application fields as industrial measurement technology, automotive engineering, and automation and production technology, they are now also found in numerous applications in transport systems, special vehicles, medical technology, aerospace, and consumer electronics.[9] Thus for example, cranes are prevented from tipping over by ascertaining the tilt angles of their booms. In electronic games, tilt sensors provide a realistic atmosphere by permitting the joysticks of game consoles to react to the movements of players. Other areas of applications include:

- Industrial manufacturing processes or medical systems (such as operating tables which can be kept level by tilt sensors)
- Antenna systems (e.g. on ships)
- Measurement or imaging systems
- Laser instruments (e.g. in metrology)
- Support for GPS navigation operations
- Security services
- Construction

While the demands of various sectors regarding the level of accuracy and the possible angular range of such sensors for numerous applications are continuously increasing, the known measurement systems for this purpose have shown some important weaknesses. A high degree of measurement accuracy generally not only requires large component size, but also places the sensors in the upper price range, of up to several hundred euros. The high price for such quality has been due to the fact that the technological problems for a reasonable production of measurement systems had yet to be mastered to the point of system maturity. And for the user too,

[9] Source: http://www.mm-applications.com/ and http://www.2e-mechatronic.de

bringing the devices into operation has often meant a complicated installation process and time-consuming calibration. Depending on the type of application, it has hitherto been necessary to decide the minimum level of precision necessary, so as to keep the cost of the sensor within limits. Thus, many applications, such as in automation and safety engineering, scientific equipment, medical technology and levelling systems, which in addition to high measurement accuracy also required the ability of installation in a very small space, have therefore not been served at all.

From this point of departure, the network partners came to be motivated to develop new tilt sensors to such a point that they would in future be able to meet high customer demands for durability, reliability, cost effectiveness and accuracy of measurement.

Object of Innovation

The objective of the present innovation project is the technological development and market introduction of novel, high-precision tilt sensors. The main task of the sensors is the measurement of the tilt angle of any object in relation to the gravitational field of the earth. In the context of this innovative project, the fundamental technological developments of a scientific institute in this area are used. These are based on the principle of the liquid-based, capacitive tilt sensor. As shown in Fig. 3.9, the space between two vertical electrodes is half filled with a dielectric liquid. If there is a tilt in the sensor, the liquid retains its position, due to the gravity, so that depending on the tilt angle, the change in the difference of the capacitance between the electrodes is measured. This difference is detected by an electronic signal output, converted to the corresponding tilt angle, and output to the user via software. The fact that the principle of this technology basically works has already been demonstrated in a previous research project at the Institute, and in scientific tests conducted by the participating industrial partners.

The particular interest in the innovation project is to validate the technological principle, particularly with regard to market-driven development. In addition to the miniaturization of the sensors, the increase in the functional density and the reliability, it is particularly the reduction in unit costs that is the crucial factor. In

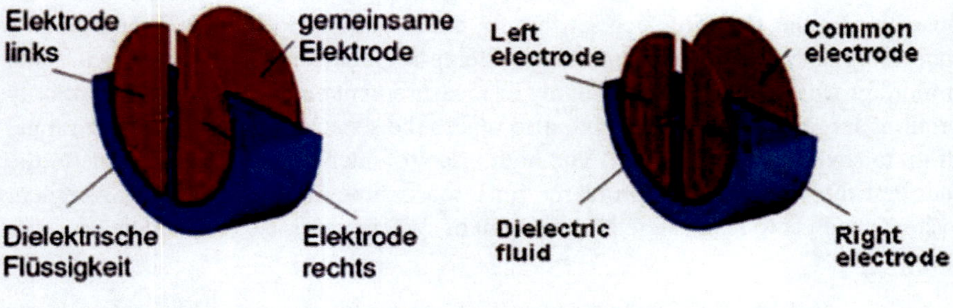

Fig. 3.9 Liquid-based capacitive tilt sensor (Source: http://www.imat.hsg-imit.de)

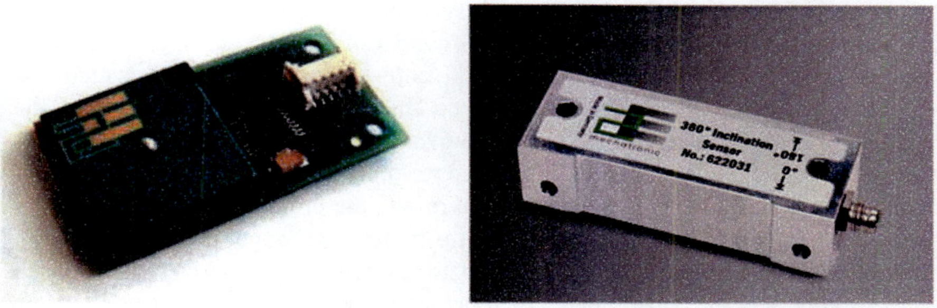

Fig. 3.10 Housing-enclosed version of a tilt sensor with low-cost manufacturing technology (*left*), and printed circuit board version (*right*) (Source: http://www.2e-mechatronic.de)

this way, the new properties of the sensor are in future to be used even in areas of application that have previously been either impossible or prohibitively expensive.

The unique feature of the innovative device described here is that the sensor quality is very high compared to competing products, and can nonetheless be produced inexpensively. In direct comparison with the competition, the cost is reduced by a factor of about five to ten. In the 360° model, the special feature by comparison to competing sensors is its very high measurement accuracy of less than 0.1° over the entire measurement range, which is thus in some cases several tenths better than the competition.

As there had yet been no technological approach to the low-cost mass production of mature liquid-based capacitive tilt sensors, the construction of a suitable production facility is a particular focus of the innovation project. Here, the modular principle is to be pursued from the outset, i.e. the ability to produce customer-specific designs quickly and easily is the goal (Fig. 3.10). This aspect of development constitutes the biggest risk in the overall project, due to lack of experience in this area.

3.7.2 *The Innovation Process*

History and Factors Influencing the Innovation Process

It was a long way from the basic idea to the ultimately successful implementation of the high-precision tilt sensor, marked by existing framework conditions and by various influencing factors. The history of the innovation process can be traced back to the 1990s, when the first contact between two of the members of the community came about. The following factors marked the structure and development of the tilt sensors innovation network particularly lastingly:

Long-term network and preliminary work in the field of sensors: The two main initiators of the network have been cooperating for over 12 years in joint preparatory work in the field of microsystems technology. The head of the participating academic institution is the co-founder of the innovation project "Innovative micro-encoders for automation and automotive technology," which has already addressed the issue. He introduced the technical prerequisites and the technological foundations for the future innovation project. This and other projects also provided the contact to a medium-sized technology company and to a technology transfer centre, which play an essential role in the present case study. The first points of contact emerged at the level of information exchange; the idea for a joint project grew, and eventually in 2006 led to an active cooperative effort in the innovation project described here. The requisite network was however not built up only in the research community; rather, the participating company, too, had already at an early stage made contact with several potential lead customers, and had discussed the project idea with them. All these relationships, some only informal, enabled the initiation of the tilt sensors innovation project in 2006.

> We knew each other, we knew what we were getting, it had stood the test in the past. (Director, Research institute, 54)

Good relations with the microsystems technology industry: Prior to the innovation projects, both the research company and the technology transfer centre have already spent many years in the microsystems technology industry, and thus have a very good overview of current technology developments, key players and trends, as well as initial market experience. These experiences were of very great value for the initiation of their own network, as the appropriate partners could be identified and incorporated without great effort.

Unfulfilled performance needs in tilt measurement technology: Generally, ever increasing demands upon measurement systems have been observable for many years in many industries, which demand increased functional integration and higher power density, along with miniaturization and weight reduction at the level of the assemblies. Particularly affected by this are the automotive industry and mechanical engineering. However, hardly any major progress has been achieved at the technological level, so that here, a potentially large market has opened up. The feasibility of the functional principle of a liquid-based, capacitive tilt sensor had been demonstrated, so the implementation of that technological principle in a marketable product was the logical conclusion for those involved.

Promoting the innovation project: From the outset, the fact that the initial idea of over a decade ago has now developed into a successful product is thanks to the extensive funding support of the innovation object. There have been specific types of support for the various stages of technological development. The development of the foundations ("The Effective Principle of the Sensor") was initially carried out at the initiators' own initiative at the Research institute, and then continued with support from the working community of industrial research associations. The results of this research project were then the foundations for the present innovation project, with the goal of achieving the practical implementation of the technology. This project was made possible with funding from the German Federal Ministry of Education and Research.

This shows how long it takes to deploy new technologies and hence to create new products. To develop not just a technology, but indeed an entirely new mode of action (Director, Research Institute, 27).

The Focal Innovation Process

The focal innovation process, which set itself the goal of developing high-quality tilt sensors to series production, refers to the period from 2006 to 2010. Special focus was placed on the interconnection of basic technological research by the participating research institute, and on strong application and business orientation and implementation for selected users by the participating technology company.

To better illustrate the process of the innovation project, it has been divided into several chronological phases. Since the technological content of the project is based directly on the preparatory work of two of the participating partners, the period of observation has been extended backwards to include the preliminary maturation phase, which was the necessary basis for the actual innovation project.

- Maturation phase: 1998–2006
- Initial and conceptual phase: 2006
- Development phase: 2003
- Realization phase: 2008–2010
- Market and diffusion phase: since 2010

As can be seen from the above history, the basic idea and the technological principle are based on the extensive preparatory work of the research institute, together with the medium-sized company. This cooperative partnership has existed for over 12 years, and is therefore seen as the starting point of the *maturation phase*. As a result of this phase, it was possible for the actors to formulate the requirements for the radical innovation with respect to market-based implementation, and thus to define the scope of the innovation project.

From my years of experience as the managing director of an initially almost insolvent company, I realized that only through active cooperation – be it in-house or in networks across company boundaries – is it possible to reap huge benefits. The fact that I had had contact with [the research institute] for a long time was what enabled such an idea phase, which then turned into real projects, to come about in the first place. (Managing Director, Company A, 90)

In 2006, the two initiators of the project idea convinced the funding agency of the importance of their innovation project, which thus enabled the official launch of the tilt sensor network. This *initial and conceptual phase* was primarily characterized by the fact that the technology transfer centre became involved for the purposes of the strategic orientation of the project. Thus was a driving force taken into the project at an early stage, which took over both the network management tasks and the active search for more partners and customers, and also the trend monitoring for the microsystems technology industry.

During the initial phase, my task was therefore… to compile the current state of the art of tilt sensors, so as to be able to support the development strategy in the network. The questions were: What's already on the market, where are there customer needs that are still open, with respect to highly accurate sensors, and which industries are potential lead markets? (Managing Director, Technology Transfer Centre, 11)

During the subsequent *development phase*, the central task was to develop the tilt sensors based on the research results, and to make it possible to produce them with series-compatible manufacturing processes. During this period, primarily the closely coordinated technical experts of the participating partners were responsible for the progress of the project.

The innovation community achieved the biggest breakthrough in technological development during the *realization phase*. During this period, it was possible to convince a lead customer of the relevance of the innovation, and to actively integrate him into the development process. Jointly, a prototype was developed, which then went into series production.

The integration of lead customers can be assessed as very positive. [The client] was very committed, showed great interest and introduced a specific need – which had previously not been able to be met. (Department Head, Company A, 25)

Thanks to the successful use of high-precision tilt sensors by the first lead customer, the focus was during the final phase, the *market and diffusion phase*, increasingly placed on the marketing of the sensor. Through the active promotion of the product at trade fairs, exhibitions and other events of the microsystems technology industry, still more customers were acquired. Today, the medium-sized Company markets the tilt sensor on a large scale, and is seeing continued high demand from interested people from a wide range of sectors.

Organizations Involved in the Innovation Process

In the present tilt sensor network, a limited number of organizations are involved in the innovation projects. All the partners are located in the southern part of the state of Baden-Württemberg, and include the following organizations:

- A research institute
- A technology-oriented medium-sized company
- A technology transfer centre
- A practice partner as a lead customer

While the medium-sized company, together with the research institute, is responsible for technology development, the Technology transfer centre provides the bridge between the technological development and the potential market for the innovation object, the tilt sensors. The practice partners provide the necessary input on customer needs and market-side requirements.

3.7.3 *Development of the Innovation Community*

The central issue in all case studies is to review the contribution of innovation communities to the success in radical innovation projects. In the following sections, we will provide detailed information on the nature and quality of the present innovation community; first however, we would like to identify all the relevant key individuals and promotors presented, who carried out various functions and roles in the tilt sensors innovation project.

Key People and Promotors

On the basis of four interviews with selected network members, the following key people listed in Table 3.21 can be assigned to the tilt sensor network. These people crucial to the network come from both the business and the research communities. In addition, another person from the area of technology transfer is involved in the innovation community. The regional distribution of the key people is limited to the southern part of the state of Baden-Württemberg.

Table 3.21 Key players in the tilt sensors innovation process (2006–2010)

	Geographic location	Key person	Promotor	Champion	Member of the innovation community
Company					
Managing director Company A	Baden-Württemberg	Yes	Power, process, relationship promotor	Yes	Yes
Department head Company A	Baden-Württemberg	Yes	Expert promotor	No	Yes
Employee Company B	Baden-Württemberg	Yes	No	No	No
Technology transfer					
Managing director, Technology transfer centre	Baden-Württemberg	Yes	Expert, process and relationship promotor	Yes	Yes
Scientific community					
Director Research institute	Baden-Württemberg	Yes	Power, expert, process, relationship promotor	Yes	Yes
Scientist Research institute	Baden-Württemberg	Yes	Expert promotor	No	No
Total		**6**	**5**	**3**	**4**

Source: Author

Of the six key people mentioned in the interviews, five can be assigned at least one promotor role, on the basis of the criteria stated in the quantitative questionnaire. All but one of the five people met all the criteria for the role of expert promotor; in two cases, this was their only promotor role. The other three people each combined several promotor roles and can hence be considered champions; all three champions are assigned both the process and the relationship promotor roles, and one covers all four possible promotor roles.

Even if the present tilt sensor network can be considered rather small, not all of these key people belong to the innovation community. The contribution of all key people is in different ways vital to the success of the innovation project, but it is in some cases limited to a single phase. Nor do the other members of the innovation community consider all key persons to be part of the core team of the network. Of the six key players in this case study, only four are members of the tilt sensors innovation community under the definition used.

Evolution: Steps and Stages of Community Development

The origins and development of the tilt sensors community depends significantly on the factors mentioned in Sect. 3.7.2, which have permanently characterized both the environment and the substantive nature of the five phases of the project presented. Table 3.22 provides an overview of the number and type of participating organizations, promotors of the innovation community and other people involved in the innovation projects throughout the five phases. Before the start of the focal innovation process (1998–2006), the long-standing cooperation between the two promotors from the scientific community and from the company was the central element. With the entry into the initial/conceptual phase (2006), the promotor from the technology transfer centre was brought in. In order to be able to demonstrate the requisite technical competence in the establishment of the production plant, the department head of the company was actively involved in the community during the subsequent development phase (2007). Subsequently, during the realization and marketing phases (since 2008), the focus was primarily on the expansion of the network with new partners from the potentially interesting markets. This description allows important conclusions:

- The close cooperation between the two champions during the maturation phase was crucial for the initiation of the network, and also for the further expansion of the innovation community.
- The promotor network is very consistent and has changed very little throughout the phases of the project. The four main actors came together at an early stage, and have played an important role in the strategic direction and substantive developments in all the phases since then. None of these four promotors left the network early.

Table 3.22 The evolution of innovation community tilt sensors

	Maturation phase (1998–2006)	Initial/conceptual phase (2006)	Development phase (2007)	Realization phase (2008)	Market & diffusion phase (2010)
Participating organizations	Research institute Technology-company	Research institute Technology-company Technology transfer centre	Research institute Technology-company Technology transfer centre	Research institute Technology-company Technology transfer centre Company B as lead customer	Research institute Technology-company Technology transfer centre Company B as lead customer Other lead customers
Innovation community	Head of research institute Managing director, Company	Head of research institute Managing director, Company Managing director, Technology transfer centre	Head of research institute Managing director, Company Department head, Company Managing director, Technology transfer centre	Head of research institute Managing director, Company Department head, Company Managing director, Technology transfer centre	Head of research institute Managing director, Company Department head, Company Managing director, Technology transfer centre
Other participating promotors during the preliminary phase or the innovation process	Other staff, Research institute	Other staff, Research institute	Other staff, Research institute	Other staff, Research institute Employee. Company, B	Other staff, Research institute Employee. Company, B

Source: Author

- By identifying potential lead customers, the promotor function of the managing director of the technology transfer centre contributed decisively to the expansion of the network during the implementation phase.

The Composition of the Innovation Community

As indicated above, the tilt sensors network includes four key people, who, with their promotor functions, constitute the innovation community. Particularly interesting in this context is the fact that three of these promotors have held multiple roles, and may thus be referred to as champions. The head of the research institute indeed combines all four possible roles in his person: those of power, expert, process and relationship promotor.

The two champions from the scientific field and the medium-sized company primarily initiated the innovation project. The functions of both include both that of power and of relationship promotor, and both had previously worked together actively in the research area. This was a great advantage, both for the creation and for the later composition of the innovation community. Both champions had, in their years spent in preparatory work, built up a very good network of connections in the microsystems technology industry, and also to providers of research funding, so that the initial recruitment of other partners for the innovation projects was completed quickly.

> This innovative project is a relatively small network, and a small network can easily be generated on the basis of an existing network or an existing community where you already know each other, and where you understand who the most important players are. (Director, Research Institute, 86)

The integration of the promotor from the technology transfer centre was a special case. At the beginning of the innovation project, the initiators were aware of the importance of linking tilt sensor technology to the requirements of potential customers and markets. This important aspect was addressed by means of the involvement of a neutral network manager, whose main responsibilities included active participation in the composition of the network, and in the interface of the technological development and the economic aspects. Again, the relationship promotor's role was of great importance, since he could contribute his experience and expertise in managing networks, and a very large "circle of friends" in the area of microsystems technology.

> But also the input or the newly identified and acquired customers of [the Technology transfer company] were very valuable in this respect, as they have yet a different perspective and insight in the world of microsystems technology, and therefore brought in a new impetus and initially unexpected partners. (Managing Director, Company A, 116)

This relationship role function represented so strongly throughout the innovation community is a major reason for the very positive assessments of the composition

Table 3.23 Responses of surveyed members on the composition of the community

Composition of the community	Members of the innovation community				All
	Champion 1	Champion 2	Champion 3	Expert promotor	
Completeness and complementarity of skills					
The group has all companies and organizations represented within that are important to the success of the innovation project	7	7	5	7	
Social capital					
The group has very good relations with actors in partner organizations that make decisions regarding the joint innovation project	7	7	6	5	
The group has excellent contacts with decision makers who are important to the joint project, even outside the partner organizations involved	7	7	6	6	
Members of the group have access to persons with relevant expertise with respect to the joint innovation projects	7	7	7	5	
Reconfiguration capability					
The group is open to new/additional partners/members	4	6	7	5	
The group was or is in a position to let go of members who are not helpful or necessary for the success of the project, or to ensure a separation from them	1	3	3	4	
Arithmetic mean for composition of the community	5.5	6.2	5.7	5.3	5.7

Source: Author. Possible answers from 1 (disagree completely) to 7 (strongly agree)

of the innovation community shown in Table 3.23. Thanks to good knowledge, all relevant members of the group are represented in the innovation project, in order to achieve the set goals. It is also very important that the members are not competing with each other, but rather that a good synergy has resulted from their expertise. With regard to social capital too, i.e., both the relationships within the promotor network and those extending its boundaries, are estimated to be very good by all the members.

The only exception to this is the aspect of the reconfiguration of the network, which was assessed at an average of 2.8 on a scale of 7. This is due to the great stability of the community that has not changed throughout the entire period. But the following quote from a member nonetheless shows that in the event of a change occurring, there would indeed be potential for reconfiguration:

> If for example conflicts should occur due to the lack of competence of a key person, I see it very much as an option to have them replaced by an appropriate shift to more competent people. (Managing Director, Company A, 95)

The Identity of the Innovation Community

To assess the identity of the promotor network, the three main aspects of "clarity of group boundaries," "functional identity" and "cohesion" between the promotors were considered in greater detail. Table 3.24 shows the results of the survey in their respective categories.

The aspect "clarity of group boundaries" is very high for the members of the community, with a mean of 6.0. Also because of the relatively small size of the group and the regional proximity of its members, the parties see themselves as a team, and are also perceived as such from the outside; over the course of the project, that has remained very stable. Only the assessment of the expert promotor is somewhat below the values of the three champions. This is due to the fact that, as an expert promotor, he is primarily responsible for moving technological developments forward, and is less involved in other areas. He therefore has few points of contact to the work of the technology transfer partner, and thus sees the group boundaries as less clearly defined than the other three members do.

Even more marked is the difference between the three champions and the expert promotor regarding the aspect of functional identity. While the first champion awarded a maximum score of 7 in all categories, the expert promotor assed it as significantly lower. Here, too, the great importance of the already long-standing collaboration between the initiators, and of strong relationship promotor roles, was demonstrated. Common goals can be much more easily formulated on this basis, and a common language had already been developed through the extensive preparatory work. The parties involved were from the start aware of how important the project was for the other members.

> It is important that the expectations and goals are clear in advance. Each partner in the network must formulate and communicate his goals and expectations, and ideally, all should stay together – and this is good. (Director, Research Institute, 86)

> The common understanding developed through many discussions and meetings, and the exchange of knowledge during the initiation phase. That allowed everyone to judge fairly well what the state of the technology was, where 'the journey' was headed, and hence where the opportunities for their company were. (Director, Research Institute, 69)

For the expert promotor, who did not to the same extent have relationships within the network and beyond, the classification of the common vision and, above all, the use of a common language, are therefore less clear.

An obvious strength of the community is reflected in the assessment of the cohesion within the promotor network. In terms of ties of the players to the innovation project, the sympathy between the members, their pride in membership, and the integration of individuals into the group, all members agreed in a consistently positive assessment.

> In particular, you notice it because information – both about knowledge and about problems – is shared openly and without a lot of formalities. Achieving the common goal is always the main focus, no partner ever worked against the goal. (Director, Research Institute, 71)

Table 3.24 Responses of surveyed members on aspects of community identity

Identity of the community	Members of the innovation community				All
	Champion 1	Champion 2	Champion 3	Expert promotor	
Clarity of group boundaries					
The group behaves as if it were a team	7	7	6	6	
The people see themselves as "insiders"	7	6	5	5	
The group is clearly seen from the outside as a unit	4	5	6	5	
All persons fit well into the group	7	7	7	6	
Each member of the group knows exactly who is a member of the group and who is not	4	7	7	5	
The group of actors is very stable; there is little membership turnover	7	7	7	6	
Arithmetic mean for clarity of group boundaries	6.0	6.5	6.3	5.3	6.0
Functional identity					
The actors agree on the importance of innovation	7	7	6	5	
The actors agree on the objectives of the joint innovation project	7	6	7	5	
The actors use the same "language" (use the same terms, technical language, etc.)	7	7	5	4	
There are no conflicts due to conceptual misunderstandings	7	6	5	5	
Arithmetic mean for identity	7.0	6.5	5.8	4.8	6.0
Cohesion					
For individuals, it is important to be part of the project	6	7	7	6	
The actors are strongly connected with the project	7	6	7	6	
All people are fully integrated into the group	7	7	7	5	
There are no personal conflicts within the group	7	7	5	4	
There is personal affinity/sympathy between the actors	7	7	6	5	
The network is characterized by a strong cohesion of its individual actors	7	7	7	6	
Each actor is proud to be part of the group	7	6	6	6	
All people feel responsible for the group, and protect it	7	6	5	4	
Arithmetic mean for cohesion	6.9	6.6	6.3	5.3	6.3
Arithmetic mean for identity of the community	6.6	6.6	6.2	5.2	6.1

Source: Author. Possible answers from 1 (disagree completely) to 7 (strongly agree)

The expert promotor did see some minor differences regarding the topic of conflict and responsibility for the group. This is due to the fact that, as the technical expert in charge, the technological hurdles which emerged and the associated time and budget-related technical changes directly involved him, and in many cases, he had to solve such problems.

Interaction in the Innovation Community

In addition to the aspects of composition and identity, the third important aspect in the evaluation of the performance of the innovation community is the question of the interaction between the promotors. Here, there are a number of key factors that have been rated by the community members as good, with an overall average of 6.1 on a scale of 7; see Table 3.25.

Favoured by the small number of members and their regional proximity, the communications between the actors, both in phone calls and at regular meetings, can be seen as very positive. The frequent coordination permits knowledge, relevant information or emerging conflicts to be quickly and openly communicated within the network, and if necessary, resolved jointly. The fact that the hierarchical position of each member within his own organization is only of secondary importance in the context of the innovation community, is shown by the following quote by the managing director of the medium-sized company:

> This is very easy to see because all those involved – regardless of their rank or function within their company – communicate on an equal basis. (Managing director, Company, 77)

The open and respectful treatment of each individual is seen as an important and very positive feature of the tilt sensors network.

> Openness and honesty are extremely important. You can often better resolve conflicts and find compromises, if you know the motivation of all involved. (Director, Research Institute, 83)

The highest value is achieved in the aspect "trust and group climate" (average of 6.5 on a scale of 7). The players agree that the group is determined by a very productive working atmosphere where everyone can bring in ideas and express them openly. This confidence and respect in dealings with one another not only allows conflicts to be resolved quickly and purposefully, but also promotes the implementation of radical innovations through the equal participation of all actors.

> Another important aspect for the quality of cooperation and the great respect in dealing with one another. Language is a point not to be underrated, since good communications skills can certainly be beneficial. (Managing Director, Company A, 84)

Performance of the Promotor Network

As a total picture for the innovation network, a value for the performance of the Tilt Sensors innovation community can now be calculated by compiling the results from

Table 3.25 Responses of surveyed members of the community interaction

Interaction among community members	Members of the innovation community				All
	Champion 1	Champion 2	Champion 3	Expert promotor	
Communications					
There is frequent communication between people	7	6	6	6	
The actors communicate frequently via spontaneous meetings/phone calls	7	7	6	5	
The actors communicate frequently directly/personally	7	6	5	6	
Project-related information is communicated openly, and shared	7	6	6	5	
Important information will not be withheld from others in certain situations	7	7	6	5	
The actors are satisfied with the information received from other members of the group	7	6	5	5	
Communication takes place without the consent/knowledge of higher authorities	7	7	4	6	
Information circulates very quickly within the group	7	7	5	6	
Arithmetic mean for communication	**7.0**	**6.5**	**5.4**	**5.5**	**6.1**
Coordination					
The sub-tasks that each of the above named people in the innovation project performs are clearly defined	6	6	7	5	
The sub-tasks that each of the above named people in the innovation project performs are well coordinated	6	7	5	5	
The performance of the tasks and the work of different people is well coordinated	6	7	6	6	
Arithmetic mean for coordination	**6.0**	**6.0**	**6.0**	**5.3**	**6.0**
Balance of contributions					
Each of the above-named people can contribute their skills fully in the common project	6	7	7	5	
Everyone in the group contributes to the achievement of the goals with their skills and capacities	7	7	7	6	
The contributions of the above-named people to the innovation projects are perceived as balanced	6	5	4	3	
Arithmetic mean for balance of contributions	**6.3**	**6.3**	**6.0**	**4.7**	**5.8**
Mutual support and conflict resolution					
The actors support each other as much as possible	7	7	6	4	
Discussions and controversies are carried out constructively	7	6	5	6	

(continued)

Table 3.25 (continued)

Interaction among community members	Members of the innovation community				All
	Champion 1	Champion 2	Champion 3	Expert promotor	
The suggestions and contributions of group members are treated with respect and discussed	7	7	6	6	
When conflicts arise, they are resolved quickly and without problems	7	6	5	4	
The group of actors is capable of reaching a consensus regarding key issues	7	6	6	5	
Arithmetic mean for support and conflict resolution	**7.0**	**6.4**	**5.6**	**5.0**	**6.0**
Effort					
Each person fully supports the innovation project	7	6	6	5	
The innovation project has a high priority for each of the people	7	7	6	7	
People make substantial efforts to move the project ahead	7	7	6	6	
The group is completely satisfied with the individual efforts of each person	6	7	5	4	
Arithmetic mean for effort	**6.8**	**6.8**	**5.8**	**5.5**	**6.2**
Trust and group climate					
A climate of trust exists between the people	7	7	6	6	
A cooperative working climate exists in the group	7	7	7	6	
All members of the group can express their views openly	7	6	7	6	
The group members share ideas without being asked, or feel free to share ideas	7	6	7	5	
Arithmetic mean for group climate	7.0	6.5	6.8	5.8	6.5
Arithmetic mean for the community interaction	**6.8**	**6.5**	**5.8**	**5.3**	**6.1**

Source: Author. Possible answers from 1 (disagree completely) to 7 (strongly agree)

the above aspects composition, identity and interaction (Table 3.26). As can be seen from the Table, the total value of 6.1 on a 7-point scale is very high, and thus supports the major thesis that this is indeed an innovation community as defined in Sect. 2.2.3.

While the three champions of the community rated all three areas very highly, the assessment of the fourth member who, in contrast to the other members, assumed only one promotor's role, was slightly lower, with an average of 5.3. This can be explained by the fact that while his obvious technical expertise permitted him to be successfully involved in the network, he contributed less to the development of the community, and did not have the same level of close contact to all other members.

Table 3.26 Performance of the tilt sensors innovation community

Performance of the community	Members of the innovation community				All
	Champion 1	Champion 2	Champion 3	Expert promotor	
Arith. mean for composition of the community	5.5	6.2	5.7	5.3	5.7
Arith. mean for identity of the community	6.6	6.6	6.2	5.2	6.1
Arith. mean for interaction in the community	6.8	6.5	5.8	5.3	6.1
Arith. mean for performance of the community	6.6	6.5	5.9	5.3	6.1

Source: Author. Possible answers from 1 (disagree completely) to 7 (strongly agree)

The fact that the excellent performance of this community is mainly due to the promotors involved is clearly confirmed by the following statement of one member:

> Active cooperation plays a crucial role: the spirit of community and the everyday togetherness in the network are the major success factors in this project. This cooperation enabled us to overcome critical phases, which might otherwise even have led to the cancellations of the project. (Managing Director, Technology Transfer Centre, 85)

Measures and Methods of Community Development

The attestation of very good performance by the members of the tilt sensors community is due in good part to the fact that from the beginning of the innovation project, a variety of measures and methods were applied (see Table 3.27). These supported the promotors not only in achieving a much more efficient interaction both within and beyond the boundaries of the innovation community, but also served to further the community's development in terms of the composition of partners and the joint formulation of goals and strategic planning for the future. In addition to such traditional instruments as project meetings, bilateral talks and workshops for good communications, the interaction with potential customers, suppliers and other members of microsystems technology sector for information creation and sharing was a very relevant aspect. Thus, it is very important for all participants to show a constant presence in the whole microsystems technology community – and not just in this network – so as to be able to pick up signals within the industry at an early stage, to maintain the exchange of information, and to be able to identify potential technological innovations in time.

> But the direct approach is not only important internally within the network, but is also essential for addressing the market and the customers. The strategic approach of my company has always been, first of all identify the market leader and speak to him. If you're successful there, it's much easier to interest additional lead customers in the project, too. (Managing Director, Company A, 113)

Table 3.27 Measures and methods in the tilt sensors innovation project

	Functional focus				Contribution to network and community development		Prerequisite
	Info. dissemi-nation & training	Info. gathering & analysis	Organization	Inter-action & en-counter	Contribution to network and community development	Contribution to which quality aspect of the community?	Methodological exertise required?
Regular status/project meetings and voting sessions	X	X	X	X	High	Identity, interaction	No
Bilateral talks	X	X		X	High	Interaction	No
Internal innovation workshops and working groups	X	X		X	High	Interaction	Partially
Trade fair visits/Presence in micro-systems community	X	X		X	High	Composition, interaction	No
Presentation series from the area of micro-systems technology	X	X		X	Medium	Composition, interaction	Yes
Discussions with experts from other industries	X	X		X	Medium	Composition, interaction	Yes
Lead user workshops		X		X	High	Composition, interaction	Yes
Trend analysis/trend monitoring		X			Medium	Composition, interaction	Yes
Internet communications	X	X		X	Low	Interaction	No
Ratings of technology potentials		X			Low		Yes
Market research/competition analysis		X			Low		Yes
Intellectual property rights analysis		X			Low		Yes
Cooperation agreement			X		High	Interaction	Yes
Number of functional focuses	6	12	2	7			

Source: Author

While there are a large number of measures and methods dedicated to the acquisition of information and interaction, very few address the organizational needs of the network. The most important point – and here, all members agree – is to draft a cooperation contract, so as to have a legally regulated basis in the event of a later dispute.

> But to get such an open and trusting culture of cooperation, it is absolutely necessary to draft a joint cooperation contract, which governs all legal foundations and prevents a drain-off of your own know-how. It is also important at the outset to set up a very precise interface definition, so that everybody knows for sure where in the project they have to work with which other partners, so as to get to their goals quickly, and to make sure that no double work is created. (Managing Director, Company A, 110)

Looking back, the interviewed members of the community saw very little need for improvements in the methodological support. In their view, only two aspects could have been implemented better:

- The possibilities for Internet communications, e.g. for faster exchange of data, were not utilized.
- To improve the efficiency of the entire project, a central project financial controlling system would have been helpful, to redirect the project again in case of deviations (in terms of budget, capacity, schedules, etc.).

3.7.4 Effects of the Innovation Community on the Project's Success

With this detailed discussion of the performance of the community and the application of the supporting methodology, it now appears logical to evaluate the success to date of the innovation project. Accordingly, the contribution of the promotor network to the success of the overall project can then be deduced from that. Since the community was at the time of the survey already at the end of the project, the interviewees were able to give very accurate estimates of the project's success.

As seen in Table 3.28, the success of the tilt sensors innovation project was rated as very high (6.2 on a scale from 1 [very low] to 7 [very high]), both in terms of effectiveness and of efficiency. The goal of developing tilt sensors to a level corresponding to the self-set goals, and to find appropriate markets and customers for them, was successfully achieved. Pilot customers from the measurement technology sector have already tested first prototypes, and the medium-sized company has begun manufacturing a sensor model.

An interesting aspect of the assessment is the higher estimate of effectiveness, by comparison with the efficiency. All four interviewees agree that the results of the project are of very high quality and that all requirements have been met. However, certain technological hurdles during the course of the project led to delays and increased the required budget, an expense largely borne by the participating partners. The clearest assessment in this respect is that of the expert promotor,

Table 3.28 Evaluation of the success of the innovation project tilt sensors

Success of the innovation project	Members of the innovation community				All
	Champion 1	Champion 2	Champion 3	Expert promotor	
Effectiveness					
To date, the joint innovation project has been very successful	7	7	7	7	
The requirements of customers or users have so far been considered systematically	7	7	6	7	
The recent results are of high quality	7	7	6	6	
The partners in the joint innovation projects are satisfied with the results	7	7	6	6	
To date, there has been little need for follow-up work	6	5	5	5	
At present, everything points to the future success of the joint innovation project	7	7	6	6	
Arithmetic mean for effectiveness	**6.8**	**6.7**	**6.0**	**6.0**	**6.4**
Efficiency					
To date, the stated objectives within the joint innovation project have been achieved	7	7	7	6	
From the perspective of the network, one can be satisfied with the progress of the innovation	7	7	7	6	
To date, the project has been carried out cost effectively	6	5	5	3	
The project is progressing on schedule	7	7	6	6	
The project is on budget	6	5	6	4	
Arithmetic mean for efficiency	6.6	6.2	6.2	5.0	6.0
Arithmetic mean for success of the innovation project	**6.7**	**6.5**	**6.1**	**5.5**	**6.2**

Source: Author. Possible answers from 1 (disagree completely) to 7 (strongly agree)

who assessed the effectiveness of the innovation project with the comparatively low value of 5.0 (arithmetic mean of all interviewees: 6.0). This is also due to the fact that he experienced these delays most clearly, due to his direct development work with the tilt sensors, and had to provide the additional effort needed to implement the necessary changes.

Apart from some small substantive and financial deviations from the original content and project plan, all the members were shown in the interviews to be very satisfied with the project's success.

> The success of the innovation project has in my view been outstanding. In the market, the sensor has very good prospects, now and in future. I myself am almost surprised by the success. For a long time, I never thought that we would come this far, together with our network partners. (Department Head, Company A, 102)

> This innovative project has been very successful, because the tilt sensor has been successfully mass-produced for a year and a half, and . . .marketed by the company. The sensor, with its accuracy and its cost-benefit ratio, has a unique market advantage over its competitors. (Director, Research Institute, 104)

> Overall, I see the innovation project as very successful, especially in view of the fact that the sales volume in our company has shown very good results with respect to the tilt sensors. The interest of the customer, for example at trade shows, continues unabated, and there are many requests coming in from many different industries. In order to handle these new requests, successful follow-on projects have also been initiated to develop the next generation of the sensor. (Managing Director, Company A, 122)

Contribution to Success of the Innovation Community

As indicated above, the members see the innovative project as very successful. In the last part of the survey, it remains to determine what contribution the cooperation of the promotors has made to the project's success. The results in Table 3.29 are clear: all four members assess the impact of the innovation community as very high. Two of the members have assigned it the highest possible rating, a 7, while the other two gave it a grade of 6. With this result, the significant impact of the innovation community to the project's success has been clearly demonstrated.

In the following, the contributions of the innovation community to project success are considered in greater detail. From the interviews conducted, eight issues were identified which contributed to the sustained success of the project:

- Many years of collaboration: Even before the start of the innovation project, there was close cooperation between the two initiators of the tilt sensors network, who worked together on other projects in the area of microsystems technology. Thus, the existing skills and preparatory work, and also the already existing very positive team spirit between the players, facilitated the initiation and development of the community. The common definition of goals and the search for potential partners and lead customers was realized very efficiently, and at a very early stage of the innovation process, the first prototype of the tilt sensor was built.
- Integration of innovative technology development actors: The right combination of actors was, from the perspective of all four interviewees, a major reason for

Table 3.29 Evaluation of the contribution of innovation community to project success

Contribution of the community to project success	Members of the innovation community				All
	Champion 1	Champion 2	Champion 3	Expert promotor	
What is your estimate of the contribution of the above named group of people to the success of the joint innovation project?	7	7	6	6	6.5

Soure: Author. Possible answers from 1 (low) to 7 (very high)

the success of the network. In this case study, all the key actors in the project complement each other along the value-added chain. Especially in the case of communities addressing radical innovations, effective operation is contingent upon the actors themselves not being directly in competition, so that the synergies are primary, and that is perceived by all. Each actor in the tilt sensors network sees a sufficiently large benefit for himself, for which it pays to actively participate in the network.

Therefore, one reason the tilt sensor was so successful was because the partners in the network proactively transferred the new and complex technology to the market, and did not follow the ideas and solutions of other competitors. (Managing Director, Company A, 121)

- Coordination by a neutral partner: The contribution of the promotor from the technology transfer centre is particularly valuable in terms of the management aspect, since with this company, all organizational tasks could be brought together and carried out by one person. But this actor was not only responsible for coordinating tasks, but also for linking the technological development to the market side requirements. And finally, through its systematic trend detection and its very good contacts to the microsystems technology industry, the company made an important contribution to network management and to the strategic direction and development of the network.

  The main focus is primarily on marshalling all available resources, personal capacities and the technical knowledge of all partners quickly and efficiently, and putting them to use. (Managing Director, Technology Transfer Centre, 12)

- Open communications and cooperation: A major factor of the community was the very open communications culture and the intense and respectful cooperation between all actors – even transcending hierarchical levels. Due to the very substantive and frequent coordination, it was possible to carry out the organizational development process without major loss of time, and thus to avoid redundancies or mistakes.

  Even the customers opened themselves up surprisingly wide, and passed sensitive information on to the network. This openness and familiarity of all involved contributed to the very lasting success. (Managing Director, Company A, 84)

- Common language: The cooperation in the promotor network was supported by the fact that regular internal workshops ensured that common goals were understood by all participants in the same way. Major errors were thus excluded from the outset.

  When you have partners from different industries who each speak a sector-specific language, it can – even though they all speak German – lead to misunderstandings: one person means something different than the other. It may also be that the world in which one actor works is far too little familiar to another – to recognize that and to solve it is very important, as is the case in any form of cooperative effort: clarity about what the other partner wants, and clarity that he has understood it. (Director, Research Institute, 86)

- Early involvement of key customers: The inclusion of a lead customer brought in during the development phase of the innovation project was the critical factor for a marketable tilt sensor. The customer introduced a very specific need that made it possible to give the development process a very goal-oriented approach. That also constituted the first step into a new market, and it was possible to acquire new customers via the existing contacts.

> Via the contacts of the lead customers and the partners in the network, new relationships are established by the snowball principle, new customers are found and new product ideas generated. (Managing Director, Company A, 33)

- A high level of self-initiative: A very large stake in the success of an innovation project is of fundamental importance for the implementation of such a project. This is particularly crucial in the case of any hurdles during the project, or in case of conflicts in the promotor network. In the present case, the network was able to overcome the technical hurdles very well, primarily because of the very high personal initiative of the partners. For the partners for example, this meant investing more of their own resources or labour power into the project than originally planned, to achieve the overall goals.
- Good contacts to the microsystems technology industry and the funding sources: During the initial phase, the existing relationship network throughout the microsystems industry was very important for the composition of the innovation network. In addition, the excellent contacts of the research institute and the technology transfer centre to the funding sources ensured the necessary public subsidies for the overall project, and thereafter a follow-up project for the production of tilt sensors.

3.7.5 Conclusion

In the context of the innovation project of the present case study, the production process and design of high precision tilt sensors ready for market-ready series production was promoted. With this radical innovation, market requirements were met which previously known sensor systems could not meet, especially with regard to the combination of quality and cost.

As the survey of the four members of the network indicates, the key players involved, the promotors, and their functions were significantly involved in the success of the network. According to the definition from Sect. 1.3 an innovation community can clearly be stated. Due to the small size of the overall network, the composition of the innovation community did not change throughout the entire duration of the innovation project. Only a slight shift in the task content was observed throughout the various phases. The promotor network was characterized from the start by a high degree of identity and very good cooperation. These characteristics allowed it to overcome technological barriers and arrive at rapid resolution of conflicts.

As the analysis shows, the innovation community had a significant impact on the success of the innovation project. Through their good cooperative effort, the partners achieved the goal of the relationships of the major players: the integration of innovative technology development actors, the coordination of a neutral partner, the open culture of communications and cooperation, the common language, the early involvement of key customers, the high level of initiative, and the good network of contacts with the microsystems technology industry and with the funding sources.

3.8 Wood Energy: Promoting Short Rotation Wood Crops for Energy Purposes

Klaus Fichter

Fact Sheet

Case Study: Wood Energy

Innovation object:	Regional agroforestry systems: Developing fast-growing wood, crop plantations for biomass and energy production
Industry sectors:	Agriculture, energy suppliers, agricultural machinery
Period analyzed:	2004–2010
Initial situation:	Biomass is considered the "sleeping giant" in energy production. Wood biomass offers several benefits that are explored in the development of "agroforestry systems" in Brandenburg, Germany.
Current situation:	Realization phase
Research highlights:	Radical innovation project, changing energy markets and policies, interesting and unusual approach to biomass and energy production issues.

Key persons involved:	15
Size of innovation community:	7
Group composition:	More than half of the members are champions
Regional distribution:	Regional (Lusatia, Germany)
Community performance:	6.2*
Contribution to project success:	5.8*
Project success:	6.8*

*On a scale from 1 (very low/poor) to 7 (very high/good).

3.8.1 Context and Object of Innovation

The importance of renewable energy sources has grown considerably since the 1990s, and will increase further in future. Climate change, the limited nature of fossil fuels and rising energy prices demand that new ways of generating energy are found. Energy from biomass is assessed as having high growth potential; indeed, it has been characterized as the "sleeping giant" of renewable energies (DLG/WWF 2006). Compared with other bioenergy processes, wood is a particularly promising source of energy, because it promises high greenhouse gas savings, more extensive plantation is possible than is the case for annual crops, and the costs of heat production are relatively low. In this context, fast-growing plantations in the form of so-called "agroforestry systems" or "short-rotation plantations" in agriculture are being seen as a new option for timber production (Landschaftspflegeverband, 2010b, 1). An "agroforestry system" is a land use system that integrates elements both of forestry (perennial woody plants) and of agriculture (annual crops, pastureland) on a farmed area (Freese 2010). Short-rotation plantations are plantations of fast growing tree species and varieties such as poplars, willows and locust trees, which provide high returns of woody biomass in short-rotation periods. In cases where the wood is used exclusively for energy, these are also called "energy forests" or "wood energy" (Landschaftspflegeverband 2010b, 1). The issue of wood energy is thus at the interface of the agriculture, forestry and energy sectors.

Object of Innovation

The object of innovation addressed in the context of this case study is the widespread introduction and establishment of agroforestry systems in the German state of Brandenburg (Fig. 3.11). These agroforestry systems are short-rotation plantations where fast-growing species, such as poplar, locust and willow are grown on usually fallow farmland or re-cultivated former lignite mining areas. These tree species, planted as cuttings, are generally harvested after 5 to 8 years, and used as biomass.

While field groves and windbreak plantations are nothing new, the innovative element in the "wood energy" concept is that the groves are planted extensively and at the same time in accordance with ecological requirements, and systematically used as part of a crop rotation process.

> What is new is that it is possible for wood to be harvested technically efficiently – I am referring here to the wood chip technology that makes it possible to introduce new tree plantations in the agricultural sector, and also makes it possible to introduce new ways of thinking into the agricultural economy. (Managing Director, Landscape Care Association, 57)

In the past, short-rotation plantations have constituted a niche in biomass use. In 2010, the area used for wood energy cultivation in Germany has been estimated to be about 3,000 ha by experts (Fig. 3.12). "We in Brandenburg are far and away

Fig. 3.11 Short-rotation plantation on a former lignite mine in Brandenburg (Source: www.tucottbus.de/multiland)

Fig. 3.12 Efficient harvesting technologies, an element of the wood energy innovation project (Source: www.energiewald.com)

number one, with some 700 ha" (Chair, Landscape Care Association 208). Taking into account the relevant sustainability and cultivation criteria, the farmland in Brandenburg potentially usable for short-rotation plantations is estimated at about 300,000 ha (Uckert 2010, 26).

The large-scale introduction of wood energy cultivation in the form of short-rotation plantations can be considered a fundamental – even radical – innovation. Although experimental plantations in Sweden and a few pioneer projects in Austria have existed since the 1990s, where the suitability of certain tree species has been explored, the pilot projects initiated in the Brandenburg region of Lusatia in 2004 are part of the ground-breaking work in the area of wood energy cultivation. From the perspective of innovators involved in Brandenburg, short-rotation plantations and their associated improvements (more efficient harvesting techniques, new types of trees, etc.) meet the essential criteria for a high degree of innovation. All those interviewed agreed that wood energy cultivation was very new, both for themselves and for their partners, that there had been little experience in this area, that wood energy cultivation led to in some cases entirely new customer benefits, and that it required new, previously non-existent collaborative efforts with partners in the market.[10]

3.8.2 *The Innovation Process*

History and Factors Influencing the Innovation Process

The innovation process of wood fuel production examined in the context of the present case study in the southern Brandenburg region of Lusatia between 2004 and 2010 has been influenced by a number of developments and factors going back to the early 1990s, and in some cases even to the period prior to the overthrow of the East German regime. In view of the history and the drivers of the innovation process, the following formative aspects can be identified, based on the interviews conducted:

A long-standing network of farmers in the region: Many of the farmers active in the context of the innovation process have known each other for decades. One of the initiators of the first pilot project for wood energy production in Lusatia is himself a trained farmer, and has excellent contacts in the region. The fact that he was able to recruit farms as business practice partners for the testing of the cultivation of wood energy for the first pilot project is due in large part to his long-standing working relationships with individual farmers.

Establishment of the Landscape Care Association 1994: As early as the beginning of the 1990s, some active people in the region understood that certain issues such as the use of renewable resources and the link between agricultural use and

[10] The evaluations of the degree of innovation ("project innovativeness") were surveyed by means of a questionnaire, using nine characteristics. Cf. Appendix in Chap. 2.

nature conservation would only be possible by means of good regional management. Such ideas led to the foundation of the Landscape Care Association in 1994, and played a key role during the subsequent introduction of wood energy cultivation. The idea of wood energy cultivation was initiated by suggestions that the chair of the Landscape Care Association brought back from the Elmia in Sweden, the largest forestry trade fair in Europe. The board of the Association was quickly won over to the idea, and the protagonists came together in a working group under the title "New Energy – New Jobs – Wood as an Energy Fuel." Marketing chains and opportunities for wood chip production were discussed, but the result was sobering: The production and supply costs were at that time still too high to be economically viable. Thus, the project ended up in the "filing cabinet" in 1999 (Schwarz and Swoboda 2005, 16).

Contacts between practitioners, scientists and political leaders: Another important factor that favoured the initiation of a pilot project for the cultivation of wood energy was the long-term contact of the Chair of the Landscape Care Association to scientists and policy-makers. His contacts with political circles in the state were the reason that in 2004, he received a request from the Brandenburg Ministry of Agriculture asking "whether I could imagine drafting a project proposal for the area of biomass, specifically, wood on set-aside farmland" (73).

Shortages of energy resources and use of set-aside farmland: The interest of the state of Brandenburg in promoting a pilot project for wood energy production was due to the co-financing available from the EU for the cultivation of wood on set-aside farmland. The scarcity of energy resources on the one hand, and the EU's set-aside policy for low-yield farmland designed to avoid over-production in Europe on the other were thus important indirect driver for wood energy cultivation in Brandenburg.

An energetic promotor with top contacts in the region: As stated above, the Chair of the Landscape Care Association played a key role in the preliminary work, the initiation and the implementation of the pilot projects for wood energy production in Lusatia. He can be characterized as the "key promotor" of the innovation process. As a skilled farmer, and on the basis of his decades of experience in the organization of companies and his excellent contacts with farmers in the region, and also with scientists and policy-makers in the state, he had since the early 1990s pushed forward regional projects, and in particular the use of renewable resources. He was effective not only in his role as Chair of the Landscape Care Association, but also in various other networking functions in the region, particularly as the initiator and manager of numerous regional projects.

The Focal Innovation Process

The focal innovation process examined in the context of the present case study includes the period from 2004 to 2010, and involves the activities to explore the possibilities of energy use of field groves, and the implementation of model projects for short-rotation plantations in the southern Brandenburg region of Lusatia. If the aforementioned history of the focal innovation process is included, the overall process can be divided into three main phases:

- The maturation phase: 1994–2003
- The initial and conceptual phase: 2004
- The development phase: 2005–2010

Innovation process research has shown that the initiation of innovative projects is usually preceded by a usually lengthy "maturation phase" (see van de Ven et al. 1999, 21) In the case of the introduction of wood energy production in Lusatia, it is possible to clearly delimit such a *maturation phase*. Based on the above-described history of the innovation process, the foundation of the Landscape Care Association can be considered the starting point of the maturation phase, since it marked the beginning of a systematic addressing of the issues of renewable resources and bioenergy. The other key aspects and factors influencing the maturation phase were presented in the previous section.

The year 2004 can be characterized as *the initial or conceptual phase*. That year, some significant framework conditions changed. While at the end of the 1990s, the Landscape Care Association had still assessed the cultivation of wood energy as uneconomical (see above), the political and economic conditions in Germany changed fundamentally in 2004, with the amendment of the Renewable Energies Act (EEG), since, with support for cogeneration, new possibilities for the economic use of wood energy were opened up. In this context, the Brandenburg state government supported the proposal to build wood energy projects on set-aside farmland, with a view towards possible subsidies, and after a preliminary evaluation of the project proposal by a panel of the Brandenburg Energy Technology Initiative. The request from the Brandenburg Ministry of Agriculture and a respected scholar from the region led to re-addressing the opportunities for woody biomass in 2004.

Applying for funding for an initial demonstration project was preceded by a series of preparatory measures. A variety of events, visits, seminars, consultations and exchanges of experience brought together a regional group of like-minded people, and motivated them to press ahead with the project (Schwarz and Swoboda 2005, 15). Their activities included trips to pioneer projects in other countries, leading to intense network relations that continue to this day.

The *development phase*, which began in 2005, continued until 2010, with a total of three different EU funded projects. The first project involved the development of a concept for short-rotation plantations in the Spree-Neisse-Bober region, building a biomass agro-network, and the development of a manual for wood energy producers (Landschaftspflegeverband 2010a). This project was co-financed under the INTERREG IIIA programme, and implemented in 2005 and 2006.[11] In the

[11] The INTERREG IIIA Programme was a regional EU German-Polish border area support programme on the territory of the German states of Hither Pomerania and Brandenburg, as well as in Poland, from 2000 to 2006.

trans-boundary projects, the Polish partners were also involved. The second project was financially supported under the EU's LEADER Plus Programme, and involved the cultivation of wood energy on demonstration plots in the Spree-Neisse-Land Convergence Region. The trees planted were fast-growing species such as willow, gray alder, black locust and poplar.

In addition to the above projects, there were numerous contacts among the participants in the focal innovation process regarding other projects that were also carried out or initiated between 2004 and 2010. These included the projects DENDROM, AGROWOOD, AGROFORST, NeBRA, INKA BB and AgroForNet. All these projects involve agroforestry systems and/or short-rotation plantations, or the use of woody biomass, and refer in full or in part to the southern Brandenburg region of Lusatia.

Organizations Involved in the Innovation Process

A wide variety of organizations are involved in the three EU-funded projects of the focal innovation process. Together, they form the "Biomass Agro-Network"; they include:

- Four farms, three in Lusatia and one in Poland
- Five research institutions, four in Brandenburg and one in Poland
- Three associations, two local and one international
- One Brandenburg state-level initiative for energy technology, and
- Two other practice partners, one regional and one foreign

The key player in the Biomass Agro-Network is the Landscape Care Association, which coordinates the project.

3.8.3 Development of the Innovation Community

The focus of the present case study is the question of how and to what extent innovation communities can contribute to the success of radical innovation projects. First, the question of which key actors and promotors were involved in the focal innovation process of wood energy in Lusatia (2004–2010) should be clarified.

Key People and Promotors

As Table 3.30 shows, a total of 15 key people can be identified in the case of the wood energy initiative. They were primarily from the Landscape Care Association, and secondarily from agricultural and forestry enterprises, and from the scientific community. Twelve of the key people are located in the state of Brandenburg, one

Table 3.30 Key people in the Lusatian Wood Energy innovation project (2004–2010)

	Geographic location	Key person	Promotor	Champion	Member of the innovation community	Member of the core community
Organization						
Chairman, Landscape management association	Brandenburg	Yes	Power, expert, process and relationship promotor	Yes	Yes	Yes
Managing Director, Landscape management association	Brandenburg	Yes	Expert and process promotor	Yes	Yes	Yes
Farming/forestry operations						
Farmer K, Farming operation	Austria	Yes	Power and expert promotor	Yes	Yes	Yes
Agricultural entrepreneur B, Farming operation	Brandenburg	Yes	Power and expert promotor	Yes	Yes	No
Agricultural entrepreneur D, Farming operation	Poland	Yes	Power and expert promotor	Yes	Yes	No
Director B, Production cooperative	Brandenburg	Yes	Power and expert promotor	Yes	No	No
Employee P, Forestry tree nursery	Brandenburg	Yes	Expert promotor	No	No	No
Farmer K, Agricultural cooperative	Brandenburg	Yes	No	No	No	No
Manager K, Farming operation	Brandenburg	Yes	No	No	No	No
Scientific community						
Scientist H, University	Brandenburg	Yes	Power and expert promotor	Yes	No	No
Scientist S, University	Brandenburg	Yes	Expert promotor	No	Yes	No
Scientist F, Institute	Brandenburg	Yes	Expert promotor	No	Yes	No
Scientist L, Research institute	Brandenburg	Yes	Expert promotor	No	No	No

(continued)

Table 3.30 (continued)

	Geographic location	Key person	Promotor	Champion	Member of the innovation community	Member of the core community
Scientist B, University	Saxony	Yes	Expert promotor	No	No	No
Scientist M, university	Brandenburg	Yes	Expert promotor	No	No	No
Number		**15**	**13**	**7**	**7**	**3**

Source: Author

in the state of Saxony, and two outside Germany. The key people are the most important participants in the Lusatian Biomass Agro-Network.

Thirteen of the key people meet the definition of a promotor. By far the most frequently encountered promotor role in the Biomass Agro-Network is that of the expert promotor. All promotors fulfil this role. Seven of the thirteen promotors can moreover be classified as "champions," because they assume two or more promotor roles simultaneously. In addition to the role of expert promotor, five people have also assumed the role of power promotor. Only two people have taken on the role of process promotor in the Lusatian wood energy focal innovation process, and only one person acts as a relationship promotor. The latter is the only person who meets the criteria for all four promotor roles: he is the chair of the Landscape Care Association.

As the case of wood energy shows, not all the promotors of an innovation process are automatically members of an innovation community. This can be explained by the fact that the promotors involved may, while supporting the innovation process in important ways, for example through the provision and operation of demonstration plots, nonetheless only cooperate with the other promotors sporadically. The other promotors therefore do not consider them part of the "core team". Hence, only seven of the thirteen promotors of the Lusatian wood energy innovation process can be considered members of the innovation community, as defined in this study. The fuel-wood case thus includes a larger, 13-person promotor network, and a seven-member innovation community embedded within it, which can be delimited from the larger promotor network by its intensive inner-community level of cooperation and strong group cohesion.

In the present case, a further need for differentiation is apparent. Thus, within the seven-member innovation community, a "hard core" can be identified which cooperates especially closely, and has a particularly strong group cohesion. For this reason, we will in this case also refer to a "core community." This includes three people: the Chair and the Managing Director of the Landscape Care Association, and a farmer; they are among the four promotors interviewed for this study.

Evolution: Steps and Stages of Community Development

The emergence and development of the Wood Energy innovation community can be described by way of the three phases of the innovation process presented above. As Table 3.31 shows, both the number of organizations involved in the innovation process and the number of members in the innovation community and the number of other promotors increases over the course of time. During the maturation phase (1994–2003), the Landscape Care Association and a university are of particular importance. The three promotors that can be identified for this phase come from these two organizations. In the initial/conceptual phase (2004), because of the extensive research and travel activities of the regional promotors, two foreign farmers are added, who, in their respective countries (Austria and Poland), have been active as pioneers in the field of short-rotation plantations since the 1990s. With the EU-funded projects that shape the development phase (2005–2010), the number of participating organizations, and the innovation community too, are expanding.

A number of important insights can be obtained from the evolution of the Wood Energy innovation community:

- The community is characterized by a high degree of continuity with respect to the “core promotors” (the Chair and the Managing Director of the Landscape Management Association). They play a key role in all three phases, both for the innovation process and for the innovation community.
- A major power/expert promotor from the scientific community played a key role in the maturation and initiation phases, but was less significant during the development phase, since, due to other commitments, he did not want to remain active in the innovation process.
- The preparation and initiation of development projects will require additional information, and the identification of potential partners. With the extensive research and travel activities during the initial phase, contact was made with other promotors, in this case, pioneers in the field of short-rotation plantations in Austria and Poland. This expanded the innovation community.
- The provision of resources (here, EU funds), and the implementation of development and realization projects required the involvement of other implementation partners and led here to a further expansion of the innovation community during the development phase.
- With the implementation of development and realization projects, the added tasks enhance not only the innovation community, i.e. the network of promotors who cooperate closely in the innovation process and have a high sense of commonality, but also the number of “loosely” associated promotors. The innovation community is thus the core of a larger promotor network.

Table 3.31 The evolution of the Wood Energy innovation community

	Maturation phase (1994–2003)	Initial/conceptual phase (2004)	Development phase (2005–2010)
Participating organizations	Landscape management association University	Landscape management association University Agricultural operation in Poland Agricultural operation in Austria	Landscape management association Two additional associations An Energy technology initiative, state of Brandenburg University Four other research institutions Agricultural operation in Poland Three agricultural operations in Brandenburg Agricultural operation in Austria Two additional practice partners
Innovation community	Chair, Landscape management association Managing director, Landscape management association Scientist H, University	Chair, Landscape management association Managing director, Landscape management association Scientist H, University Farmer K, Agricultural operation Agricultural entrepreneur D, Agricultural operation	Chair, Landscape management association Managing director, Landscape management association Scientist D, University Scientist F, University Farmer K, Agricultural operation Agricultural entrepreneur D, Agricultural operation Agricultural entrepreneur B, Agricultural operation
Other participating promotors during the preliminary phase or the innovation process	Scientist B, University Scientist M, University	Scientist B, University Scientist M, University Scientist L, Research institute	Scientist H, University Scientist B, University Scientist M, University Scientist L, Research institute Director B, Production cooperative Employee P, Forestry tree nursery

Source: Author

The Composition of the Innovation Community

As Table 3.30 shows, the innovative community includes seven promotors, five of whom assume multiple promotor roles, and can therefore be classified as "champions." With the Chair of the Landscape Care Association, the Wood Energy innovation community has a champion who assumes all four promotor roles. For this reason, and in view of the fact that he is both one of the initiators of the innovation process and the one who was the central driving force during the entire maturation, initiation and development phases, he can be characterized as the "top champion." A community member describes his role as follows:

> He ... really is a focal point, and has contacts in many areas ... to politicians, to the Ministry, of course, and I also think he is a kind of engine. (Agricultural Entrepreneur B, Agricultural Operation, 57)

He is the only member of the network who assumes the role of relationship promotor. As chair of the Landscape Care Association and in various professional roles, he has many contacts, both to farmers and to political decision-makers and scientists. He therefore has great social capital; this also applies to the innovation community as a whole. As the responses of the surveyed community members show (Table 3.32), these people see their group as having a large number of good relationships with policy makers, and as being equipped with a high level of expertise.

From the perspective of the surveyed members, who all assume several promotor roles, and can thus be classified as "champions," the innovation community also largely has the necessary cooperation partners who are critical to success. On a scale of 1 (disagree completely) to 7 (strongly agree), three of the respondents gave a 6. Only one community member gave a 4, and evidently still sees the need to strengthen the composition of the community with people critical to success, although he did not say so specifically during the interview. This community member, designated here as a "Champion 4," is in general different from the other three surveyed members in terms of his evaluation of the community. This can be explained by the fact that contacts between him and the other community members developed relatively late, in 2008–2009, so that he has only recently been perceived as part of the community. Moreover, the interaction between him and the rest of the community is largely confined to two of its seven members– the chair of the Landscape Management Association ("Champion 3") and a farmer ("Champion 2"). Due to this intense, but only sporadic contact with the innovation community, Champion 4 is here referred to as the "satellite."

The fact that the community has since its formation in the maturation phase been in a position to attract new members shows its capacity for reconfiguration. A member of the community observes that the persons involved:

> ... move around... depending on the project. The partners are there; they're the contacts in the network. If, let's say, I need seed, I'll talk to those people and say: What's happening with you? Is something developing, or not? (Managing Director, Landscape Care Association, 176)

Table 3.32 Responses of members surveyed on community composition

Composition of the community	Members of the core community			"Satellite"	All
	Champion 1	Champion 2	Champion 3	Champion 4	
Completeness and complementarity of skills					
The group has all companies and organizations represented within that are important to the success of the innovation project	6	6	6	4	
Social capital					
The group has very good relations with actors in partner organizations that make decisions regarding the joint innovation project	6	6	6	6	
The group has excellent contacts with decision makers who are important to the joint project, even outside the partner organizations involved	5	6	7	6	
Members of the group have access to persons with relevant expertise with respect to the joint innovation projects	7	6	7	7	
Reconfiguration					
The group is open to new/additional partners/members	6	7	7	7	
The group was or is in a position to let go of members who are not helpful or necessary for the success of the project, or to ensure a separation from them	6	7	6	5	
Arithmetic mean for composition of the community	**6.0**	**6.3**	**6.5**	**5.8**	**6.2**

Source: Author. Possible answers from 1 (disagree completely) to 7 (strongly agree)

The Identity of the Innovation Community

The identity of an innovation community is determined by the clarity of group boundaries, the degree of functional identity and the intensity of group cohesion. As Table 3.33 shows, three of the four surveyed members of the Wood Energy innovation community consider the group's limits very clear, with values of 6 or 7. Champion 4 assumes a special position. The fact that he considers it less clear who belongs to the community and who does not, can be explained by the fact that he only joined the community quite late (2008–2009), and the interaction between him and the rest of the community is largely limited to two of its seven members.

As Table 3.33 shows, the functional identity of the Wood Energy community is very highly developed. The members are very united with respect to the focus on innovation, wood energy and the objectives of the common innovation project, and speak "the same language." Only Champion 4 sees conflicts due to conceptual

Table 3.33 Responses of community members surveyed on aspects of community identity

Identity of the community	Members of the core community			"Satellite"	All
	Champion 1	Champion 2	Champion 3	Champion 4	
Clarity of group boundaries					
The group behaves as if it were a team	6	6	6	3	
The people see themselves as "insiders"	7	7	7	6	
The group is clearly seen from the outside as a unit	7	6	6	4	
All persons fit well into the group	6	6	6	6	
Each member of the group knows exactly who is a member of the group and who is not	7	6	6	3	
The group of actors is very stable; there is little membership turnover	7	6	7	3	
Arithmetic mean for clarity of group boundaries	**6.7**	**6.2**	**6.3**	**4.2**	**5.8**
Functional identity					
The actors agree on the importance of innovation	7	7	7	7	
The actors agree on the objectives of the joint innovation project	7	7	7	7	
The actors use the same "language" (use the same terms, technical language, etc.)	7	6	7	7	
There are no conflicts due to conceptual misunderstandings	6	6	7	1	
Arithmetic mean for identity	**6.8**	**6.5**	**7.0**	**5.5**	**6.4**
Cohesion					
For individuals, it is important to be part of the project	7	6	6	6	
The actors are strongly connected with the project	7	7	6	6	
All people are fully integrated into the group	7	7	6	6	
There are no personal conflicts within the group	7	6	6	7	
There is personal affinity/sympathy between the actors	6	7	7	6	
The network is characterized by a strong cohesion of its individual actors	6	6	7	6	
Each actor is proud to be part of the group	6	6	6	4	
All people feel responsible for the group, and protect it	6	7	6	3	
Arithmetic mean for cohesion	**6.5**	**6.5**	**6.3**	**5.5**	**6.2**
Arithmetic mean for identity of the community	**6.6**	**6.4**	**6.4**	**5.1**	**6.1**

Source: Author. Possible answers from 1 (disagree completely) to 7 (strongly agree)

misunderstandings; the other respondents do not see this at all. This special position can be explained by Champion 4's late entry into the innovation community. One community member describes the very high degree of functional identity as follows:

> Biomass and wood energy are a growth market. This is the fundamental tie that binds... to establish wood cultivation here in the region, that's what we all want. (Managing Director, Landscape Care Association, 200)

Table 3.33 moreover shows that the cohesion between those actively promoting wood energy cultivation in the region is extremely strong. For all community members, being part of the projects is important. They see themselves as closely connected with them, and as fully integrated into the group. They see little or no personal conflict; personal sympathy values are very high – 6 or 7. Only in the sense of responsibility for the group does Champion 4 differ from the others. The fact that the Wood Energy community is both a community of purpose and a group with a very high level of personal cohesion is also supported by the interviews. As an example of this, we quote the following statement by a community member:

> However, sometimes we are really purely a community of purpose, which is also okay. But sometimes we are just a community, where no one has to say a word, you can just feel it. (Farmer K, Agricultural Operation, 140)

As the chair of the Landscape Care Association stresses, the feeling of community is created only if something is actively done for it:

> Of course, that only works well if it grows out of personal bonds. If he comes up here with his wife, then I'll take my wife, and we'll all drive through the Spree Forest, so that they can get to know a different kind of cultural landscape, and it's the same the other way around, if I'm down there. (Chairman, Landscape Care Association, 165)

Interaction in the Innovation Community

The construct of the performance of innovation communities is based on the assumption that the contribution of a network of promotors to the success of an innovation project increases with the quality of interaction within. The case of the Wood Energy community then raises the question as to how good the communication, coordination, individual effort, and balance between contributions of community members, as well as the mutual support and trust and the group climate are.

As Table 3.34 shows, the quality of interaction within the Wood Energy community is extremely good, from the perspective of its members. With an arithmetic mean of 6.3 for all aspects of interaction and for all the members surveyed, the quality is very close to the maximum value of 7. The balance of the contributions of the various community members is assessed particularly positively (an arithmetic mean of 6.7), as are trust and group climate (6.6), as well as mutual support and conflict resolution (6.4). The assessments of the community members surveyed are largely identical with respect to these aspects of interaction, and vary by a maximum of one point on the scale of 1–7. Only the question of coordination draws a

Table 3.34 Responses of community members surveyed on community interaction

Interaction among community members	Members of the core community			"Satellite"	All
	Champion 1	Champion 2	Champion 3	Champion 4	
Communications					
There is frequent communication between people	7	5	5	5	
The actors communicate frequently via spontaneous meetings/phone calls	7	6	6	6	
The actors communicate frequently directly/personally	5	6	6	5	
Project-related information is communicated openly, and shared	6	7	7	6	
Important information will not be withheld from others in certain situations	6	7	7	6	
The actors are satisfied with the information received from other members of the group	6	6	6	6	
Communication takes place without the consent/knowledge of higher authorities	7	6	7	7	
Information circulates very quickly within the group	7	7	7	6	
Arithmetic mean for communication	**6.4**	**6.3**	**6.4**	**5.9**	**6.2**
Coordination					
The sub-tasks that each of the above named people in the innovation project performs are clearly defined.	7	6	6	4	
The sub-tasks that each of the above named people in the innovation project performs are well coordinated	7	6	5	4	
The performance of the tasks and the work of different people is well coordinated	7	6	6	4	
Arithmetic mean for coordination	**7.0**	**6.0**	**5.7**	**4.0**	**5.7**
Balance of contributions					
Each of the above-named people can contribute their skills fully in the common project	7	7	7	7	
Everyone in the group contributes to the achievement of the goals with their skills and capacities	7	7	7	7	
The contributions of the above-named people to the innovation projects are perceived as balanced	7	6	6	5	
Arithmetic mean for balance of contributions	**7.0**	**6.7**	**6.7**	**6.3**	**6.7**
Mutual support and conflict resolution					
The actors support each other as much as possible	6	7	6	7	
Discussions and controversies are carried out constructively	7	6	6	6	
The suggestions and contributions of group members are treated with respect and discussed	7	6	7	6	

When conflicts arise, they are resolved quickly and without problems	6	7	6	6	
The group of actors is capable of reaching a consensus regarding key issues	7	6	6	6	
Arithmetic mean for support and conflict resolution	**6.6**	**6.4**	**6.2**	**6.2**	**6.4**
Effort					
Each person fully supports the innovation project	7	6	6	6	
The innovation project has a high priority for each of the people	6	6	6	5	
People make substantial efforts to move the project ahead		6	7		
The group is completely satisfied with the individual efforts of each person	6	6	7	5	
Arithmetic mean for effort	**6.3**	**6.0**	**6.5**	**5.3**	**6.0**
Trust and group climate					
A climate of trust exists between the people	7	7	6	6	
A cooperative working climate exists in the group	7	6	6	6	
All members of the group can express their views openly	7	7	6	7	
The group members share ideas without being asked, or feel free to share ideas	7	7	7	7	
Arithmetic mean for group climate	**7.0**	**6.8**	**6.3**	**6.5**	**6.6**
Arithmetic mean for the community interaction	**6.7**	**6.3**	**6.3**	**5.8**	**6.3**

Source: Author. Possible answers from 1 (disagree completely) to 7 (strongly agree)

range of responses varying by two points. The aspect of trust and good group climate is of particular importance in innovation communities, because these are informal network relationships lacking formal safety and sanction mechanisms. In addition to personal contacts and cooperation in concrete projects, openness between members is an essential factor in the development of trust. The managing director of the Landscape Care Association attributes the team spirit and the good group climate in the Wood Energy community to mutual openness and honesty:

> The sense of community is, in my view, the result of the openness that exists between us, and the honesty with which we treat each other; that's what ultimately makes the difference. (Managing Director, Landscape Care Association, 228)

Another member of the Wood Energy community also stresses the importance of mutual openness:

> And if I notice that there's something going on under the negotiating table, I say: 'Let's please get it on the table!' (Farmer K, Agricultural Operation, 111)

An essential basis for the high quality of interaction among the Wood Energy promotors is the functional communication, which obtains means score of 6.2, extending across all the different aspects of communication and across all four surveyed community members. This value would be even higher if the members were to assess the frequency of communication as very high. However, three of the four respondents gave this aspect a score of only 5 on a 7-point scale. This may be because not all community members are located in Lusatia, which does indeed make more frequent personal communication difficult. As the example of the Wood Energy community described shows, a very high frequency of communication does not necessarily seem to be a prerequisite for ensuring that information will if necessary be passed on rapidly, and that the members will be satisfied with the information they receive. This suggests that communication frequency is not always very high, but high enough to meet the needs in particular situations. The intensity of the exchange within the community is described by one of the interviewed members:

> I don't think there's any group that sends each other more e-mails about God knows what and everything about it, including the details, than this group. That has grown up over many years in our conversations, that everybody should send everyone e-mails. So this morning we met, and then the first ones showed up, and said, "You know, I'm so glad you send so many e-mails, they're so informative." And then I had to say, thank God I got this or that from you. (Farmer K, Agricultural Operation, 148)

The only aspect of interaction that was assessed somewhat lower by one of the community members surveyed is coordination. The assessment with an average value of 4 by Champion 4 may be due to his special role as a relatively young member of the community, and be focused on his interaction with two other members. The slightly lower rating by Champion 4 could be related to the fact that although he is active in wood energy production and in close consultation with the other community members, he has yet to take part any of the formal projects in the region (see above) which have been carried out by the rest of the community.

Table 3.35 Performance of the Wood Energy innovation community

Performance of the community	Members of the core community			"Satellite"	All
	Champion 1	Champion 2	Champion 3	Champion 4	
Arith. mean for composition of the community	6.0	6.3	6.5	5.8	6.2
Arith. mean for identity of the community	6.6	6.4	6.4	5.1	6.1
Arith. mean for interaction in the community	6.7	6.3	6.3	5.8	6.3
Arith. mean for performance of the community	**6.6**	**6.4**	**6.4**	**5.5**	**6.2**

Source: Author. Possible answers from 1 (disagree completely) to 7 (strongly agree)

The Performance of the Promotor Network

Based on the data of the promotors surveyed and the other information sources used for this case study, the performance of the Lusatian Wood Energy promotor network can be considered very high. On a 7-point scale, the average of the values of all the surveyed community members yields a value of approximately 6, both for the aspect of composition and that of identity and interaction of group members (Table 3.35). Champion 4 assumes a special position; as a relatively young member of the community with no previous involvement in the funded projects of the Biomass Agro-Network, he gives slightly lower values on average than do his counterparts in the community. This is particularly evident with regard to the aspect of community identity.

These results clearly show that the wood energy promotor network is an innovation community as defined in Chap. 1 (cf. Sect. 1.3.2).

Measures and Methods of Community Development

The methods and measures identified by the members of the Lusatian Wood Energy innovation community interviewed which have been used within the focal innovation process to date are shown in Table 3.36. The focus is on those measures and methods that primarily support the purpose of interaction and exchange of the actors. Eleven of the seventeen measures and methods identified have their functional focus in that area. Various forms of information dissemination and training are of particular importance in the innovation process. Such events as the annual "Brandenburg Wood Energy Day," held since 2005, or the "Wood Energy Producer's Manual," published by the network members, are examples of forms of information transfer and training, of which a total of ten different types of measures are stated. Forms of targeted information gathering and analysis, and of organization, are stated much less frequently (5 each). In particular, the use of

Table 3.36 Measures and methods in the Lusatian Wood Energy innovation project

	Functional focus				Contribution to network and community development		Prerequisite
	Info. dissemination & training	Info. gathering/ analysis	Organization	Inter-action, encounter	Contribution to network and community development	Contribution to which quality aspect of the community?	Methodological expertise required?
Personal conversations (e.g., "back room")	X	X		X	High	Interaction	No
Internet platform	X	X		X	Medium	Interaction	Partially
Events (e.g. Wood Energy Day)	X	X		X	High	Interaction	No
Presentations ("travelling preacher", "word-of-mouth")	X			X	High	Composition, interaction	No
Experience, information (infotainment): e.g. energy bike path	X			X	Medium	Interaction	Partially
Photo contest/photo exhibition	X				Medium	Interaction	No
Demonstration projects (e.g., pilot plots, model energy farm)	X		X	X	Medium	Identity, interaction	Yes
Learning together in projects	X				High	Identity, interaction	No
Informational materials (e.g., "Wood Energy Cultivation" Info.blatt No. 6)	X				Medium	Identity (for author)	Yes
Wood energy producer's manual	X				Medium	Identity (for author)	Yes
Excursions		X		X	High	Composition, identity, interaction	No
Trend analysis		X			Medium	Composition, interaction	Yes
Project meeting	X	X	X	X	High	Identity, interaction	No
Projects (for example, EU-funded projects)			X	X	High	Identity, interaction	Yes
Networks & associations			X	X	High	Identity, interaction	Partially
Working groups (e.g., "Renewable Resources," "Producer's Guide"			X	X	High	Identity, interaction	No
Producer group			X		High	Identity	Yes
Number of functional focuses	**10**	**5**	**5**	**11**			

Source: Author

analytical instruments is obviously very little developed in the Biomass Agro-Network. Here, the instrument of trend analysis was stated only once.

As Table 3.36 also shows, it is in particular the measures and methods of interaction and organization management that contribute to the development of the network and the community.

3.8.4 Effects of the Innovation Community on the Project's Success

Before considering, in the following, the contribution of the Wood Eenergy innovation community to the success of the innovation project, it must first be ascertained whether and to what extent the focal Lusatian Wood Energy innovation process has been successful to date.

According to the four promotors/champions interviewed, the effort to move forward wood energy production in the region of Lusatia and in the state of Brandenburg has been fairly successful. In the arithmetic mean of all success aspects examined and of all community members surveyed, the aspect success achieved a value of 5.7 on a scale of 1–7 (see Table 3.37). It is striking that here too, Champion 4 gave lower scores than the other community members surveyed. This too can be attributed to his special status as a relatively new community member and to the lack of involvement to date in cooperation projects. Moreover, it is striking that the effectiveness of the project is evaluated more highly than its efficiency. According to three of the four champions surveyed, the project is to date not entirely on schedule or on budget. This appears to be the only real shortcoming of the innovation process to date. It is also striking that one champion evaluated the success of the project considerably more highly than did his community counterparts. This may be partially due to the fact that this is the manager of the Landscape Care Association, who, in his capacity as the coordinator of the EU-funded projects, naturally has a special interest in ensuring that the projects are successful, and carried out efficiently. The same is true for Champion 3, who was responsible for the project in his capacity as chair of the Landscape Care Association.

In the interviews, community members showed a high degree of satisfaction with the course of the Wood Energy innovation project. Examples of the responses to the question "How do you rate your success to date?" included the following comments:

> Now, qualitatively, we have clearly made progress in recent years, through exchange, by trial, and networking and exchange of experience. (Agricultural Entrepreneur B, Agricultural Operation, 109)

> The bottom line is that I am satisfied, both in terms of market development and in terms of the pace. That suits me, because a massively forced pace, where the actors can't keep up, doesn't make sense. It's going in the right direction. So it has been successful, yes. (Agricultural Entrepreneur B, Agricultural Operation, 111)

Table 3.37 Evaluation of the success of the Lusatian Wood Energy innovation project

Success of the innovation project	Members of the core community			"Satellite"	All
	Champion 1	Champion 2	Champion 3	Champion 4	
Effectiveness					
To date, the joint innovation project has been very successful	7	6	6	5	
The requirements of customers or users have so far been considered systematically	6	6	6	5	
The recent results are of high quality	7	6	7	5	
The partners in the joint innovation projects are satisfied with the results	6	6	6	5	
To date, there has been little need for follow-up work	7	2	6	5	
At present, everything points to the future success of the joint innovation project	7	6	7	5	
Arithmetic mean for effectiveness	**6.7**	**5.3**	**6.3**	**5.0**	**5.8**
Efficiency					
To date, the stated objectives within the joint innovation project have been achieved	7	6	6	5	
From the perspective of the network, one can be satisfied with the progress of the innovation	6	6	6	6	
To date, the project has been carried out cost effectively	7	6	6	3	
The project is progressing on schedule	7	4	5	3	
The project is on budget	7	5	5	3	
Arithmetic mean for efficiency	**6.8**	**5.4**	**5.6**	**4.0**	**5.5**
Arithmetic mean for success of the innovation project	**6.7**	**5.4**	**6.0**	**4.5**	**5.7**

Source: Author. Possible answers from 1 (disagree completely) to 7 (strongly agree)

> You're never satisfied. But the success can be measured in the change of opinion. In the beginning, with the farmers, there was rejection and scepticism. That's been reversed in recent years. The success is that, now we say, 'I'll send so-and-so over, you have to take a look at this!' And it gets carried on. And in time, we've now got to the point that people are talking about it, and I see that as a success. (Managing Director, Landscape Care Association 262)

Contribution to Success by the Innovation Community

The promotors and champions interviewed see the contribution of the innovation community to the success of the Wood Energy joint innovation project as very high. Three of the community members give a score of 7, on a scale of 1 (very low contribution) to 7 (very high contribution), the other respondent gave a 6. From

Table 3.38 Evaluation of the contribution of the innovation community to project success

Contribution of the community to project success	Members of the core community			"Satellite"	All
	Champion 1	Champion 2	Champion 3	Champion 4	
What is your estimate of the contribution of the above named group of people to the success of the joint innovation project?	7	7	6	7	6.8

Source: Author. Possible answers from 1 (low) to 7 (very high)

these responses and on the basis of the information in the interviews, it can be concluded very clearly that the success achieved would not have been possible without the close cooperation of the promotors (Table 3.38).

The question also arises: what is the specific contribution of the innovation community to that success? Based on the interviews, six major contributions of the promotor network have been identified:

1. The "midwife" function: As was shown in the description of the focal innovation process, the development phase from 2005 through 2010 was preceded by a maturation phase from 1994 through 2003 and an initial/conceptual phase in 2004. A central feature of the cooperation of the promotors during the maturation phase was to lay the technical foundations for the initiation of development projects. Thus, the promotors during the maturation phase, e.g., the Renewable Resources Working Group of the Landscape Care Association, concerned themselves with the possibilities and challenges of agroforestry systems. It was here that the major technical foundations for future EU-funded projects developed. In addition, during the maturation phase and then especially during the initial phase, vital contacts with potential cooperation partners in agriculture and in the scientific community, as well as with regard to government subsidies, were established. With the cooperation of the promotors, the innovative idea of wood energy cultivation in Brandenburg was then transferred from an informal phase to a formal organization (EU-funded projects).
2. Exchange of knowledge and experience: A very important feature of the innovation community in both the initial and the development phases was its knowledge and experience exchange. Examples include numerous excursions during the initial phase, and the local exchange of experience with pioneers of wood energy cultivation, e.g., in Poland and Austria. Another important locus of knowledge and experience is the Brandenburg Wood Energy Day, organized by the promotors of the network since 2005. These activities both helped develop and strengthen the innovation community and provided the exchange of experience, scientific expertise and political know-how necessary to ensure the success of wood energy crop cultivation. The innovation community secured the attainment of knowledge and experience resources critical to that success.

> And this experience – that's the next highlight. Whether in the seminar or on the charabanc, we bring people together and get great conversations. (Chair, Landscape Care Association, 110)

3. Obtaining financial resources critical to success: Another key contribution of the Wood Energy innovation community was the acquisition of the funds necessary for the implementation of pilot and development projects. Key contributing factors here included not only the good relations of individual promotors to potential donors, such as subsidy providers, but also the fact that the promotors combined their respective contacts. A central feature of innovation communities is thus the pooling of social capital to obtain resources critical to success.
4. Successful integration of key actors critical to success: Due to the close cooperation between the promotors of different organizations and groups, it was possible in the case of Lusatian wood energy to concentrate the resources of such relevant actors as farmers (fields for experimentation and cultivation), tree nurseries (suppliers of cuttings), the scientific community (scientific basis; environmental impact assessment), regional associations (organization, public relations), nature conservation groups (monitoring of environmental impact, public acceptance), or government institutions (subsidies, biomass plans, etc.) in the context of projects and activities, so that the cause of short-rotation plantations in Brandenburg could be advanced considerably.
5. Mitigation or resolution of conflicts endangering innovation: As was clearly shown in the description of the project's success, there was at the beginning of the project work, a high scepticism, and in some cases even rejection of wood energy cultivation, both among farmers and among authorities and the nature conservation groups. Demonstration projects and the joint advocacy network of promotors largely succeeded within a few years in eliminating this scepticism and rejection, which was a danger to the innovation project.
6. Influence upon the general conditions: Cooperation in the promotor network also enables influence on the political context, as shown by the following statement by a community member:

> Then you also have to try to get into the commissions at the state level, and to think ahead. That means, we in the Landscape Care Association and other associations are now involved in the development of the 'Biomass Action Plan 2020' of the Brandenburg Ministry of Agriculture. (Chair, Landscape Care Association, 65)

3.8.5 Conclusions

The focal innovation process examined in the context of this case study includes the activities for the exploration of the possibilities of energy use of field groves and the implementation of model projects for short-rotation plantations in the southern Brandenburg region of Lusatia. The large-scale introduction of wood energy cultivation in the form of short-rotation plantations can be seen as a fundamental (radical) innovation.

As the case analysis shows, the innovation process is influenced by a number of promotors and the close cooperation between them. The existence of an innovation community could be proven, which has grown steadily from the maturation phase through the initial phase to the development phase; its composition with respect to central "core promotors" shows a high degree of continuity, but has also undergone phase and task-specific changes. The innovation community has a competent composition, a high degree of identity and very good quality of interaction between its members. The investigation demonstrated that the success of innovation processes was essentially due to the activities of the promotors and their close cooperation.

The case study supports the hypothesis that the success of a radical innovation project depends largely on the existence and performance of an innovation community. The central contribution of the community to the success of wood energy was that it helped move an innovative idea from the informal stage to a formal project organization (the "midwife" function), provided for the exchange of knowledge and experience, supported the acquisition of resources, such as subsidies which were critical to success, integrated success-relevant actors, and ensured improved acceptance of the innovation idea among key contacts, while improving the overall political context.

3.9 Carbon Nanotubes: Developing a Lead Market for Nano-materials

Stefanie Bunzel

Fact Sheet

Case Study: Carbon Nanotubes

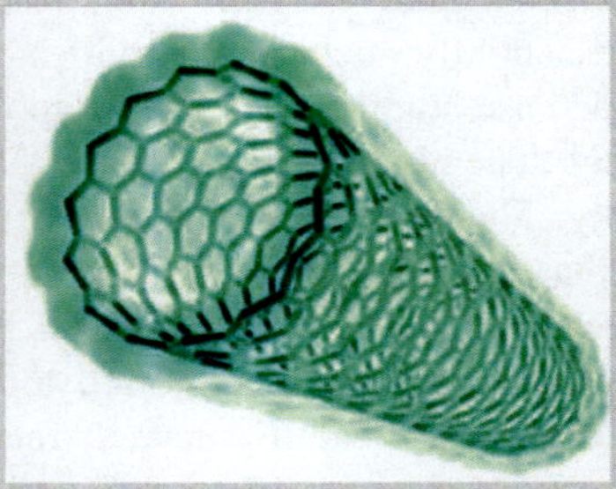

Innovation object:	Bringing Carbon Nanotubes for a variety of different applications to market
Industry sectors:	Chemistry, engineering, electronics
Period analyzed:	2006–2011
Initial situation:	The remarkable chemical and physical properties of nano-scale carbon structures like nanotubes (CNTs) have been known for some time, and while CNTs are still largely at a research stage, initial applications in the areas of energy, transport, construction, materials and electronics are emerging
Current phase:	Development phase
Research highlights:	Radical innovation project, stunning range of future applications and potential market opportunities, broad cooperation of specialist industries, producers and reseachers required

Key persons involved:	14
Size of innovation community.	6
Group composition:	More than half of the members are champions
Regional distribution:	National (Germany)
Community performance:	5.4*
Contribution to project success:	5.8*
Project success:	5.4*

*On a scale from 1 (very low/poor) to 7 (very high/good).

3.9.1 *Context and Object of Innovation*

Since the Japanese physicist Sumio Iijima helped carbon nanotubes (CNT) to succeed in 1991, these tubular structures have experienced an extraordinary world-wide development. Both nanotechnology and materials technology are considered key technologies for the twenty-first century (BMBF 2010), and it is precisely at this interface, i.e. in the area of nano-materials, that carbon nanotubes are located.

The distinctiveness of this particular class of nano-materials is its multifarious range of excellent properties, which open the way to entirely new products and markets. Their special atomic structure and geometric dimensions and wide range of mechanical, chemical, thermal, optical, electro-optical and electronic properties make CNTs interesting for a variety of commercial applications. There is almost no industry that cannot benefit from the potentials of CNTs, and although they are still largely in the research stage, initial applications in the areas of energy, transport, construction, electronics, and sports are emerging (Fig. 3.13).

In one and a half decades of research, great progress in production and high concentration has been made, so that today, high-quality CNTs can be produced in large quantities. Intensive research is being conducted into the use of CNTs for such applications as field emission sources, electrochemical electrodes, components in composite materials, field-effect transistors, and sensor elements. Even today, CNTs are being used in various technical applications, such as sports equipment, medical devices and lithium-ion batteries. For example, the plastic of tennis rackets is mixed with CNTs, resulting in a composite material with increased stiffness and stability, together with very low weight.

Research activities in the area of CNTs have been going on for a number of years, in various of countries, including Germany. Most of the successes have been achieved in the U.S. and Japan, since these countries discovered very early the importance of this field, and have moved it ahead, in joint projects with government support. Now Germany too has recognized the enormous potential of this technology, and formed a national innovation alliance, in which all major German key

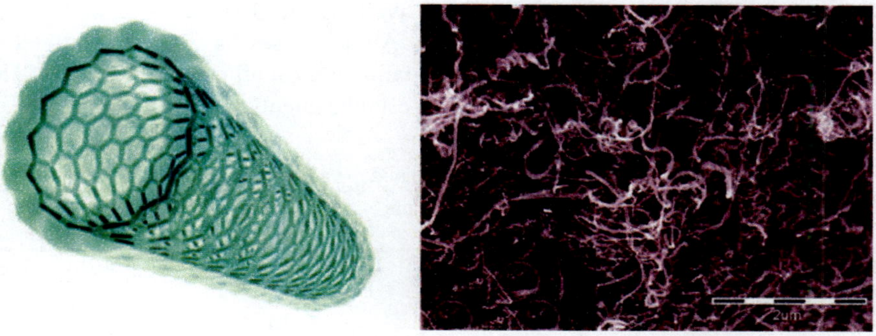

Fig. 3.13 *Left*: 3-D model of a CNT; *right*: CNT agglomerate (Source: *Left*: Denis Nikolenko - Fotolia.com; *right*: Bayer MaterialScience)

CNT players are involved. Bringing together this expertise for application and product development is to result in a rapid market entry, so as to ensure that German research is not left behind by the activities in the U.S. and in Japan.

The research needed to obtain market maturity in the relevant application areas is however still enormous, so that this is certainly a highly complex or radical technology. The major challenge consists in a rapid establishment and development of CNT technology leadership in Germany, in order to be able to participate in the rapidly growing CNT market and achieve a global profile.

Object of Innovation

From this initial state has emerged the motivation of the innovation alliance to establish a lead market for CNT production and implementation in Germany, via a close link between technology and application development. The objective of the present project is the development to market maturity of CNTs as an innovation object, for a variety of fields of application and industries.

CNTs are characterized by a combination of extremely high tensile strength and low weight. The electrical conductivity of CNTs is outstanding. Current densities of 109 A/cm^2 can be achieved. This is about 1,000 times higher than that provided by copper or silver wires. CNTs at room temperature have a specific thermal conductivity of 6,000 W/(m*K), almost three times as high as the specific heat conductivity of diamonds (2,300 W/[m*K]), and almost 14 times higher than that of silver (430 W/[m*K]) (ZTC 2009). Other mechanical properties are shown in Table 3.39.

Because of their many features and functionalities, there are numerous potential areas of application in almost all industries and products. The resulting requirements are directly reflected in basic research, i.e. the production, functionalization and dispersion of CNTs. Basically, the task of the innovation alliance is the responsible research and development of basic technologies and applications for CNT-based products. By bringing together all the essential CNT skills, the alliance is to enable the creation of the basis for building a sustainable lead market in Germany. CNT production technology is available in Germany at a world-class level, with a high potential for expansion. Already, uses have been identified in many applications with high market potential, in almost all economic sectors. The long-range planning for the sustainable establishment of CNT technology as

Table 3.39 Mechanical properties of CNTs compared to carbon fibre and steel

Material	Density (g/cm^3)	E-module (GPa)	Tensile strength (GPa)	TS/weight (standardized)
Single-walled CNT	1.4	542	65	462
Multi-walled CNT	1.8	400	2.7	15
Carbon fibre	1.6	152	2.1	13
Steel	7.6	207	0.7	1

Source: ZTC (2009)

a promising technology of the future in the German market is therefore being very strongly supported by the federal government.

> We will not see for several years whether or not today's efforts pay off. And with the ever-present safety discussions, the Alliance also continually has to fight for a good public image. (Department Head, Company B, 25)

There is hardly any other area of technology in which such a development task requires such a high level of interdisciplinary cooperation, as is currently the case in the research area of nano-materials. Not only do the disciplines of chemistry, physics, biology and engineering sciences flow together here; a very long value-added chain must also be covered, so as to actually arrive at marketable products. A major chemical corporation which is at the very top of the value-added chain in the production of CNTs recognized this fact very early, and gave the impetus for the creation of the present innovation project. In the network, all basic technological and all essential manufacturing, functionalization and dispersion steps are covered, and promising fields of application with the appropriate component, system, and finishing processes are integrated and networked. Various issues along the value-added chain are addressed – from basic research ("how can CNTs be manufactured, modified and incorporated into a material?") to concrete applications in the areas of energy and environmental technology, mobility, lightweight construction and electronics.

For all users and manufacturers of CNTs and CNT-containing materials, the project is a unique opportunity to explore the potential of this class of materials, and to identify the parameters in which work will in future necessarily proceed. Whether or not all objectives set can be achieved is still open, but the continuous research is causing the use of CNTs to grow steadily.

> CNT technology is very complex and also novel, and for most applications, it is still in its infancy. In particular, the very long value-added chains – from producers of the particles through system integration – make development very expensive. While there are already the first CNT products on the market, research is still very far from mass deployment. (Department and Project Head, Company A, 19)

Because of the complexity of the innovation project, there are now 90 partners working in 27 network projects with the overall objective of implementing this practical CNT application. None of the partners could possibly carry out such a project alone, neither technically nor financially.

3.9.2 The Innovation Process

History and Factors Influencing the Innovation Process

Research and development work in the field of carbon nanotubes has already been underway for several years in Germany, but to date mostly only sporadically, and with a very low level of application focus. Based on the history of the innovation process under consideration, it is possible to identify the following factors which have contributed essentially to the construction and development of the innovation project as it exists today and which – as will be explicitly demonstrated below – have served as a guide for this innovation community:

Impending danger of falling back behind the level of global CNT research: In just a few years, application-oriented research in the U.S. and Asia has developed so rapidly that Germany threatened to fall behind internationally in the field of carbon nanotubes. While in those countries, the networking of skills – even across organizational boundaries – had been recognized as a very effective measure for development projects with a radical character, this process of re-thinking only started somewhat later in Germany. By now however, Germany has been able to catch up again in some areas, thanks to the innovation alliance.

> As already mentioned, value-added chains in CNT development are very long, and not even major corporations such as [Company A] can handle all the process steps, from the preparation of the tubes to their use in the products themselves. The examples from the U.S. or Japan show that this is only possible in the context of networks. These countries recognized this principle well before Germany, and have therefore already pulled dangerously far ahead on the world market. In order for Germany not to fall behind completely, the innovation alliance was founded. (Department and Project Head, Company A, 23)

Breakthrough in CNT production: The participating major corporation, which can also be considered the main initiator of this innovative project, was already addressing the manufacturing issues related to production of carbon nanotubes long before the start of the innovation process. A few years ago, the corporation achieved a technological breakthrough by building a prototype system for the large-scale production of CNTs. With this facility, it succeeded in producing CNTs on a scale of tonnes in very good quality – a production capacity that was outstanding worldwide. This milestone is an essential element for all the activities and development work planned in the innovation project, because the CNTs can now be made available within the network in sufficient quantities.

Existing key contacts to the CNT community: The central actor in the innovation project had already been in contact with other actors and members of the CNT community in Germany and internationally for many years prior to the creation of the alliance. These contacts contributed greatly to supporting the maturation phase, in which the structure and the substantive orientation of the network first took shape. Especially in informal meetings or discussions, the joint needs were revealed, and solutions for implementation discussed.

> And then we thought about what you would have to do to commercialize that clear international lead in terms of nanotechnology research. And so we thought, "well, we'd have to …". So, that's how that happened. (Department Head, Company E, 39)

Very good contacts with the Federal Ministry of Education and Research (BMBF) and an extensive research program in the field of materials innovations: The very good contact between the central major corporation and the BMBF/ funding agency also proved to be particularly important, since that was what made the funding for the overall project possible in the first place. With the very good insight into the processes and requirements of the BMBF, some decision-making processes could be accomplished much faster, and the projects officially launched in 2008, in spite of the enormous size and number of partners – after only a year and a half of preparation.

The Focal Innovation Process

The focal innovation process in the present case study covers all relevant development steps necessary to move carbon nanotubes from the basic technologies to market-ready products in various fields of application. The specificity of the focal innovation process is the high degree of interconnection between all areas (production, functionalization, dispersion, parts/components, finished products), which is a significant factor due to the very long supply chains and the many partners involved. The focal innovation process considered here includes the period from 2006 to 2011, and can, with the above-described history, be broken down into three phases:

- Maturation phase, from the beginning of the decade to 2006
- Initial and conceptual phase, from 2006 to 2008
- Development phase, from 2008 to about 2012

The official launch of the innovation alliance was preceded by a long *maturation phase*, which was characterized mainly by basic technological research, but much less application orientation. However, the application focus grew with the first CNT-based products that conquered the market, and initiated the first in-house discussion on the sustainability and further development intensity of CNTs.

> CNTs have been an issue for a very long time in our company, and there have been numerous in-house discussions in recent years on their potentials and their development possibilities. (Department and Project Head, Company A, 8)

As the technological advantage of Asia and the U.S. became ever more clearly visible, the pressure to catch up also grew for the companies in this area of technology, since there was a risk of not being able to profit any more from the existing market potential of CNTs in the long term. Since going it alone was out of the question from the start, because of the complexity of the technology, the first talks were initiated between the future partners of the carbon nanotubes innovation alliance.

> So we knew the people involved, and we also knew their "personal networks", and we maintained them. (Department Head, Company E, 28)

These initial inter-company talks marked the start of the second phase, the *initial and conceptual phase*, which extended for over a year and a half, from 2006 to 2008. The idea rapidly emerged to build a nation-wide network in Germany, and to pool all relevant skills into a large overall project. As early as the beginning of the phase, contact was sought with the BMBF and the funding agency, to discuss the financing of such a project.

Until the official start of the project, and the start of *development phase* in 2008, the initiators of the joint innovation project were able to recruit over 80 industrial partners and more than 30 other partners from the research community to work on the project. Within the alliance, some 20 networked and concurrent single projects were set up to address the fundamental issues of both basic CNT issues and concrete CNT applications. The results of each project are continuously fed back to the other projects, which ensure a high degree of networking.

According to current plans, the focal innovation process is to run until 2012, and has over the course of the development phase integrated several new application projects, and new partners, in the focal process of innovation, due to emerging trends and market developments. However, due to the sheer size of the network, some partners have inevitably left the project again – for example, because of the temporary economic crisis of 2008–2009.

Organizations Involved in the Innovation Process

The 27 now publicly funded single projects in the innovation alliance each involve a variety of organizations. A total of 90 partners from the industrial and research communities are together addressing the various development issues. Together, they constitute the overall network of the Inno.CNT innovation alliance. However, not all organizations play an equally important role in terms of the structure, and of the strategic and substantive development of the overall network. The following organizations have been seen as particularly important in the focal innovation process in the context of the case study survey:

- Two major chemical industry corporations
- A medium-sized technology company
- A major chemical and energy corporation
- A technology development company with a European focus
- A large corporation with a focus on electrical, drive transmission, medical and power plant technology

The central player in the alliance for innovation is one of the two major chemical industry corporations, since it was actively involved in both the maturation and initiation phases, and has also assumed the central network management tasks.

3.9.3 Development of the Innovation Community

The goal of this case study is to answer the question of the extent to which the present network of key actors has contributed to the success of the overall project. As a first step, a detailed analysis of all key players and promotors of the innovation alliance involved in the focal innovation process between 2006 and 2011 is necessary.

Key People and Promotors

In the described case study, 14 key people can be identified who in various ways played an essential role in the focal innovation process (see Table 3.40). As we can see, these people come almost exclusively from medium-sized companies and large corporations, and are distributed throughout Germany. Two of the key people mentioned in the interviews come from the funding agency which is responsible for the public funding of the overall project.

Of the 14 key people, 12 fulfil a promotor's role. The most frequently represented role is that of power promotor (seven people), followed closely by expert promotor (six people) and relationship promotor (five people), and the role of process promotor is represented by only three people. Six of the promotors are champions, in that they assume two or more promotor roles. However, there is no person who meets the criteria for all four types. One important factor is that not all key people have a promotor function (two people), and not all promotors, or even all champions, are members of the innovation community. While their contribution to the network is seen as important, their interaction with other community members is rather low or sporadic, so that they are not perceived as part of the "core team".

For the further consideration of the case study, the focus is therefore on the composition, identity and interaction of the six members of the innovation community. Interviews were conducted with four of these people.

Evolution: Steps and Stages of Community Development

As described in the focal innovation process, the emergence and development of the carbon nanotubes community up through the time of the survey consisted of three phases. In order to take the chronological aspect into account, Table 3.41 shows the changes during this phase with regard to the participating organizations and the members of the network. It is clear that the network has grown throughout the phases, both in terms of the number of partners and of the members of the network and other key people involved. During the maturation phase, major corporation A played a crucial role as the initiator; nonetheless, no one of its staff, including those active during this phase as key people, is seen as part of the innovation community. Moreover, they were also no longer actively involved in the later focal innovation process. Then, during the initial and conceptual phase, the medium-sized company and the major corporations C, D, E and F, which were the essential organizations for the substantive orientation of the network, joined the network with the corresponding promotors and other key people. Also of interest during this phase is the effect of the two employees of the funding agency, who played a major role primarily prior to the formal establishment of the network. With its formal establishment and the launch of the first single projects, the development phase began, during which the overall network grew to its current strength of 90 organizations

Table 3.40 Key people in the CNT innovation process, 2006–2011

	Geographic location	Key personnel	Promotor	Champion	Member of the innovation community
Company					
Department and project head, Company A	Central Germany	Yes	Power, process and relationship promotor	Yes	Yes
Project manager, Company A	Central Germany	Yes	Process and relationship promotor	Yes	Yes
Department head, Company A	Central Germany	Yes	Power promotor	No	No
Employee 1, Company A	Central Germany	Yes	Power promotor	No	No
Employee 2, Company A	Central Germany	Yes	Expert promotor	No	No
Managing director, Medium-sized company B	Southern Germany	Yes	Power promotor	No	No
Department head, Medium-sized company B	Southern Germany	Yes	Expert and Process promotor	Yes	Yes
Employee 1, Company C	Southern Germany	Yes	Power and relationship promotor	Yes	No
Employee 2, Company C	Southern Germany	Yes	Expert promotor	No	No
Section chief, Company D	Central Germany	Yes	Expert promotor	No	No
Team chief, Company D	Central Germany	Yes	Expert promotor	No	Yes
Department head, Company E	Southern Germany	Yes	Power and relationship promotor	Yes	Yes
Project chief, Company F	Southern Germany	Yes	Power and expert promotor	Yes	Yes
Public institution					
Employee 1, Funding agency	Northern Germany	Yes	No promotor role	No	No
Employee 2, Funding agency	Northern Germany	Yes	No promotor role	No	No
Total		**14**	**12**	**6**	**6**

Source: Author

Table 3.41 The evolution of the CNT innovation community

	Maturation phase (2000–2006)	Initial and conceptual phase (2006)	Development phase (2008)
Participating organizations	Company A	Company A Medium-sized company B Company C Company D Company E Company F Funding agency	Company A Medium-sized company B Company C Company D Company E Company F Entry of all others, approx. 90 partners from research community, business
Innovation community		Department & project head, Company A Department head, Medium-sized company B Department head, Company E Project head, Company F	Department & project head, Company A Project manager, Company A Department head, Medium-sized company B Team chief, Company D Department head, Company E Project head, Company F
Other promotors participating during the preliminary phase or the innovation process	Department head, Employees 1 & 2, Company A	Department head, Employees 1 & 2, Company A Managing director, Medium-sized company B Employees 1 & 2, Company C Section chief, Company D Employees 1 & 2, Funding agency	Managing director, Medium-sized company B Employees 1 & 2, Company C All 27 single project heads

Source: Author

from the business and scientific communities. That also caused the innovation community to grow by another two people. During the development phase, another group important for the innovation projects was also formed, composed of the managers of all single projects, which supported the vital exchange of substantive information between the projects.

This evolution of the community already provided the first interesting findings:

- Independent of the size of the overall network, after the maturation phase, the innovation community is characterized by a relatively high level of staff continuity. The six members identified had a significant share in the organization and development of the project.
- Company A and its key people, as the initiator and driving force during all three phases, played a crucial role. However, the key people who participated during the maturation phase were no longer involved by the time of the formal launch of the network.
- The relationship promotor's role was of great significance during the initial and conceptual phase, since suitable cooperation partners and skills for the numerous single projects had to be found in Germany, and tied into the project. This process also continued during the development phase.
- The active involvement of the funding agency in the initial and conceptual phase permitted a very targeted development of the innovation project up to the point of formal establishment in 2008. This allowed the needs of the network partners to be brought into line with the requirements of the funding agency at an early date.

The Composition of the Innovation Community

In order to be able to assess the performance of the network of promotors, we will, in the next three sections, examine more closely the assessments of its composition, identity and interaction made by the six innovation community members in the interviews.

Table 3.42 shows that all four promotors surveyed see the composition of the community itself as very positive, with an average value of 6.1 on a scale of 1–7. For the aspect of completeness of competences, three of the four promotors have awarded a 6, while only one champion, in whose scope of responsibility the identification of new partners fell, gave the slightly lower rating of 5.

> There are occasionally situations where it appears that a project consortium is not adequately occupied by partners, so that I then have the task of finding another suitable partner, so as to close the competency gaps. (Department and Project Head, Company A, 131)

The social capital of the innovation community is assessed as even higher, with most respondents awarding a 7. The role of relationship promotor, occupied by several members, allowed for an excellently managed network of relationships, both internally and across network boundaries. This aspect was already crucial during the preparatory period, and also later in the network.

> The composition of key people during the founding period was largely present before the official launch of the network. Since the major players already knew each other beforehand through the still small CNT community, the contact and the subsequent discussions of ideas was very easy to establish. Then came the cooperation by the snowball method – each of the key people brought more people from their respective spheres of activity into the network. (Department and Project Head, Company A, 61)

Table 3.42 Responses of members surveyed on community composition

Composition of the group of the above persons	Members of the innovation community				All
	Champion 1	Champion 2	Expert promotor	Champion 3	
Completeness and complementarity of skills					
The group has all companies and organizations represented within that are important to the success of the innovation project	5	6	6	6	
Social capital					
The group has very good relations with actors in partner organizations that make decisions regarding the joint innovation project	7	6	6	6	
The group has excellent contacts with decision makers who are important to the joint project, even outside the partner organizations involved	7	7	7	7	
Members of the group have access to persons who have relevant (technical) expertise with respect to the joint innovation projects	7	7	7	7	
Reconfiguration capacity					
The group is open to new/additional partners/members	6	5	6	6	
The group was or is in a position to let go of members who are not helpful or necessary for the success of the project, or to ensure a separation from them	4	6	5	5	
Arithmetic mean for composition of the community	**6.0**	**6.2**	**6.2**	**6.2**	**6.1**

Source: Author. Possible answers from 1 (disagree completely) to 7 (strongly agree)

The reconfiguration of the community was assessed slightly lower, particularly with regard to the aspect of excluding less useful or necessary members from the project. This may be because no such procedure has yet been necessary. New members, on the other hand, are incorporated into the community as needed, although the formal framework of the publicly funded project often makes such an accession by a new partner more difficult.

The Identity of the Innovation Community

To assess the identity of the community, the aspects of "clarity of group boundaries," "functional identity" and "cohesion" were queried in the interviews. Here, the arithmetic mean of responses was 5.2 on a 7-point scale (see Table 3.43). Due to the enormous size of the network of 90 partners, who began work in their

Table 3.43 Responses of members surveyed on aspects of community identity

Identity of the community	Members of the innovation community				All
	Champion 1	Champion 2	Expert promotor	Champion 3	
Clarity of group boundaries					
The group behaves as if it were a team	6	6	5	4	
The people see themselves as "insiders"	5	5	4	4	
The group is clearly seen from the outside as a unit	2	3	4	4	
All people fit very well into the group	6	7	6	6	
Each member of the group knows exactly who is a member of the group and who is not	5	5	5	5	
The group of actors is very stable, little membership turnover	6	6	5	6	
Arithmetic mean for clarity of group boundaries	5.0	5.3	4.8	4.8	5.0
Functional identity					
The actors agree on the importance of innovation	5	6	5	6	
The actors agree on the objectives of the joint innovation project	5	6	6	5	
The actors use the same "language" (use the same terms, technical language, etc.)	6	4	4	4	
There are no conflicts due to conceptual misunderstandings	6	4	5	5	
Arithmetic mean for identity	5.5	5.0	5.0	5.0	5.1
Cohesion					
For these people, it is important to be part of the project	5	6	4	5	
The actors are strongly connected with the project	7	6	5	5	
All the people are fully integrated into the group	7	5	6	5	
There are no personal conflicts within the group	6	5	5	5	
There is personal affinity/sympathy between the actors	5	6	5	6	
The network is characterized by a strong cohesion of its individual actors	6	6	4	6	
Each actor is proud to be part of the group	5	5	4	6	
All the people feel responsible for the group, and protect it	5	4	4	5	
Arithmetic mean for cohesion	5.8	5.4	4.6	5.4	5.3
Arithmetic mean for identity of the community	**5.4**	**5.3**	**4.8**	**5.1**	**5.2**

Source: Author. Possible answers from 1 (disagree completely) to 7 (strongly agree)

respective single projects at different times, the assessments of the clarity of group boundaries differ widely. While the group of members has remained relatively stable over the course of the project, and they have been very well integrated into the group, the public image as an entity receives a very low rating, with values of between 2 and 4.

The assessments with respect to functional identity are somewhat more homogeneous. Although each community members understandably wishes to draw benefits for his own company from his work in the network, all members agree on the role and objectives of the overall project, assigning values of 5 and 6. The very long value-added chains and the associated comprehensive expertise in all disciplines involved also explain the slightly lower values of between 4 and 6 for the assessment of the common language and the resulting misunderstandings.

> There are, in such large collaborative projects, always a few exceptions, inasmuch as some people cannot fully integrate themselves into the team structure. Recently for example, some milestones were coming up, and some of them couldn't be met in the projects, there are indeed a few coordinators who may go against the flow, or there's a certain partiality, and the goals need to be re-discussed. And of course there's always need for discussion, but I believe that what works well on the other hand is that you always get back together again, thanks to the frequent meetings. (Team Chief, Company D, 82)

The feeling of togetherness too can be seen as positive, with an average of 5.3 on the scale. Once again, the size of the overall project had a significant effect, so that the pride of being part of this community, or even of building a sense of responsibility, was often somewhat lower. Compared to the three champions, the views of the expert promotor was lower, at 4.6, which can also be explained by the fact that this promotor has only been a member of the community since the beginning of the development phase in 2008.

> One standard for a successful project, by your definition, is just the number of people who are on a first name basis with one another. The percentage is quite high for us, and we call each other by our first names because we all have the feeling of having the same family name, I'd almost say. We feel like a big family of researchers, no matter what company name is on your business card. [...]. That's really important; you go have a drink together and think out loud – together. And then one idea leads to another and something comes of it. (Department Head, Company E, 40)

Interaction in the Innovation Community

The last aspect of the performance of the innovation community refers to the communication, coordination, balance of contributions, mutual support, effort and trust, and group climate. Again, the arithmetic mean of 5.4 indicates a very good overall quality of interaction (see Table 3.44).

> Only through the constant exchange of requirements, levels of development and problems, between the single projects, can the overall goal of the alliance, the development of new CNT markets, be achieved. (Department Head, Company B, 14)

Table 3.44 Responses of members surveyed on community interaction

Interaction among community members	Members of the innovation community				All
	Champion 1	Champion 2	Expert promotor	Champion 3	
Communications					
There is frequent communication between the people (actors)	7	6	6	7	
The actors communicate frequently via spontaneous meetings/phone calls	7	6	7	7	
The actors communicate frequently directly/personally	7	5	6	6	
Project-related information is communicated openly, and shared	4	5	5	4	
Important information will not be withheld from others in certain situations	4	5	4	4	
The actors are satisfied with the information received from other members of the group	5	4	5	4	
Communication takes place without the consent/knowledge of higher authorities	5	6	6	6	
Information circulates very quickly within the group of the people	5	5	5	5	
Arithmetic mean for communication	**5.5**	**5.3**	**5.5**	**5.4**	**5.4**
Coordination					
The sub-tasks that each of the above named people in the innovation project performs are clearly defined	6	6	5	6	
The sub-tasks that each of the above named people in the innovation project performs are well coordinated	5	4	4	3	
The performance of the tasks and work of different people is well coordinated	5	6	6	4	
Arithmetic mean for coordination	**5.3**	**5.3**	**5.0**	**4.3**	**5.0**
Balance of contributions					
Each of the above-named people can contribute their skills fully in the common project	6	6	7	7	
Everyone in the group contributes to the achievement of the goals with their skills and capacities	5	5	6	5	
The contributions of the above-named people to the innovation projects are perceived as balanced	4	5	5	5	
Arithmetic average for balance of contributions	**5.0**	**5.3**	**6.0**	**5.7**	**5.5**
Mutual support and conflict resolution					
The actors support each other as much as possible	5	6	5	5	

(continued)

Table 3.44 (continued)

Interaction among community members	Members of the innovation community				All
	Champion 1	Champion 2	Expert promotor	Champion 3	
Discussions and controversies are carried out constructively	6	6	6	6	
The suggestions and contributions of group members are treated with respect and discussed	6	6	5	6	
When conflicts arise, they are resolved quickly and without problems	5	5	5	5	
The group of actors is capable of reaching a consensus regarding key issues	5	6	5	6	
Arithmetic mean for mutual support and conflict resolution	**5.4**	**5.8**	**5.2**	**5.6**	**5.5**
Effort					
Each of above-named people fully supports the innovation project	6	5	6	5	
The innovation project has a high priority for each of the people	4	5	4	6	
The group of above-named people make substantial efforts to move the project ahead	5	6	5	6	
The group is completely satisfied with the individual efforts of each person	5	5	4	4	
Arithmetic mean for effort	**5.0**	**5.3**	**4.8**	**5.3**	**5.1**
Trust and group climate					
A climate of trust exists between the people	6	5	6	5	
A cooperative working climate exists in the group	6	6	6	7	
All members of the group can express their views openly	6	6	6	7	
The group members share ideas without being asked, or feel free to share ideas	5	6	6	7	
Arithmetic mean for group climate	**5.8**	**5.8**	**6.0**	**6.5**	**6.0**
Arithmetic mean for community interaction	**5.4**	**5.4**	**5.4**	**5.5**	**5.4**

Source: Author. Possible answers from 1 (disagree completely) to 7 (strongly agree)

The members see the area of trust and group climate as particularly positive (6.0). One member expressed the following about this aspect:

> The absolute focus is on learning to value the other person, getting to know them and respect them as a person. That to me is extremely important. (Department Head, Company E, 18)

This explains the very good mutual support and conflict resolution. This good and trusting interaction not only allows for a quick resolution of existing conflicts, but also identifies potential problems early and prevents their possible escalation even from the outset. Indispensable for this point is good communications (5.4), which, through a large number of project meetings at several levels, provides for the necessary space for interaction.

> And the cooperation is considerably simplified by having regular meetings with a great flow of information. For example, if two projects are working on the same problems, and we didn't know about it in advance, or haven't detected it until now, during the on-going projects, it will certainly come to light at the next meeting, when we're all sitting at the table and talking about what's new, what problems are being handled, and then someone will say, 'yes, we've addressed that problem too, from a different angle', and then you can look at it bilaterally, to see how to solve the problem together. (Team Chief, Company D, 73)

The relatively lowest assessments, with values of 5.0 and 5.1, can be observed for the balance of contributions and effort. Here too, the reasons involve the size of the overall project and the breakdown into single projects, which make a real overview of activities and contributions extremely difficult.

The Performance of the Promotor Network

The fact that the present promotor network is indeed an innovation community as defined in Sect. 1.1 can be shown by on analysis of Table 3.45. There, the arithmetic mean of the performance of the community, consisting of the above separate values for composition, identity and interaction of the group, is shown with a very good score of 5.4 on the scale of 1–7. Here, all four interviewees are very much in agreement in their assessments, so that none differs significantly from the mean.

Measures and Methods of Community Development

In the context of the interviews, the four promotors identified a number of methods and measures that actively supported the building and development of the community. These 13 measures listed in Table 3.46 each address a different quality aspect of the community, in accordance with functional focus, and in some cases require

Table 3.45 Performance of the CNT innovation community

Performance of the community	Members of the innovation community				All
	Champion 1	Champion 2	Expert promotor	Champion 3	
Arithmetic mean for composition of the community	6.0	6.2	6.2	6.2	6.1
Arithmetic mean for identity of the community	5.4	5.3	4.8	5.1	5.2
Arithmetic mean for community interaction	5.4	5.4	5.4	5.5	5.4
Arithmetic mean for performance of the community	**5.5**	**5.5**	**5.3**	**5.4**	**5.4**

Source: Author. Possible answers from 1 (disagree completely) to 7 (strongly agree)

Table 3.46 Measures and methods in the CNT innovation process

	Functional focus				Contribution to network and community development		Prerequisite
	Info. Dissemination & training	Info. Gathering & analysis	Organi-zation	Interaction & encounter	Contribution to network and community development	Contribution to which quality aspect of the community?	Methodological expertise required?
Regular status/project meetings and coordinating sessions	X	X	X	X	High	Composition, identity, interaction	No
Project-internal and cross-innovation workshops	X	X		X	High	Interaction	Partially
Annual Meeting for the overall project	X			X	High	Interaction, identity	No
Internet platform/Wiki	X	X		X	Medium	Interaction	Partially
Newsletter	X			X	Low	Interaction	No
IP coordination for patent applications	X	X			Medium	Interaction	Yes
Contracts (under contract, cooperation agreement, contract R & D . . .)			X		High	Interaction	Yes
Literature/patent watch	X	X			Low		Partially
Trend detection and analysis		X			Medium	Composition, interaction	Yes
Identification of potential partners		X	X	X	High	Composition, identity	Partially
Market research		X			Low		Yes
Task force on communications for the external representation	X			X	High	Identity, interaction	Partially
Scientific community with external experts		X		X	High	Interaction	No
Number of functional focus	**8**	**9**	**4**	**6**			

Source: Author. Possible answers from 1 (disagree completely) to 7 (strongly agree)

methodological expertise, which can for example be introduced by an external moderator in workshops, an IP expert for monitoring patents, or marketing experts for public relations.

Most methods are focused on the provision (8) and gathering (9) of information, which is quite understandable and meaningful in terms of network size and the absolute necessity of networking the single projects. Slightly smaller is the proportion of the methods and measures to support the interaction and encounter, in which case the extensive project meetings are seen as sufficient measures. In support of the organizational aspects of the network, only four measures, hence considerably fewer, were stated.

3.9.4 Effects of the Innovation Community on the Project's Success

In order to determine conclusively whether the innovation community of the carbon nanotubes innovation alliance contributed to the success of the innovation project, we would first like to examine the success of the recent focal innovation process more closely.

The Success to Date of the Innovation Project

> At the end of the day, the alliance will be measured by the achievement of goals – and that will have to be determined in the coming months. (Department and Project Head, Company A, 136)

Although the development work on innovative projects is still in progress, and the success cannot actually be measured until the end of the funding period in 2012, all four respondent members of the innovation community are very confident that the major milestones of the overall project will be achieved. There are already initial demonstrators, but the foundation for the development of truly large markets has not yet been laid. It is still too early in the course of the project for this, especially since some application-oriented single projects have only recently been launched.

> For all users and manufacturers of CNTs and CNT-containing materials, Inno.CNT is a unique opportunity to explore the potentials of CNTs, and to get to know the parameters on which we will be able to, and will have to, work in future. Perhaps we will not achieve every goal, but we will have learned the points at which we have to move forward to use all the characteristics of CNTs in an ideal way. (Department Head, Company B, 38)

Table 3.47 gives an overview of the assessments of promotors with regard to the effectiveness and efficiency of project success. Both mean values certify the innovation project a very good process to date. The value for efficiency, 5.5, is slightly higher than that for effectiveness (5.3), which among other things speaks for a very good network management, since it means that the targets set are both on

Table 3.47 Evaluation of the success of the CNT innovation project

Success of the innovation project	Members of the innovation community				All
	Champion 1	Champion 2	Expert promotor	Champion 3	
Effectiveness					
To date, the joint innovation project has been very successful	5	5	6	6	
The requirements of customers or users have so far been considered systematically	5	6	5	5	
The recent results are of high quality	3	6	6	6	
The partners in the joint innovation projects are satisfied with the result	5	6	6	5	
To date, there has been little need for follow-up work	5	4	5	5	
At present, everything points to the future success of the joint innovation project	5	5	6	6	
Arithmetic mean for effectiveness	**4.7**	**5.3**	**5.7**	**5.5**	**5.3**
Efficiency					
To date, the stated objectives within the joint innovation project have been achieved	5	5	5	5	
From the perspective of the network, one can be satisfied with the progress of the innovation	5	6	6	6	
To date, the project has been carried out cost effectively	5	6	6	5	
The project is progressing on schedule	5	6	5	6	
The project is on budget	5	6	6	5	
Arithmetic mean for efficiency	5.0	5.8	5.6	5.4	5.5
Arithmetic mean for success of the innovation project	**4.8**	**5.5**	**5.6**	**5.5**	**5.4**

Source: Author. Possible answers from 1 (disagree completely) to 7 (strongly agree)

schedule and on budget. The assessment of the effectiveness is slightly less homogeneous. In assessing the quality of the results, champion 1 awards only a 3, while the other three respondents awarded a 6. The reason for this may be related to the cross-network role of champion 1, who has better insight into whether these results can contribute usefully to other single projects, or will be viable on the market. Champions 2 and 3 and the expert promotor are involved in specific individual projects, and primarily assessed their own results.

Contribution to Success by the Innovation Community

Table 3.48 clearly shows that all members of the innovation community assess the contribution to the success of the project promotors as very high. On a scale of 1–7,

Table 3.48 Evaluation of the contribution of innovation community to project success

Contribution of the community to project success	Members of the innovation community				All
	Champion 1	Champion 2	Expert promotor	Champion 3	
What is your estimate of the contribution of the above-named group of people to the success of the joint innovation project?	5	6	5	7	5.8

Source: Author. Possible answers from 1 (disagree completely) to 7 (strongly agree)

two of the promotors give a value 5, another a 6 and champion 3 even awards the highest ranking, a 7. Without the active cooperation of the promotors, the current success would not have been possible.

> I am convinced that while you can provide networks, frameworks, and the entire environment, but ultimately it's all only worth something if it is then filled with life by individuals in the network. For this reason, I value the role of individuals very highly. Framework conditions are often necessary conditions, but not sufficient ones. And in order to assure project success within the framework conditions, you always need the respective key individuals. (Department and Project Head, Company A, 83)

In the last section of the present case study, the contribution of the community to project success shown in Table 3.48 will be examined, and to essential aspects will be ascertained. The following key points can be summarized from the interviews:

- *Bringing together CNT expertise*: The members of the innovation community agree that CNTs can be considered a real innovation driver for a variety of application fields. Due to the extremely long value-added chains, even the large corporations involved cannot carry this innovation from basic development to the market on their own. The participating promotors already ascertained this at an early stage, and therefore brought together all available German expertise to a joint innovation project, to promote research jointly and interactively. To date, in Germany, the innovation alliance is clearly a unique joint project in terms of structure and scope, bringing together partners from business and the scientific community, including many small and medium-sized companies. This composition could make it possible to enable a breakthrough for the promising CNT technology in Germany as well, and thus to realize a variety of applications in a number of different industries and markets. Of course, with such a large number of research partners, a certain competitive situation is inevitable, but due to the diversity of applications, each party has so far found his own sphere of activity. Cooperation agreements, etc. within the innovation project also avoid a non-controlled knowledge drain.

> So in the direction [of the funding agency], I would say that it is definitely a high-performance network, because people came together who were previously not together. Hence, the personal networks, with their different characteristics of individuals, have grown. And the Innovation Alliance has definitely contributed to strengthening Germany as an industrial location. (Department Head, Company E, 15)

- *Intensive and trusting cooperation along the value-added chain*: The success of the overall project is directly dependent on the intensive exchange of information between the individual projects along the value-added chain. Without the specific requirements from applications, the basic research work cannot be goal-oriented in the production, functionalization and dispersion of carbon nanotubes. By the same token, the end users also require precise information about the performance and operating conditions of CNTs for their implementation as marketable products. Through the organizational structure, but especially through the strong networking and collaboration of promotors from the respective single projects and at the overarching level, this aspect must be taken into account.
- *The relationship with the BMBF, government and other international alliances*: Especially during the maturation and initiation phases, the most important relationships were between the promotors and the funding agency. Not until the BMBF had realized the need for funding did the network come together in the present form. The major impetus for the work in the network came through the relationships with international alliances, which are structured similarly to the present carbon nanotubes innovation alliance. For example, at upcoming annual meetings, other networks will be invited, for example those consisting of partners from the European or even global area, which are essentially pursuing the same goals as the present projects. This measure allows the promotors to network their own innovation communities with other similar innovation communities at an overarching level.
- *Continuous substantive further development*: Trends are of major importance in the alliance, since the innovation community has the intent of continually developing itself further substantively. During the development phase of the network for example, nine more application-oriented single projects were added, thanks to the fact that trends, potentials and opportunities have been recognized which had not yet not been identified in these fields at the outset of the alliance effort, or which had at that time not yet been mature enough to be implemented. Both internal and external experts have the permanent task of collecting all possibly relevant ideas from around the world, analysing them, and preparing them for their own community. Even if a "threat" were to arise (e.g. by the discovery of an alternative technology), the key people could come together very quickly and discuss the implications for their own projects, and take appropriate action.
- *Intensive public relations*: Another important aspect arose only during the course of the innovation project: the relationship with the public in Germany. Several times, technologies have – regardless of their actual performance – not been successfully implementable because their risks outweighed their benefits, in the public perception. A well-known example is genetic technology. To some extent, the assessment of the risks is based on the fact that the advantages and disadvantages of new technology are not widely known, and there has been no public discourse about it. To prevent such a situation, members of the innovation community are undertaking major efforts to inform the public by means of educational campaigns about the opportunities, risks and potentials of CNTs.

> A news story, say, for example, that CNTs acted like asbestos, would be an absolute innovation killer. (Department Head, Company B, 72)

3.9.5 Conclusion

In the present case study, the focal innovation process of the carbon nanotubes innovation alliance was considered in detail and the role of key persons analyzed. The aim of the project is to build a national community across value-added chains, and the application-oriented development of this CNT technology, which can be considered radical. This network is designed to ensure that Germany not fall behind in the worldwide progress of CNT research.

In conclusion, the interviews conducted with four key members of the network permits the assessment that the innovation process was largely determined by the intensive cooperation of six promotors. These promotors, who constituted the innovation community, have come together since the beginning of the maturation phase, and throughout the initial phase through to the development phase, and show a fairly high degree of personal continuity. Both in the areas of composition and identity and in the quality of interaction, the innovation community shows an overall high level of performance.

On this basis, it was shown in the case study that the current success of the innovation project clearly depends on the existence and effectiveness of the existing innovation community. The contribution to success of the promotor network in the carbon nanotubes innovation alliance is shown in the bringing together of CNT competence from across Germany, the intensive and trusting cooperation along the value-added chains, the very good relations with the project promotors, government and other international alliances, in the continuous substantive development, and in the intensive public relations.

3.10 Conclusions: How Do Innovation Communities Evolve, What Makes Them Successful?

Klaus Fichter

After presenting the results of each single case study, we will draw some conclusions by pulling together and comparing the cases. We will do this by referring to the guiding questions that have been articulated as part of the research design:

- Evolution: How does the size and structure of innovation communities change over time?
- Performance: How do innovation communities contribute to the success of innovation projects?
- Quality: Which features make innovation communities successful?
- Methods: Which measures and methods constitute and improve the performance of innovation communities?

The first three questions are discussed below. The question of suitable methods and measures is treated in Sect. 5.2.

3.10.1 Evolution: How Does Size and Structure of Communities Change?

As the observations in Sects. 3.5–3.9 have shown, in all of the cases examined, several promotors can be identified that work together closely (interaction quality), perceive one another as a team (identity) and in this respect, fulfill the characteristics of an innovation community as defined in Sect. 3.3. The number of members of the community innovation ranges from four in the case of Connected Energy and Tilt Sensors, to seven members in the case of Wood Energy (see Table 3.49). This result is supported by the findings from the large-number survey of innovation networks that is detailed in Chap. 2 (see Table 2.14 in Chap. 2), as well as by earlier case studies (Fichter 2009b) that found that innovation communities usually have a rather small number of members. This is understood to be due to the fact that innovation communities are informal promotor networks that require personal contacts, a group feeling and a certain degree of trust between community members, all of which can generally only be ensured by a smaller group size.

The innovation communities studied all had at least two champions: individuals playing two or more professional "promotor" roles. In four out of five cases, the proportion of champions within the entire community was 50% or higher. Only in the case of Dye-sensitized Solar Cells, the share of champions was smaller (two out of five community members in total – the other three members only performed one promotor role).

Table 3.49 Key data from the five case studies

Focal innovation process			Number of ... involved in focal innovation process				
Case	Region	Period analyzed	Key persons[a]	Supporters	Promotors[b]	Champions[c]	Members of the innovation community
Connected Energy	Northern Germany	2009–2011	8	3	5	2	4
Dye-sensitized Solar Cells	Germany/ Switzerland	2008–2011	8	2	6	2	5
Tilt Sensors	Baden-Württemberg	2006–2010	6	1	5	3	4
Wood Energy	Brandenburg/ International	2004–2010	15	2	13	7	7
Carbon Nanotubes	Germany	2006–2011	14	2	12	6	6

Source: Author
[a] *Key persons* includes supporters and promotors
[b] *Promotors* includes promotors with a single role (power, expert, process, relationship promotors) as well as promotors with two or more roles (universal promotor = champions)
[c] *Champions* are persons with two or more promotor roles

The examined cases have also shown that not all promotors within a particular innovation process actually take part in the innovation community. This has become particularly clear in the case of Wood Energy and Carbon Nanotubes, where not even half the promotors are in fact members of the innovation community – possibly because some promotors support the innovation process professionally and technically, but do not cooperate that closely with other promotors and are therefore not perceived as part of the community. This aspect highlights the need to clearly distinguish between the identification of the promotors of a focal innovation process, and the separate question whether these promotors collaborate closely, see themselves as part of a team, and can in this sense be counted as a part of the innovation community.

Another important insight from the innovation case studies is that, in addition to the promotors, other key persons who support the enterprise through their activities, competence or decision-making power also have notable influence on the initiation, progress and even overall success of the innovation project (Table 3.49). Their "drive" in the innovation process may not be as continuous or as straightforward as the promotors', so they are not defined as such. But even so, their selective support is of great relevance for the overall success of the innovation project. To distinguish them from the main promotors, they are here referred to as "supporters." The group of key persons hence comprises promotors and supporters. As laid out above, all examined cases feature promotors as well as supporters.

The results of the case studies reveal that all innovation communities work across different levels of the innovation system (cf. Fig. 1.2 in Chap. 1). They all involve promotors from the company level, the value chain, and the framing and linking level. Despite the fact that all innovation communities are cross-level networks of promotors, most of their activities relate to the value chain level. While the concept presented in Chap. 1 (cf. Fig. 1.2) assumes that it is sufficient to differentiate three levels of the innovation system, the five cases analyzed in the InnoCo-project indicate that there is a need for further differentiation on the value chain level. As the five cases display, the main level of activities of innovation communities on the value chain level can either relate to (cf. Fig. 4.1 in Sect. 4.1.1):

1. A broader field of a basic technology, like e.g. carbon nanotubes, or a comprehensive usage system, like e.g. the growing of wood crops and the whole value chain system of using wood crops for energy purposes
2. A specific product or application of a new technology or solution

The first sub-category is obviously more complex and comprises a broad range of different value chains, products or applications. An example for an innovation community that mainly works and interacts on this first sub-level is the network of promotors in the Wood Energy case. It comprises promotors from all levels of the innovation system, but its focus is to develop and promote agroforestry systems and usage systems for short-rotation wood crops as a whole. Their work is not limited to single projects or specific products, but covers a broad range of projects, timber products, harvesting methods and forms of biomass energy use in the German federal state of Brandenburg.

The work and interaction of the innovation community in the case of Carbon Nanotubes is also mainly related to the first sub-category of the value chain level. The network of promotors in this case tries to build up a nation-wide network in Germany, and to develop a lead market for a basic new technology with a high number of different value chains, products and applications.

The innovation communities in the cases of Connected Energy, Dye-sensitized Solar Cells and Tilt Sensors are all focused on a specific application or product of a new technology. Thus the work and interaction of these innovation communities is related to the second sub-category of the value chain level.

As Table 3.50 indicates, the number of key persons and promotors involved in the more complex innovation cases of Wood Energy and Carbon Nanotubes is significantly higher than in those innovation cases that focus on one specific product or application. Also, the number of members in the innovation community is higher in communities that relate their work to comprehensive production and usage systems (Wood Energy), or broader fields of basic technologies (CNT). This is plausible since more complex innovation endeavours require the integration of more skills, resources and actors than those aimed at promoting a specific product or application.

As Table 3.51 shows, each promotor role is represented at least once in all of the innovation communities analyzed, i.e. the role of power promotor, expert promotor, process promotor and relationship promotor is present in every network. However, there are some differences in the frequency of these roles. While all communities have at least three members playing a expert promotor role, the process promotor role, for instance is present only once: in the case of Connected Energy. Also, the relationship promotor role occurs less frequently than the others. The results of these case studies correspond with the empirical findings from the initial large-scale survey, which were presented in Chap. 2 (see Table 2.7 in Chap. 2).

Table 3.50 Main level of activity of innovation communities in the five cases analyzed

Focal innovation process Case	Number of . . . involved in focal innovation process: Key persons[a]	Supporters	Promotors[b]	Champions[c]	Members of the innovation community	Main level of activity of innovation community: Level and sub-level of the innovation system
Wood Energy	15	2	13	7	7	Value chain level: comprehensive production and usage system
Carbon Nanotubes	14	2	12	6	6	Value chain level: comprehensive field of a basic technology
Dye-sensitized Solar Cells	8	2	6	2	5	Value chain level: specific product- and application-focus
Connected Energy	8	3	5	2	4	Value chain level: specific application-focus
Tilt Sensors	6	1	5	3	4	Value chain level: specific product-focus

Source: Author

[a] *Key persons* includes supporters and promotors

[b] *Promotors* includes promotors with a single role (power, expert, process, relationship promotors) as well as promotors with two or more roles (universal promotor = champions)

[c] *Champions* are persons with two or more promotor roles

Table 3.51 Members of the innovation communities in the five cases analyzed

Case	Promotors (Number of *persons*)	Number of . . . promotor *roles* in the innovation community: Power	Expert	Process	Relationship	Champions (Number of *persons*)	Champions with . . . two roles	three roles	four roles
Connected Energy	4	2	3	1	1	2	1	1	0
Dye-sensitized Solar Cells	5	3	3	2	2	2	0	2	0
Tilt Sensors	4	2	3	3	3	3	0	2	1
Wood Energy	7	4	7	2	1	5	4	0	1
Carbon Nanotubes	6	3	3	3	3	5	4	1	0

Source: Author

Table 3.51 also shows that all innovation communities have at least two champions – members who assume two or more promotor roles. Every promotor network also has at least one champion playing three or even four parts

Table 3.52 Evolution of innovation communities: number of members in each phase (permanent members)

Case	Phase of the innovation process					
	Maturation	Initiation	Concept	Development	Realization	Market/ diffusion
Connected Energy		2009: **3 (3)**	2009/10: **4 (3)**	2010/2011 **4 (3)**		
Dye-sensitized Solar Cells		2006–2007 **4 (3)**	2007–2008 **3 (3)**	2008–2011 **5 (3)**		
Tilt Sensors	1998–2006 **2 (2)**	2006 **3 (3)**	2006 **3 (3)**	2007 **4 (4)**	2008–2010 **4 (4)**	Since 2010 **4 (4)**
Wood Energy	1994–2003 **3 (2)**	2004 **5 (2)**	2004 **5 (4)**	2005–2010 **7 (4)**		
Carbon Nanotubes	2000–2006 **0**	2006–2008 **4 (4)**	2006–2008 **4 (4)**	2008–2011 **6 (4)**		

Source: Author

simultaneously. Hence we can deduce that, for the functioning of communities in radical innovation projects, innovation champions play a central role.

One of the lead questions of the case studies was: How does the size and structure of the innovation communities change over time? To answer this question, the innovation processes analyzed were divided into different phases. The innovation process and its respective phases are unequal in length depending on the case – however, all cases covered include an initiation, a conceptualization and a development phase, so that it is possible to study the evolution of the innovation community.

As Table 3.52 shows, the number of community members steadily increases from the initiation or planning phase to the final development phase. This is plausible in the sense that when nearing implementation, the number of organizations involved increases (see Sects. 3.5–3.9). Since the entire innovation network is growing and implementing partners, e.g. "pilot customers," become more and more involved, the need for representation in the innovation community through promotors from participating organizations also increases.

The five cases examined further confirm the initial observation from earlier case studies (Beucker 2007; Fichter 2006; Springer 2007), which are characterized by the simultaneity of personnel continuity on the one hand and turnover on the other. In all five communities we can identify promotors that are members from the beginning to the end, or continuously remain a member upon entry into the community. The number of continuous members is displayed in parentheses in Table 3.52. In the Wood Energy case, for example, in the maturation phase, the community is initiated by two promotors. They are supported by a power promotor who retires from the innovation process, and community after the initiation phase, however, due to other work commitments. In the initiation phase, three new promotors join the community, two of which will also remain continuous members. The third leaves after the initiation phase, so that the concept phase is carried out with four members. These can be referred to as "core promotors." In the following development phase, three additional promotors that are actively engaged in the development and become part of the innovation community join these four core promotors.

As Table 3.52 shows, all innovation communities have permanent members (number in parentheses). These numbers remain stable from the initiation to the development phase in the Connected Energy, Solar Cells and Carbon Nanotubes case studies. In the other two cases, they even increase. At the same time, a certain degree of turnover can be observed in four of five cases, as discussed above in the example of the Wood Energy innovation community. The solar cells group began to shrink temporarily and then regrow, due to the task situation in individual phases, and the resulting need for reconfiguration.

3.10.2 *Performance: How Do Innovation Communities Contribute to Success?*

As seen in Tables 3.49 and 3.53, all the networks of promotors examined meet the requirements for an innovation community as defined in Sect. 3.3. On a scale from 1 (very low) to 7 (very high), all analyzed sub-constructs and constructs in the arithmetic mean of all respondents were measured with 5 or above. From the

Table 3.53 Performance of innovation community, project success and contribution of innovation communities

Case	Performance of innovation community				Project success and contribution of innovation community (IC)	
	Mean value[d] community composition[a]	Mean value[d] community identity[a]	Mean value[d] community interaction[a]	Mean value[d] community performance[a]	Mean value[d] project success[b]	Mean value[d] success contribution of IC[c]
Connected Energy	5.3	5.6	5.7	5.6	5.3	6.0
Dye-sensitized Solar Cells	5.8	5.3	5.1	5.2	5.3	5.8
Tilt Sensors	5.7	6.1	6.1	6.1	6.2	6.5
Wood Energy	6.2	6.1	6.3	6.2	5.7	6.8
Carbon Nanotubes	6.1	5.2	5.4	5.4	5.4	5.8

Source: Author

[a] Possible answers for each item of the multi-item construct from 1 (disagree completely) to 7 (strongly agree)

[b] Possible answers from 1 (not successful at all) to 7 (very successful)

[c] The question was: What is your estimate of the contribution of the above named group of people to the success of the joint innovation project? Possible answers from 1 (low) to 7 (very high)

[d] Mean value: All constructs were measured using multiple items (see Sects. 3.5–3.9). All items offered a seven-point answer scale. The multi-item scales were each aggregated by calculating the arithmetic mean, and treated as observed indicators. The mean value given here is the arithmetic mean per case of all four responding members of the innovation community.

perspective of the community members that were interviewed (4 per community), each promotor network has high values for community identity (5.2 or above) and for interaction quality (5.1 or above). Also, the respondents hold their community to be well or very well composed (5.3 or above). This results in high to very high performance (5.3–6.2) for all investigated innovation communities.

As Table 3.53 further shows, the high performance of innovation communities corresponds with good to excellent success in terms of the joint innovation project (project success). The average for the project's success as estimated by all community members interviewed is 5.3 (Connected Energy, Dye-sensitized Solar Cells) or above. In all cases studied, the contribution of the community to the project's success is rated as high to very high (5.8–6.8). This further supports the findings from the large-scale survey of innovation networks, as presented in Chap. 2: that the community surrounding a radical innovation project is a key factor for the project's success.

In contrast to the large-scale survey (see Chap. 2), where only one person was interviewed in every innovation network (usually the network coordinator), in the case studies, four members of each innovation community were interviewed and asked to fill out a questionnaire, permitting the examination of several different perspectives and estimates from within the innovation project regarding the project itself and the level of cooperation. The absolute difference between the highest and lowest value from the four community members interviewed is here designated as *span*. The results show that the respective estimates of the surveyed constructs were very homogeneous (a span of 1 or less) or homogeneous (2 to 1). Of the 30 spans listed in Table 3.54, 12 have difference values (spans) of 1 or below, and 11 have values between 1 and 2. Only six spans are between 2 and 3, indicating a more heterogeneous valuation of that construct by the respondents. Only one value is above 3, indicating a very heterogeneous assessment. The latter concerns the solar cell community, where one member evaluated community coordination and balance of contributions substantially lower than all other community members (see Sect. 3.6).

It is noteworthy that all community members estimated the contribution of the innovation community to the success of their joint innovation project as high to very high, and that these estimates were very consistent within each responding community. Even if it can be assumed that the community members have an interest in assuming their cooperation has a positive or very positive impact on the success of their joint innovation project, and the absolute values therefore should be treated with caution, the homogeneity of responses does indicate that the innovation community is in fact making a real contribution to the success of the project (Table 3.54).

The central question regarding the contribution of innovation communities to project success is: How do innovation communities contribute to the success of innovation projects? Underlying the question of "how" are features and functions that promotor networks can fulfil in view of successful collaborative innovation processes.

As Table 3.55 shows, innovation communities contribute to the success of joint innovation projects through their composition and identity as well as through the

Table 3.54 Performance of innovation community: Assessments by different community members

Case	Lowest – highest mean value IC composition	Span highest – lowest value IC composition	Lowest – highest mean value IC identity	Span highest – lowest value IC identity	Lowest – highest mean value IC interaction	Span highest – lowest value IC interaction	Lowest – highest mean value IC performance	Span highest – lowest value IC performance	Lowest – highest mean value project success	Span highest – lowest value project success	Lowest– highest mean value success contribution of IC	Span highest – lowest value success conribution
Connected Energy	4.5–6.5	2.0	4.4–6.8	2.4	4.6–6.9	2.3	4.6–6.8	2.2	4.3–6.5	2.2	5–7	2
Dye-sensitized Solar Cells	5.5–6.0	0.5	4.3–6.2	1.9	3.1–6.2	3.1	3.9–6.0	2.1	4.4–5.8	1.4	5–6	1
Tilt Sensors	5.3–6.2	0.9	5.2–6.6	1.4	5.1–6.6	1.5	5.2–6.4	1.2	5.5–6.7	1.2	6–7	1
Wood Energy	5.8–6.5	0.7	5.1–6.6	1.5	5.8–6.7	0.9	5.5–6.6	1.1	4.5–6.7	2.2	6–7	1
Carbon Nanotubes	6.0–6.2	0.2	4.8–5.4	0.6	5.4–5.5	0.1	5.3–5.5	0.2	4.8–5.6	0.8	5–7	2

Source: Author

Table 3.55 How do innovation communities contribute to innovation success?

Case	Community composition	Community identity	Community interaction	Other
Connected Energy	Creating and improving business contacts (sponsors, funding agencies), good performance due to manageable size and complementary skills	Reinforcing social cohesion of the group	Increases conflict resolution capacities	
Dye-sensitized Solar Cells	Bringing project critics together with business contacts	Helping define clear common goals, positioning and planning further innovation activities	Organizing effective knowledge exchange, building trust and supporting effective conflict management.	
Tilt Sensors	Creating a powerful innovation network through long-term cooperation of key promotors, through complementary skills, early-on integration of pilot customers, and through good contacts and funding possibilities in the industry	Providing a common "language" and understanding of shared objectives	Optimizing coordination by involving a neutral project partner (Technology transfer centre), open communication culture and respectful collaboration, High motivation and intiative of members	
Wood Energy	Acquiring critical financial support by concentrating social capital (relationships) and integrating success relevant players (clients, environmental organizations)		Effectively exchanging knowledge and experience, mitigating or solving project-threatening conflicts, influencing favourable general requirements through frequent and decisive collaboration of promotors	Midwife function: Promoting innovation ideas into a formal organizational structure
Carbon Nanotubes	Concentration of skills within the CNT technology sector through different professional promotors, good relations to funding institutions, politics, international alliances		Frequent, trustful collaboration	Extensive publicity work

Source: Author

quality of their interaction. The composition of the community, encompassing promotors from different organizations and departments, increases the social capital of the innovation project. The individual promotors bring along their own contacts and relationships to the community. This integration can give access to donors and partners who can provide critical resources, such as financing, technological know-how, reputation, for the further development and implementation of an innovation. An innovation community also creates or improves a communal spirit for otherwise dispersed promotors. By strengthening team spirit and sense of commonality, the individual promotors are supported in their work for the innovation project. This is particularly critical in phases and situations where an innovation project is under pressure regarding time, funding or performance, or where there is strong opposition to the goals of the innovation project.

The community can also ensure that the objectives of a joint innovation project are further clarified and formulated. It also can contribute to the development of a mutual "language" and common understanding within the innovation project, established by the close cooperation of the promotors within the community.

An innovation community also supports effective knowledge transfer and sharing through its well-functioning communication and interaction features. If the organization allows for confident and respectful cooperation, it will be able to address conflicts arising within the network, and help solve them. Successful innovation communities can therefore increase the conflict resolution capabilities of the network, enhance dialogue with external stakeholders, critics or interest groups, and improve the overall acceptance of the innovation project.

Finally, the investigated cases also show that innovation communities play a decisive part in early phases of innovation and in project initiation. Due to the informal nature of the promotors' cooperation, it becomes feasible to acquire project funds and move an innovative idea from an informal stage to a more formal project phase. Promotor networks therefore contribute to the institutionalization of an innovation process, and to the allocation of critical resources.

Therefore, the case studies support the assumption expressed in Sect. 3.2 that innovation communities, through their collaboration, contribute to a project's success by:

- Helping to promote an innovative idea from an informal stage into a formal organizational stage or project phase.
- Motivating promotors and strengthening their role in overcoming innovation barriers.
- Improving the social capital of innovation networks, allowing for the acquisition of performance-critical resources, such as funding, knowledge, reputation.
- Helping solve or mitigate project-threatening conflicts.

According to the features of innovation communities described, we can conclude that innovation communities are especially important when:

1. Institutionalization: ... an innovative idea needs to be taken from an informal stage to a formal organization or project phase ("midwife" function).

2. Resource acquisition: ... resources that the individual organizations cannot provide or could not or only with considerable effort attain alone, like money, knowledge, skills, reputation, etc., are required.
3. Conflict resolution: ... conflicts or acceptance problems that individual partners could not or only with considerable effort address need to be resolved.

3.10.3 Quality: Which Features Make Innovation Communities Successful?

Finally, the third key question that the case studies were founded on should be addressed: Which features make innovation communities successful?

Taking the five case studies and the overall insights into the evolution and the success of innovation communities presented above into account, the following properties can be described as the specific power of innovation communities, creating or improving their contribution to the success of a joint radical innovation project:

Composition

- *Champions*: The performance of a particular innovation community is created and reinforced through the involvement of champions. In particular, long-term close relations between champions who have initiated an innovative project help improve the probability of success.
- *All promotor roles occupied*: In an innovation community, all promotor roles should be represented. This is not only the case for the power and expert promotors, but also for process and relationship promotors.
- *Core group (continuity)*: Due to growing and changing challenges in innovative projects, the community's composition usually changes in size and composition from the initial and maturation stage to the development and implementation phase. A core group of promotors and champions that remain in the community from the beginning to the end of the project seems to be crucial to overall stability and performance over time.
- *Manageable community size*: Since innovation communities are informal networks of promotors that require personal contacts, a team spirit and a sense of trust between all community members to function, a manageable size of the group seems to improve performance or enable it in the first place. The ideal size certainly depends on the complexity of the innovation project and on the number of partners required for the success of a project. In general, this number will however be team-oriented, lying between three and seven.
- *Promotors from all performance-related organizations*: Innovation network research has shown that the complementarity of expertise between participating

partners is of special importance to the success of a collaborative innovation project, and that all project-relevant organizations are involved. Regarding the performance and efficiency of innovation communities, it is important that as many relevant organizations as possible are represented in the community as promotors.

- *Social capital of members*: As the innovation community cases examined have shown, one measure of success is the degree in which they combine the social capital of individual promotors and thus facilitate access to decision-makers, "resource controllers" and stakeholders relevant to project success (e.g., investors, sponsors, key clients). The innovation performance of a community to a great extent depends on the contacts and relationships individual promotors provide, and on their willingness to share them with the rest of the promotor network.
- *Reconfiguration capacity*: Since project tasks can change during the innovation process and the demand for individual partners may change over time depending on the stage of the project, it is not only the association of innovating organizations that needs to check its composition regularly, but also the associated innovation community itself. Reconfiguration may also be necessary in certain situations in which participating promotors realize that they no longer "fit together" due to increasingly divergent interests and want to go their own way.

Identity

- *Functional identity*: The case studies have demonstrated that the functional identity of an innovation community is a key condition of successful collaboration. The performance of a community depends, therefore, on its capability to develop common targets and perspectives for the shared project and a common "language" and basic understanding of project goals and future options.
- *Group cohesion*: Most of the community members interviewed very clearly and unanimously displayed a "team spirit" between community members, and further indicated that their promotor network displayed good to excellent cohesion. Network performance to this extent largely seems to depend on the way in which its promotors manage to create this group cohesion.

Interaction

- *Mutual trust*: Of 20 promotors that were interviewed in the five case studies, all but one rated personal trust and social climate in their innovation community as good or excellent. This again underpins the assumption that merely informal groups like innovation communities in particular depend on trust to function, since there are no formal security measures or sanctions. Creating trust among community members, which is likely to enhance performance, is therefore one of the key tasks with regard to the functioning of the innovation community.

- *Intensive and respectful cooperation*: Trust is usually only gained through personal encounters and experience in practical cooperation. The interviewed community members, in fact, formulated intense and respectful cooperation as an important precondition for the functioning of their community. The same applies to personal initiative or motivation as well as to a high level of interest in the joint project, which promotors explicitly expect of one other.
- *Conflict management*: The particular strength of innovation communities is their capacity for close and confidential cooperation, which includes addressing and resolving project-related conflicts. This requires, however, the willingness and ability to resolve conflicts at the individual promotors' level. As the case studies have shown, resolving personal or professional conflicts within an innovation project is facilitated when the innovation community is managed or advised by a neutral partner (an innovation coach, technology transfer facility, or similar).
- *Coordination and coaching*: Some of the cases studied indicate that an innovation coach or intermediary (e.g., a technology transfer institution), managing the innovation network and community as a neutral moderator and referee, can contribute to the conflict resolution within the innovation network. In addition to their conflict management skills, innovation coaches or intermediaries can also assume the role of independent process or relationship promotors. In this function, they then even become a member of the innovation community in question (see Sects. 3.6, 3.7 and 5.4) what again may significantly improve their performance.
- *Effective exchange of knowledge and experience*: The performance of an innovation community ultimately depends on the degree to which it is able to organize an effective exchange of knowledge and experience, e.g. in the form of workshops, conferences or field trips.

References

Alderfer CP (1987) An intergroup perspective on group dynamics. In: Lorsch JW (ed) Handbook of organizational behaviour. Prentice-Hall, Englewood Cliffs, pp 190–222

Aulinger A (2005) Entrepreneurship und soziales Kapital. Netzwerke als Erfolgsfaktor wissensinten-siver Dienstleistungsunternehmen. Metropolis, Marburg

Backhaus K, Erichson B, Plinke W, Weiber R (2008) Multivariate Analysemethoden: Eine anwendungsorientierte Einführung. Springer, Berlin

Beucker S (2007) Die Innovation Solon-Mover der Solon AG. Fraunhofer IRB, Stuttgart

Beucker S, Fichter K (2007) Trends und Rahmenbedingungen für das Innovationssystem Farbstoffsolarzelle – Perspektive 2020. Retrieved 25 Apr 2011 from http://www.borderstep.de/pdf/P-Beucker-Fichter-Trends_und_Rahmenbedingungen_fuer_das_Innovationssystem_-Farbstoffsolarzelle_Perspektive_ 2020-2007.pdf.

BITKOM (2008) Studienreihe zur Heimvernetzung: Gesellschaftlicher Nutzen der Heimvernetzung, Band 2: Ergebnisse der Arbeitsgruppe 8 "Service- und verbraucherfreundliche IT" zum dritten nationalen IT-Gipfel 2008 [Study series on home networking: societal benefits of home networking (vol 2): Results of working group 8 "Service and consumer-friendly IT", for the third

National IT summit 2008]. http://www.bitkom.org/files/documents/Studie_Gesellschaftlicher_Nutzen_Heimvernetzung.pdf. Accessed 21 Mar 2011

Bortz J, Döring N (2006) Forschungsmethoden und evaluation, 4th edn. Springer, Heidelberg

Bourdieu P (1983) Ökonomisches Kapital, kulturelles Kapital, soziales Kapital. In: Kreckel R (ed) Soziale Ungleichheiten. Schwartz, Göttingen, pp 183–198

Coleman JS (1990) Foundations of social theory. Harvard University Press, Cambrigde, MA

Dayan M, Di Benedetto CA (2009) Antecedents and consequences of team-work quality in new product development projects: an empirical investigation. Eur J Innovat Manage 12(1): 129–155

DLG e.V. und Umweltstiftung WWF Deutschland (Hrsg) (2006) Nachhaltiger Anbau und energetische Verwertung von Biomasse. Empfehlungen der AG Landwirtschaft und Naturschutz von DLG und WWF. Retrieved 3 Mar 2011 from http://www.wwf.de/fileadmin/fm-wwf/pdf_neu/Nachhaltiger_Anbau_und_energetische_Verwertung_von_Biomasse_WWF-Flyer.pdf

Duschek S (1998) Kooperative Kernkompetenzen: Zum Management einzigartiger Netzwerkressourcen. Führung+Organisation 67(4):230–236

Easley RF, Devaraj S, Crant JM (2003) Relating collaborative technology use to teamwork quality and performance: an empirical analysis. J Manage Inform Syst 19(4):247–268

Eberl P (2003) Vertrauen und Management. Schäffer-Poeschel, Stuttgart

Eisenhardt KM (2007) Theory building from cases: opportunities and challenges. Acad Manage J 50(1):25–32

Federal Ministry of Education and Research (BMBF) (2010) Ideas. Innovation. Growth. High-Tech-Strategy 2020 for Germany. Federal Ministry of Education and Research (BMBF), Bonn

Federal Statistical Office (2008) Economics and statistics 2/2008 (corrected version, Apr 2008). Pro-fessional Series 5, no. 1

Fichter K (2006) Das "e-place"-Konzept der IBM Deutschland. Fraunhofer IRB, Stuttgart

Fichter K (2008) Netzwerkbildung und Netzwerkmanagement als Instrument des kommunalen Klimaschutzes: Theoriegrundlagen und empirische Befunde. Berlin

Fichter K (2009a) Grundlagen des Innovationsmanagements. Carl-von-Ossietzky-Universität Oldenburg, Oldenburg

Fichter K (2009b) Innovation communities: the role of networks of promotors in open innovation. R&D Manage 39(4):357–371

Fichter K, Beucker S (2008) Innovation communities: Promotorennetzwerke als Erfolgsfaktor bei radikalen Innovationen. Fraunhofer IRB, Stuttgart

Fichter K, Beucker S, Noack T, Springer S (2007) Entstehungspfade von Nachhaltigkeitsinnovationen: Fallstudien und Szenarien zu Einflussfaktoren, Schlüsselakteuren und Internetunterstützung. Fraunhofer IRB, Stuttgart

Freese D (2010) Agroforstsysteme im Klimawandel. Presentation to the Sixth Brandenburg Wood Energy Day. Presentation foils. Retrieved 3 Mar 2011 from http://www.bioenergie-portal.info

Friedlander F (1987) The ecology of work groups. In: Lorsch JW (ed) Handbook of organizational behavior. Prentice-Hall, Englewood Cliffs, pp 301–314

Future Technology Consulting (ZTC) of the VDI Technology Center GmbH (2009) Carbon nanotubes: a potential new class of materials for Germany – technology integration, Future technologies, No. 79. Dusseldorf

Gemünden HG, Salomo S, Hölzle K (2007) Role models for radical innovations in times of open innovation. Creativity Innovat Manage 16(4):408–421

Gerybadze A (2003) Gruppendynamik und Verstehen in Innovation Communities. In: Herstatt C, Verworn B (eds) Management der frühen Innovationsphasen. Springer, Wiesbaden, pp 145–160

Gerybadze A (2004) Technologie- und Innovationsmanagement: Strategie, Organisation und Implementierung. Vahlen, München

Hauschildt J (1999) Promotors and champions in innovations – development of a research paradigm. In: Brockhoff K, Chakrabarti AK, Hauschildt J (eds) The dynamics of innovation: strategic and managerial implications. Springer, Berlin, pp 167–182

Hausschild J, Gemünden HG (1999) Promotoren, Champions der Innovation (2. erw. Auflage). Gabler, Wiesbaden

Högl M, Gemünden HG (2001) Teamwork quality and the success of innovative projects: a theoretical concept and empirical evidence. Organ Sci 12:400–449

Högl M, Gemünden HG (2005) Teamarbeit in Innovationsprojekten. In: Högl M, Gemünden HG (eds) Management von Teams: Theoretische Konzepte und empirische Befunde, 3rd edn. Deutscher Universitätsverlag, Wiesbaden, pp 93–126

Högl M, Proserpio L (2004) Team member proximity and teamwork in innovative projects. Res Pol 33:1153–1165

Högl M, Ernst H, Prosperio L (2007) How teamwork matters more as team member dispersion increases. Prod Innovat Manage 24:156–165

Landschaftspflegeverband Spree-Neiße e.V. (Hrsg) (2010a) Erzeugerhandbuch BioEnergieregion Lausitz. Cottbus

Landschaftspflegeverband Spree-Neiße e.V (2010b) Energieholzanbau: Kurzumtriebsplantagen (KUP) und Agroforestrysysteme (AFS). Informationsblatt No. 6

Mullen B, Cooper C (1994) The relation between group cohesiveness and performance: an integration. Psychol Bull 115(2):210–227

Nahapiet J, Goshal S (1998) Social capital, intellectual capital, and the organizational advantage. Acad Manage Rev 23(2):242–266

Putnam RD (2000) Bowling alone: the collapse and revival of American community. Simon & Schuster, New York

Rese A, Sand N, Baier D (2009) Relevance and contribution of promotor networks in cross-company innovation projects: an empirical investigation. 16th International Product Development Management Conference, Twente, 7-9 June 2009, p 20

Ripperger T (1998) Ökonomik des Vertrauens: Analyse eines Organisationsprinzips. Tübingen

Rost K, Hölzle K, Gemünden HG (2007) Promotors or champions? Pros and cons of role specialisation for economic progress. Schmalenbach Bus Rev 59:340–363

Salomo S (2003) Konzept und Messung des Innovationsgrades: Ergebnisse einer empirischen Studie zu innovativen Entwicklungsvorhaben. In: Schwaiger M, Harhoff D (eds) Empirie und Betriebswirtschaftslehre: Entwicklungen und Perspektiven. Schäffer-Poeschel, Stuttgart, pp 399–427

Salomo S, Weise J, Gemünden HG (2007) NPD planning activities and innovation performance: the mediating role of process management and the moderating effect of product innovativeness. J Prod Innovat Manage 24(4):285–302

Sand N, Rese A (2011) Comparing promotor networks in radical and incremental innovation projects: roles and interaction as success factors. In: Baier D, Gaul W, Rese A, Würth R (eds) The entrepreneurship – innovation – marketing – interface. Proceedings of the 3rd symposium in Cottbus. Swiridoff, Künzelsau, S. 233–250

Scholl W (2005) Grundprobleme der Teamarbeit und ihrer Bewältigung. In: Högl M, Gemünden HG (eds) Management von Teams: Theoretische Konzepte und empirische Befunde. Deutscher Universitätsverlag, Wiesbaden

Schwarz K, Swoboda J (2005) Mit Biomasse heizen – Energieholzanbau in der Lausitz. LEADER Forum 3:15–16

Springer S (2007) Von der Faser bis zum Druck: Das Coated-Coldset-Netzwerk. Fraunhofer IRB, Stuttgart

Sydow J (2006) Management von Netzwerkorganisationen (4. aktualisierte und erweiterte Auflage). Gabler, Wiesbaden

Sydow J, Duschek S, Möllering G, Rometsch M (2003) Kompetenzentwicklung in Netzwerken: Ein typologische Studie. Gabler, Wiesbaden

Trappmann M, Humell Hj, Sodeur W (2005) Strukturanalyse sozialer Netzwerke: Konzepte, Modelle, *Methoden*. VS Verlag für Sozialwissenschaften, Wiesbaden
Uckert G (2010) Motivation zur Produktion von Energieholz in landwirtschaftlichen Betrieben – Ergebnisse einer Befragung in Brandenburg. Presentation to the Sixth Brandenburg Wood Energy Day, 26 Aug 2010. Retrieved 3 Mar 2011 from http://www.bioenergie-portal.info
Van de Ven AH, Polley DE, Garud R, Venkataraman S (1999) The innovation journey. Oxford University Press, New York, Oxford
Weyer J (ed) (2000) Soziale Netzwerke, Konzepte und Methoden der sozialwissenschaftlichen Netzwerkforschung. Oldenbourg, München
Yin RK (2003) Case study research: design and methods, 3rd edn. Sage, Thousand Oaks

Chapter 4
Managing Innovation Communities: Best Practice Reports

Abstract At the core of this chapter is the question whether and how innovation communities can be actively supported and managed, e.g. by steering or influencing their configuration or offering structures and methodological background.

While the case studies in the proceeding Chap. 3 have given insight into five innovation communities and into the effects and interaction observed between the members of those communities, the following three practice fields have a different scope. They were selected and accompanied to gather experience with the practical aspects of managing innovation communities. The three practice fields overlap with three of the case studies, so that interesting insights can be gained from the comparison of the analytical background of the innovation community and the experience in practical management. Thus, all of the practice fields focus on radical innovations that take place in various sectors of high-tech industries (e.g., information and communication technologies, renewable energies and material sciences).

To avoid conflicting interests between the theoretical analysis and the research on managerial support of innovation communities, the practice fields were accompanied by different researchers than those who analyzed the case studies.

While the empirical findings presented in Chaps. 2 (survey) and 3 (case studies) made clear that innovation communities have a significantly positive influence on the success of innovation projects, the following Chap. 4 will show that the levels of the innovation system the community touches upon, as well as the development phase the community is in, will require different measures to build up, support and organize a community. Hence, there is no standard recipe for the management of innovation communities, merely some recommendations on measures and activities one could take into account. Finally, as the communities change over time, constant adaptation of the applied measures and management tasks is required.

K. Fichter and S. Beucker (eds.), *Innovation Communities*,
DOI 10.1007/978-3-642-22128-6_4, © Springer-Verlag Berlin Heidelberg 2012

4.1 The Management of Innovation Communities: Possibilities and Limitations

Severin Beucker

Taking the results from the previous chapter into account, it can be assumed that innovation communities profit from active support and measures that change the composition and thereby influence the performance and the quality of the community. This also implies that innovation communities can profit from the management of these factors. Clarifying the options and limitations for such a managerial support was one of the objectives of the project InnoCo. For this purpose, three practice fields within the project will serve as experimental fields to observe and test the initiation and setup and the active support of innovation communities. The practice fields were selected according to similar criteria as those applied for the selection of the case studies (see Sect. 3.3):

- Coverage of different phases of radical innovation processes: The three practice fields should allow insights into different stages of radical innovation projects and the corresponding development of innovation communities.
- Coverage of different degrees of complexity in innovation networks: Complexity (topical or technical complexity, number of partners etc.) is an important factor that allows insights into the options and the limitations of managerial support of innovation communities.
- Geographical/regional distribution and further distinguishing features: The geographical and regional distribution of the community members are, together with the industry sector and the size of the involved organizations, important elements that provide postulations for the management of innovation communities.
- Availability and access to the practice field: Finally, the availability and accessibility of the innovation network was the most critical factor for identifying the practice fields. The willingness to take part in the endeavour of creating an innovation community and being part of an experimental test field for managerial support requires strong commitment and cooperation of the individuals in the community.

According to these criteria, the following three practice fields were selected at the beginning of the project. The innovation communities belonging to these fields have been analyzed in detail in the case studies (see Sects. 3.5, 3.6 and 3.9):

- Information and Communication Technologies: This practice field focuses on the initiation and development of an innovation community that is establishing a project and network in the field of sustainable ICT innovations. The project is developing smart home technology and marketable concepts and products for the building and property sector that connect to the Smart Grid development in the energy sector. This practice field required a complete setup and development of the innovation community. It is of particular interest with regard to possible implications on the initiation and setup of innovation communities.

- New Materials: The practice field of new materials focuses on an existing large research and development network for carbon nano tubes (CNTs) with the objective of analyzing and defining precise cooperative structures for the creation of applicable products from this new nanotechnological material class. Because of its complexity and size, this practice field can provide insights into the performance and quality management of existing innovation networks and into the development and improvement of innovation communities.
- Renewable Energies: In this practice field, a partially existing network on a new photovoltaic technology was reconfigured and extended to build up a productive and stable innovation community for value creation and user integration. This field had to deal with the transition of already existing relations between organizations and individuals and is therefore especially interesting to analyze in view of the evolutionary development of innovation communities.

The above table summarizes the practice fields that will be analyzed in the following chapters, and their characteristics and possible implications for the management of innovation communities (Table 4.1):

With regard to the results from the analysis of the innovation communities in Sect. 3.10, conclusions for the initiation and management of innovation communities can be drawn. It becomes evident that the communities differ in two important aspects: the level of the innovation system the community is embedded in, and the specific development phase the actual community is in. It therefore can be assumed that the level of the innovation system and the phases of the evolution and development of the innovation community (see Sect. 3.10) require different measures and methods for its management. Both factors will be further explored in the following chapters.

Table 4.1 Characteristics and possible implications of the practice fields

Practice field	Main focus	Possible implications for the management of IC's
Information and communication technologies	Complete set-up and development of a new innovation community	Initiation and development of innovation communities
New materials	Analysis, organization and arrangement of cooperative structures	Improving the performance and quality of networks and innovation communities
Renewable energies	Reconfiguration and extension of partially existing networks and communities	Evolutionary development of innovation communities

Source: Author

4.1.1 *Level Dependency of the Management of Innovation Communities*

An important factor for the management of innovation communities seems to be the level of the innovation system the community is primarily acting on. The description of the innovation system levels (see Fig. 2.1 in Chap. 1) indicates that networks of promotors can be cross-organizational or cross-level social networks, which make management of the communities a challenging task. As promotors' activities and their integration in broader innovation networks relate to various levels of an innovation system, community management requires specific tasks and measures. The analysis of the practice fields will look into the following aspects:

- Framing and linking level: In order to provide new technologies and innovation projects with the appropriate framing and funding from within a national or international innovation system, it is necessary to align the goals of innovation activities with political goals and vice versa, and to obtain public funding and support for the development of new technologies and applications. Moreover, the development of technical standards is a key issue on the framing level. Another important role play networking and communication measures on a political level that allow the varying interest groups to interact and to seek for the settlement of conflicts and the development of joint strategies.
- Value chain level: The management of the innovation communities on the value chain level can be differentiated into two subcategories according to their scope of activities. As the results in Sect. 3.10 show, innovation communities are either primarily related to broader fields of technology (e.g., carbon nanotubes) and usage systems (e.g., smart home), or have a stronger product orientation and are focused on value chains for specific products or applications. In the case of a broader technology field or usage system, the community can be organized according to potential applications and possible value chains of a technology, for instance. Here, the challenge for the innovation community is to provide orientation, to structure the field into applicable sub-structures, and to align and organize partners along future value chains. In the case of product-focused communities, the challenge is to evaluate specific value chains and future markets, integrate customers, and extend the network with a producer-user community. The focus in this case lies on the creation and adaptation of a powerful network to support the new product.
- Company level: On the company level, a key issue is how to convince decision makers and resource controllers of a specific innovation idea or project, and integrate internal promotors. Another important task is to provide measures for information and knowledge transfer and raise the awareness for a new product, service or application field of a technology. Often the promotors from a company can use the measures to further involve and integrate key actors from their own organization into the innovation community and thus broaden the innovation network.

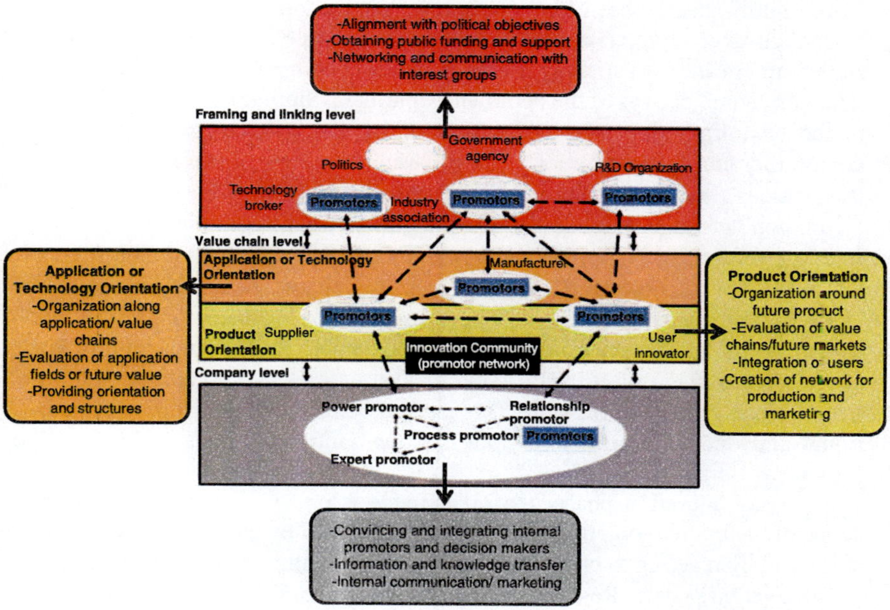

Fig. 4.1 Levels of innovation systems and implications for management (Source: Author)

The above figure sums up the options and implications for managerial support of innovation communities on different levels of the innovation system (Fig. 4.1).

4.1.2 *Phase Dependency of the Management of Innovation Communities*

In addition to the level of the innovation system, the development phase of the innovation community is an important factor in community management. The analysis of the case studies in Sect. 3.10 indicated that the composition of the community changes over time and that its development is an evolutionary and cyclical rather than a linear process. Three major phases of community development can be distinguished: the initial phase in which the community is initiated and established, the developing phase in which the focus is on the development and improvement of a community's performance (see Sect. 3.2), and finally the reconfiguration phase in which the community is fundamentally enlarged or changed according to new tasks or challenges, e.g. when transitioning to a new phase of the innovation process. It can be concluded from the experience and results in the practice fields that managerial support in these phases of community development should consider the following aspects:

- In the initial phase, the focus of management lies on analytical, communicational and interactive tasks, such as finding a suitable organizational form, the right partners and best strategy for the innovation community, and clarifying the interests and objectives of the community partners. Similar tasks can be required in the reconfiguration phase of a community e.g. if a reorganization of the community and the integration of new partner becomes necessary. The aim of this phase is also to reach a mutual understanding that allows the development of a common language, shared ideas and definitions in the community.
- In the development phase, the support shifts towards tasks that define steps for the realization of objectives. Typically, in this phase, the management of the community focuses on the execution and realization of the community's goals by transforming them into applicable tasks. Another important task in this phase is the establishment of efficient workflows, structures and processes that allow the community to perform well. This phase is often characterized by organizational and interactive measures like meetings and workshops within specific projects.
- In the reconfiguration phase, the main focus is on extending the impact and scope of the innovation project. This can be achieved by integrating the required skills and by reacting to challenges like changed conditions on various levels of the innovation system. Re-examining strategies and objectives of the community also belongs in this phase. If necessary, these can be modified or adapted. In some cases, after fundamental reconfiguration phases, the innovation community may again enter a new initial phase.

The following figure summarizes the observed phases of the community development and the implications for managerial support. The applied measures and methods will be further discussed in Chap. 5 (Fig. 4.2).

In practice, the phases of community development cannot always be clearly distinguished from each other, so that the following recommendations for innovation community management in specific phases should be seen as "rules of thumb" rather than as a binding framework.

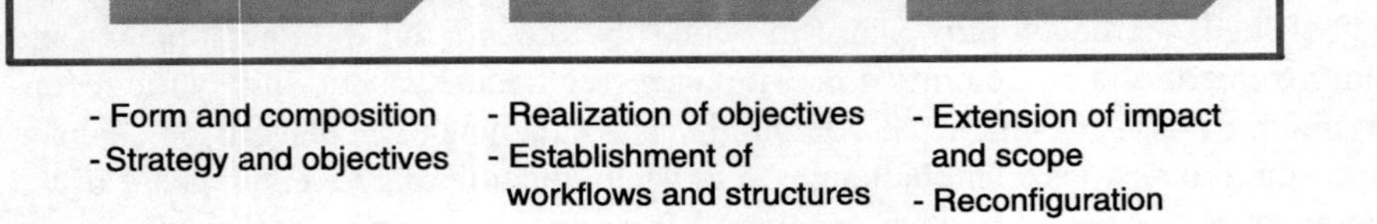

Fig. 4.2 Phases of community development and implications for management (Source: Author)

4.2 Information and Communication Technologies: Initiating and Developing the Connected Energy Community

Severin Beucker

The information and communication technology (ICT) sector is typified by its high R&D intensity, short product cycles and dynamic and competitive markets, especially in the telecommunication and Internet services field. Large ICT companies in particular are expected to provide not only efficient products and services, but also contribute to important social challenges of the future, e.g. climate change, work-life balance, or the informational self-determination of social groups. Against this background, there are two key challenges for innovation processes in the ICT industry:

1. Corporate innovation processes are to be opened, facilitating regular exchange and cooperation with R&D partners, customers and stakeholders. This is an important aspect expressed in the new paradigms of "Open Innovation" (Chesbrough 2003; Rohrbeck et al. 2007) and "Interactive Innovation" (Fichter 2005).
2. Customer and user demands as well as social requirements need to be considered early on in the innovation process. The new challenge lies in the directed founding of innovation communities that provide additional performance-related knowledge at the beginning, and support the development and launch phase subsequently.

Against this background, the accompaniment of practice fields of sustainable ICT innovation by the project InnoCo has shown to be a promising approach for gaining new insights in an environment of radical change, increasing interactivity and sustainability of innovation processes.

4.2.1 Origin of the Practice Field "Sustainable ICT Innovations"

The practice field "Sustainable ICT Innovations" originally arose from a contact between T-Labs (German Telekom's research department) and the Borderstep Institute through a previous research project. In an initial talk in 2008, both presented the InnoCo project objectives and assessed the research and innovation fields that seemed particularly promising for the future and relied most heavily on supportive innovation communities.

Amongst others, Web 2.0 applications, e.g. for social networking via Internet and other ICT, for online collaborations, or for distribution of user-generated content (blogs, videos, etc.), were shortlisted as possible topics. In addition to these user-centred web services (C2C or consumer-to-consumer applications), in terms of future Web X.0 innovations, new business opportunities are already opening up for telecommunication and Internet companies through so-called M2M applications.[1] The potential of ICT for optimizing all levels of the value chain in a given field can, for instance, greatly benefit the energy sector (e.g., power generation, control, management and consumption). One aim of the German Telekom has been to increase applications of ICT in the smart grid and smart energy sector, helping to improve sustainability and resource-, cost- and eco-efficiency of this industry.

Next to these activities, T-Labs and one of its partners, the Laboratory for Distributed Artificial Intelligence (DAI) at the Technical University of Berlin, discussed the possibility of continuing research work already begun in a joint research project for intelligent home networks, SerCHo or Service-Centric Home. Here, the objective was to integrate the latest developments in software engineering for smart home networks into practice-and market-relevant products and services. These are product or service innovations that meet the demands of end consumers, and at the same time help solve future social issues like energy efficiency and technical convenience. Smart home networking and automation technologies offer considerable potential here, due to the great share of households of total energy consumption. In Germany, rented properties and estates are of great relevance to energy and climate policy, since they make up around 60% of German housing and are responsible for significant fractions of national energy consumption. If the dormant energy saving potentials were tapped, this could make a considerable contribution to energy efficiency and climate protection.

One key issue here is how to make the energy used in heating, ventilation and power supplies in rented properties transparent to residents. It will become mandatory to present concrete opportunities and options of influencing and controlling energy consumption, as well as create incentives for energy saving. The mobilisation of energy efficiency potentials in rented estates through smart controls of energy consumption and energy storage is one possible approach examined in the project.

Intelligent building automation technologies provide great opportunities for this goal, since they can draw from well-approved technical components such as sensors, actuators and control devices. In rental units, central functions like heating or lighting controls can be optimized in terms of energy efficiency or cost reduction for a large number of consumers. The main objectives and issues of the practice field "Sustainable ICT Innovations" lie in identifying these potentials in smart home technology and automation, and in developing marketable concepts and products for the building and property sector. Moreover, the integration of smart home technologies into powerful data and electricity networks, the "Smart Grid", is to be further encouraged.

[1] Machine-to-Machine (M2M) refers to automatic information exchange between terminals such as (industrial) machines, automats, vehicles, or containers, with one central control centre. This enables remote control, monitoring and maintenance of machines, plants and systems.

The innovation community this practice field is based on has been analyzed in detail in a case study in Sect. 3.5. In addition to the findings of the case study, the following chapters focus on experience and conclusions drawn from the work in the practice field, as well as on the active participation in the evolution of innovation community. As the Connected Energy community was literally built from scratch, a special interest lies on factors that helped the community to evolve and develop. The level and phase dependency (see Sect. 4.1) of the community development is also examined.

4.2.2 *Development and Management of the Connected Energy Community*

The founding of both the Connected Living association and the complementary working group Connected Energy, dealing in particular with energy management in smart home networks, was accompanied by the practice field "Sustainable ICT Innovations" within the research project InnoCo. The Borderstep Institute supported the development of an innovation community "Connected Energy" through different means and methods. The process leading to the development of the innovation community is described in the section below.

Founding of the Connected Living Association

As mentioned in Sect. 3.5, the practice field Sustainable ICT Innovations originated from an already terminated research project on intelligent home networking, SerCHo – Service Centric Home, which was funded by the German Federal Ministry of Economics and Technology (BMWi) from 2005 to 2008. At the same time as the preliminary exchanges between T-Labs and Borderstep in 2008 were taking place, the network began developing an innovation centre "Connected Life". This centre was supposed to enlarge the knowledge originally gained in the SerCHo project, with the goal of further improving intelligent home networks. One important influence in this phase was that the BMWi was continuously prepared to promote and support the topic, all with higher intentions of positioning Germany as a leader in the field in the long term.

The potential organizational structure of the innovation centre was discussed in several conversations between the DAI Labs, German Telekom's T-Labs and the Borderstep Institute. Research and development on home networks, and the need for standardization were determined as the central objectives of the innovation centre which, in combination with the proposed funding of the BMWi, made the foundation of a non-profit organization appear appropriate.

At this stage, the Borderstep Institute's contribution to the process consisted of the following activities:

- Consulting regarding the choice of organization (non-profit vs. economic association), preparation of proposals for association structures and for functions of the advisory board.
- Analysis of existing networks and initiatives for smart home networks on a national and international scale, in order to determine their most distinguishing features.
- Identification of further potential partners for the association and the working group "Connected Energy".

In the end, the innovation centre was established as the non-profit association "Connected Living". Founding members include the DAI Lab Deutsche Telekom's T-Labs, and the companies Riedel Automation Systems, and Prosyst.

Based on the results from the project SerCHo, the formation of topical working groups was proposed as a means of internally structuring the association. After multiple re-organization and restructuring processes, the following working groups have been in place as substructures of Connected Living since 2011: Connected Communication, Connected Energy, Connected Health, Connected Media, and Connected Safety. In addition, there are two cross-sectional working groups discussing user interfaces and technical requirements for home networks as a whole: Connected Usability and Connected Technology (Fig. 4.3).

The association has by now come to include some 40 research institutions and enterprises. The German Federal Ministry of Economics and Technology (BMWi) supports Connected Living, endorsing the opportunity to continue to further develop the area of smart home networks. With the support of the BMWi, German research institutions and companies are to be further encouraged in their efforts in establishing marketable smart home network solutions.

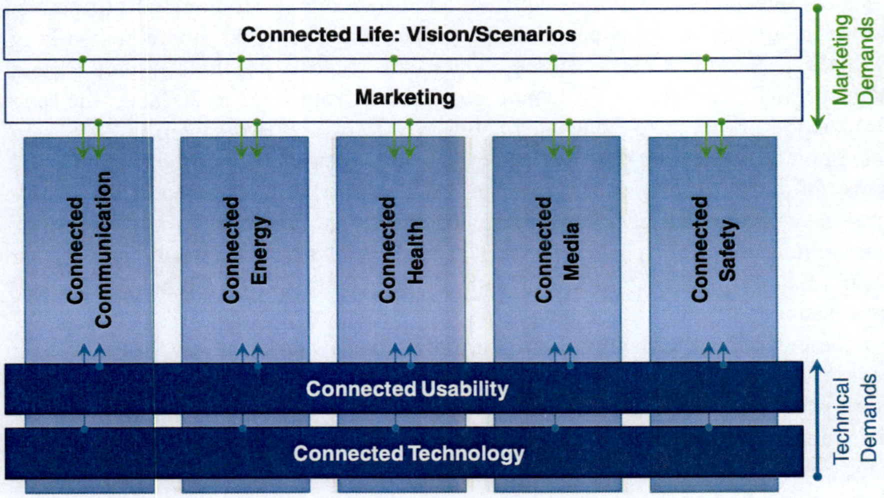

Fig. 4.3 Structure of working groups in the Connected Living association (Source: Connected Living 2011. Copyright Next Generation Media 2007)

Development of the Connected Energy Community

The innovation community Connected Energy has its origins in the working group of the same name. The latter was formed in the founding phase of the association Connected Living in 2009. Both former SerCHo project partners and new members of the Connected Living association are represented in the working group.

Due to the topical similarities, the Borderstep Institute became a member of this group. The working group management was in the hands of an energy supplier at the time of foundation. Once the energy company had withdrawn from the project coordination due to lack of staff, the Borderstep Institute took over the management of the working group. By now, ten organizations belong to the group, five of which are in turn members of the innovation community with its own communication and exchange structures. Partners in this innovation community include: the DAI Lab, Orga Systems Riedel Automation Systems the energy suppliers Vattenfall Europe and EWE and the Borderstep Institute.

Central to the formation of the innovation community was the development of a joint project proposal, SHAPE: Service-oriented home automation platform for energy efficiency – aimed at developing a decentralized energy management system for rented properties as an integral part of an intelligent home network. The proposed energy management system:

1. Provides greater transparency, influence and control possibilities in private energy consumption of end consumers, who are thereby assisted in minimizing their energy consumption through synchronized power consumption according to time- or load-dependent electricity rates.
2. Allows for the utilization of new services, located at the consumer/energy provider interface, that bundle energy-saving efficiency potentials and peak-reducing shift potentials.
3. By involving service providers, helps energy suppliers and network operators include flexible electricity rates and smart meters for remote power and energy demand management of households.

The project proposal was first submitted to the BMWi in January 2010, further revised, then handed in as a project application, and approved as a funded project in mid 2010. The joint research project that all the partners of the innovation community take part in has now been running since July 2010. It represents a complex innovation process that consists of technical products as well as complementary services. For the development of this community, different skills from a number of fields and professions were required, e.g. building services and appliances, ICT and telecommunications, energy business, and innovation and technology management. This in turn is reflected in the variety of the community members.

As described in Sect. 3.5, the innovation community consists of four promotors. An important insight from the innovation case study is that, in addition to the promotors, other key persons who support the enterprise through their activities, competence or decision-making power also have notable influence on the initiation, progress and even overall success of the innovation project. Their "drive" in the innovation process may not be as continuous or as straightforward as the promotors', so they are not defined as such. But even so, their selective support is of great relevance for the overall success of the innovation project. To distinguish them from the main promotors, they are here referred to as "supporters" (cf. Sec. 3.5). In addition to the four promotors of the Connected Energy community, another four supporters belong to the greater network in which the innovation community is embedded.

As described in Chap. 1, promotors often belong to different levels of an innovation system (framing and linking, value chain and company level). In the case of the best practice field, the key actors (cf. the analysis of promotors and supporters in Sect. 3.5) can be attributed to different levels of the innovation system, indicated by the white and grey coding in the following figure (Fig. 4.4).

The figure shows that the community involves promotors and supporters from the framing and linking level as well as from the value chain level which in this case

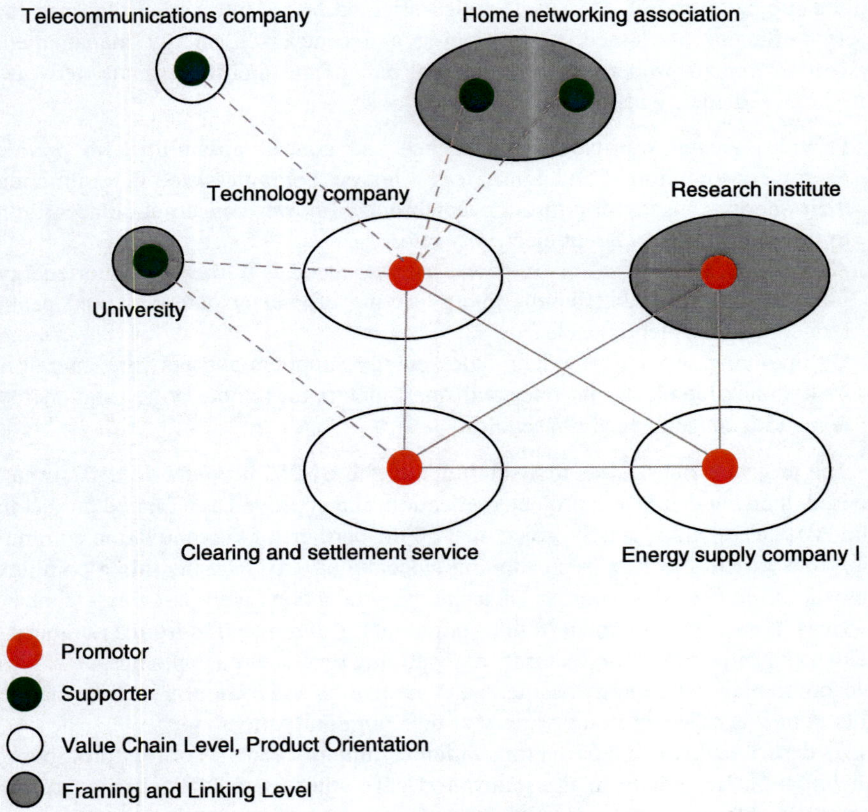

Fig. 4.4 Key actors of the Connected Energy community and their integration into the innovation system (Source: Author)

is product-oriented. From accompanying work in the best practice field, it is known that additional actors are also involved and important at a company level. The energy supply company, for instance, is actively trying to involve individuals from various internal departments into the innovation process and the community too. They are making sure that they can promote ideas for new products and services for energy management systems in their own organization, and further develop and strengthen the innovation network by incorporating more partners from their own organizations.

4.2.3 Phases of Community Development

The following chapter will take a closer look at the phases of the community development and the steps that were taken to form and organize the network. As mentioned above, the Connected Energy community evolved from scratch and the formation process from July 2009 until today went through notable phases. The starting point was in July 2009 when the association's working group was founded, where the key persons met in a workshop and developed a common understanding and targets for the innovation process. Within the working group, the four promotors (see chapter above) soon formed a small team that started developing ideas for a joint funded research project. Important for the formation of the innovation community were the events and phases listed in (Table 4.2)

The table clearly indicates that the formation of the innovation community as it presently exists was the result of a number of smaller steps that led to increased understanding and the constitution of joint objectives. For this process, regular communication and the development of scenarios, documents and finally, a project proposal were important steps. The table also shows that all three phases of community development are influenced by measures of community development.

4.2.4 Measures and Methods for Community Development

From the steps and phases of community development described in the chapter above, it is evident that a number of measures and methods have been applied in the community's development phase and its management. One reason for the successful implementation of these measures and practices was the Borderstep Institute's scientific and methodological experience in coaching similar innovation projects. The steps and applications that the institute played a decisive role in are summarized below:

- Workshops: During several workshops, the group developed a joint conceptual understanding and drafted project ideas and sketches, notably a 2-day strategy workshop organized by the association Connected Living and the DAI Laboratory,

Table 4.2 Development phases of the Connected Energy innovation community

	Event	Influence on community development
Initiation phase		
July 2009	General Assembly and 2-day strategy workshop of Connected Living association: • Establishment of work groups • Development of common understanding regarding topics and objectives • First definition of scenarios for roadmaps and scenarios for distributed energy management architecture	Initiation of the Innovation Community and development of a common understanding of topics and objectives
July 2009–Nov. 2009	Joint development of a project idea, based on first scenarios, in a "living document", coordinated by Borderstep	Systematic development of conceptual understanding and objectives of the IC
Development phase		
Nov. 2009	General Assembly of the Connected Living association with a first presentation of ideas for distributed energy management in rented estates (SECCO) in front of representatives of the German ministry of economics BMWi	Enhanced group perception through joint public presentation of the project ideas
Dec. 2009	Working group meeting, further members joined, team leadership adopted by Borderstep	Strengthening of group competencies, redefinition of group boundaries
Nov. 2009–Jan. 2010	Renaming of the project idea to "SHAPE", first draft for a project proposal for a BMWi funded project – a longer process accompanied by phone conferences	Refined understanding of concept, goals, partners and their respective tasks
Jan. 2010	Consultations with the project executing agent, DLR, on the project outline for SHAPE	Enhanced perception as a group through joint public presentations of project ideas
Feb. 2010	Working group meeting, review of project ideas and scenarios	Reassurance of mutual understanding and project targets
Feb. 2010–March 2010	Development of the project outline	Refined understanding of concept, partners and their respective tasks
March 2010	Presentation of the project outline in front of a review committee appointed by the BMWi	Enhanced perception as a group through joint public presentation of project ideas

(continued)

Table 4.2 (continued)

	Event	Influence on community development
From April 2010	After a first positive assessment in April 2010, the project outline is converted into a project proposal, one additional partner (Telekom) is included	Redefining group boundaries, redefining project objectives and roles
Stabilisation phase June 2010	Covenant for the project SHAPE by the executing agency, DLR	Institutionalization of the Innovation Community as a project team
Since July 2010	Ongoing project SHAPE since July 1, 2010	Redefined working plan and project objectives

Source: Author

held in July 2009. Here the members were dispatched to various thematic work groups and first topical discussions initiated in the working groups. Independent of their respective contents, these workshops served to facilitate communication between community members and the surrounding network.

- Working group meetings: In addition to the workshops, several group meetings and telephone conferences were held by the Connected Energy working group with different objectives. These meetings were mostly moderated by Borderstep and served as a communication medium for community members as well. They also enabled professional and personal exchange between members and therefore played an important role in team building.
- Scenario Technologies and Roadmapping: The most important result of the first strategy workshop was the development of scenarios and roadmaps by the working groups, creating a common understanding of objectives and defining specific areas of application for smart home networks and the research partners' respective roles. The roadmap was developed primarily as a guideline to describe the long-term perspectives of each work group, and to distil the distinct strategies and interests of individual groups into one common "mission statement" for the entire Connected Living project.
- Living Document: As a shared knowledge base of the Connected Energy working group, the "Living Document" had a central function in conceptual debate and helped to establish common understanding on the topic of decentralized energy management. It also served as a foundation for the development of the project outline for SHAPE.

4.2.5 Lessons Learned

The case of the Connected Energy community proves that both the origin of the innovation and the systematic application and use of measures and methods of community development play a central role in the foundation, organization and management of innovation communities. Since here, an entirely new community was formed by members of the Connected Energy working group, there was great demand for coordination in developing a common understanding of topics, objectives, roles and functions, and on that basis, defining and promoting ideas for a joint innovation project.

In this procedure, the Borderstep Institute's scientific monitoring and the practical measures and methods described in the chapter above both played an important supportive role in the development and preservation of the innovation community, and accordingly, for community composition, communication design, strategic development and long-term planning in general. The value of this technical and methodological support is recognized by the community members who continue to use these techniques for the development of new project ideas, for involving new partners, or for systematic planning of topics and processes.

The community has presently conquered the initiation and development phases, but not quite reached the reconfiguration phase yet. In the ongoing working process, it has become apparent that more partners from the property and telecommunication sectors are needed in order to successfully develop products and services as well as the community itself. Furthermore, by some of its members, the community is seen as a means of extending the network to the company level of the organizations involved as well, thereby aiming to extend and strengthen the innovation network and its impact at a framing and linking level.

4.3 Renewable Energies: Coaching Innovation Communities for Next Generation Solar Cells

Willy Bierter

Renewable energies, especially photovoltaics, are a main thematic emphasis of German and European research funding and energy policies. Despite diminishing funding for solar electricity in Germany, photovoltaics is regarded as a great potential to accomplish the goal of an energy supply based on renewable energies. Both politics and industry acknowledge that the costs for producing electricity from photovoltaics need to be reduced in order to be competitive. These are important framework conditions for the development of so-called dye-sensitized solar cells, as, due to specific material features, the technology has great cost reducing potentials and opens up many new application possibilities (e.g., in architecture). Research funding has an important function in the innovation process, as it reduces investment risks, helps cooperating partners to position themselves in the scientific and market surroundings, and positively influences pre-competitive structures and competitive partnerships.

Worldwide research on dye-sensitized solar cells (DSSC) has already been conducted for more than 15 years. DSSCs are considered promising candidates for the next generation of solar cells. Compared to conventional silicon solar cells, they have the great advantage of lower costs and lower energy consumption in production (for a description of DSSC technology see Sect. 3.6). Moreover, transparency and colour selection, features other types of solar cells do not display, provide greater freedom of design. For further cost reduction, monolithic, serially interconnected modules are the most promising. The fact that no complex processes like bonding or wiring individual solar cells are involved in module manufacturing contributes to the cost reduction and helps paving the way for mass production. The technology is expected to be marketable around 2012. Before that, a number of important preconditions have to be accounted for, since the DSSC technology necessitates setting up completely new value creation networks and developing new markets.

4.3.1 Origin of the Practice Field "Renewable Energies"

The practice field of DSSCs is of particular interest and value for the analysis of managerial support and the development of innovation communities for the following reasons:

1. The development of DSSC has the character of a radical innovation, as it concerns the development of new businesses and product lines with an entirely new set of performance features and with significant cost reduction potentials,

which can transform existing markets and create new ones and therefore requires administrative support.
2. The DSSC technology is still in an early innovation phase, as is the innovation network itself. The organization and coaching of innovation communities is a suitable approach to tackle the various complex tasks accompanying it.
3. Many knowledge fields, experiences and competencies need to be brought together coherently and further developed in order to bring the technology successfully to market.

The practice field partly relies on earlier experiences from the research project ColorSol (see Sect. 3.6). This project had three main objectives: (1) the cooperative technological development of glass-based DSSCs in selected application fields; (2) the support of technology and market developments using technology and product roadmaps, life-cycle analysis and success factors for the solar industry; and (3) transferring the results to the solar industry.

The ColorSol project consortium consisted of application-oriented research institutes, enterprises from the glass und chemical industry, and a planning office for photovoltaics. In this phase, productive research networks and development organizations as well as potential lead users of the DSSC technology were identified. Detailed enquiries were made into the DSSC technology and emerging technological, economic and social trends surrounding it. The information compiled was used for the exchange of technical knowledge and the discussion of consumer product specifications. Furthermore, market potentials, costs, investments and global marketing structures were assessed and then debated with potential production and marketing partners. Towards the end of the project, meetings were also arranged with venture capital groups, to discuss the foundation of an enterprise to produce and commercialize glass-based DSSCs.

Towards the end of the project it became clear that a more selective development and differentiation of DSSCs had become necessary in order to successfully position the technology in specific application fields and markets. Against this background, the DSSC practice project was initiated from within the common project InnoCo – Innovation Communities, aimed at realigning the research topics and promoting the setup of a corresponding innovation community. At the beginning of the DSSC practice project, technological varieties of DSSCs as well as the resulting application fields and potential constellations of partners were assessed with research partners from the earlier ColorSol project.

The focal innovation process in this DSSC practice project includes the development of a monolithic DSSC and the corresponding production technology, as well as the establishment of a productive and efficient innovation community supporting this process. These new and complex requirements and tasks made it necessary to look for new partners, to merge the competencies distributed among the different partners involved "along the line", and to define and implement specific projects. In order to direct all this properly, sensible external coaching can help to guide the innovation community in a productive direction and enhance its success (see Sect. 5.3 on coaching innovation communities).

The specific innovation community of this practice field has been analyzed in detail in the case study summarized in Sect. 3.6. In addition to the findings described there, the focus in the following chapter lies on practical steps of reorganizing and accompanying the innovation community over the course of the research project InnoCo.

A special interest of the work in this practice field lies on measures that advance the setup and reorganization of an innovation community, and support its development over time and over various levels of the innovation system (see Chap. 1).

4.3.2 Main Goals and Tasks of the DSSC Innovation Community

The main goals of the DSSC practice project were:

1. The development of a productive and stable innovation community consisting of manufacturers and suppliers, with the aim of establishing a new value creation network for DSSCs (value-network community).
2. The inclusion of trendsetting users and decision-makers like solar architects, façade planners, builders etc. (manufacturer-user community).

One central goal when setting up the value creation network was to gain additional partners and suppliers for the existing consortium. These were meant to expand the required profile of competencies, so that individual system and consumer solutions could be offered, as well as DSSC modules in larger, industrial-scale numbers. Therefore a partner who was to warrant the marketability of DSSC products with his knowledge of target markets and international distribution structures was selected for distribution and marketing. Moreover, it was essential to find partners and suppliers with appropriate skills and experience, as well as the necessary machines and equipment, for manufacturing. Not only did new partners need to have professional and market know-how, entrepreneurial skills were also required.

Apart from the search for suitable partners and suppliers, the most important tasks were:

1. Clarifying converging and diverging interests and attitudes with regard to cooperation.
2. Developing a draft programme for the value creation community.
3. Learning with and from task-specific projects in the pilot phase of setting up the innovation community.
4. Establishing the value creation-community.

For the realization of these tasks, the method of systemic innovation and network coaching was employed. For market and technological developments (competitors, existing products, etc.) and future customer needs, Internet-based expert surveys and technological and market trend assessments were used.

Setting up a manufacturer-user community is primarily about convincing trend-setting users and key clients, and fostering longer-term relations with important market partners. For that purpose, measures like innovation workshops with lead users were used to evaluate expectations on future products and services and to derive possible future value chains.

4.3.3 Setting up the DSSC Innovation Community

The development of competitive DSSC products is a complex task because technologies from different scientific disciplines such as physics, chemistry and engineering have to be combined and blended. In order to develop a technology of this complexity into a marketable product and simultaneously prepare the corresponding markets, a network or innovation community of potential producers, system or component suppliers, and future users and customers is essential. Therefore, the innovation process not only comprises the technological development, but also the preparation and opening of the market.

The success of DSSC clearly depends on the development of an efficient innovation community, consisting of the most important actors in the value creation network, as well as an innovation community of producers and users. Correspondingly, at the centre of the project was:

1. The development of a competitive and stable innovation community of producers and suppliers with the aim of setting up a new value-creation network for DSSCs (value-creation community).
2. The inclusion of future trendsetting users of DSSCs and DSSC products, which at present are primarily seen in small applications like consumer electronics and lighting or are integrated into the façades of buildings.
3. The strengthening of the German and European innovation community in the face of competing Asian research groups.

The aim in setting up and enlarging the value-creation community was to find additional partners to supplement and expand the profile of competencies, whether as:

- Suppliers skilled in upscaling the production of components for DSSC modules like pastes, dyes and transparent conductive oxide (TCO) glass, and experienced with production and the required machinery and equipment.
- Partners in the field of marketing, sales and distribution who can ensure the successful marketing and sales of DSSC products with their knowledge of target markets and international distribution structures.
- Pilot customers prepared to experiment with pilot DSSC products and share their experiences.

New partners should not only have competencies in terms of technology and markets, but also bring in entrepreneurial spirit and qualifications.

The origin of the innovation community lies in the research project ColorSol (see Sect. 3.6). By the end of this project it had become evident that the existing network needed a reorientation and reconfiguration phase to successfully develop the DSSC technology and corresponding marketable products.

At the beginning of the work in the practice field in July 2008 the DSSC innovation community therefore consisted only of a small chemical company and the Borderstep Institute. The chemical company was founded in May 2003. It is 100% privately owned and completely independent. Its core competences are in one of the basic materials important for DSSCs. It is a component supplier for a number of solar applications like DSSCs or organic photovoltaic cells (OPVCs) (see Sect. 3.6).

A further strategic partner, another small company from the sector of specialty chemicals from Switzerland, joined the DSSC innovation community shortly afterwards. This company was founded in 1993 and is a pioneer in the field of DSSC research. One of its founders is a former PhD student of Michael Grätzel of the École Polytéchnique Fédérale de Lausanne (EPFL), the discoverer and most important researcher in the field of DSSCs. The company is world leader in DSSC specialty chemicals, such as nanoscale titanium oxide particles, ruthenium-based dyes, electrolytes and ionic liquids. It is in the course of installing pilot assembly line for smaller DSSC modules (30 $\times$30 cm), and is an important manufacturer of plasma light lamps for photovoltaic testing and certification purposes. Until today, it is the only company in the DSSC field that is economically self-supporting, financing its investments solely from its cash-flow. Moreover, the company holds all the essential property rights and patents itself, which is of great importance for the commercialization of the DSSC technology.

The activities of the two companies from the chemical sector partly overlap but are also complementary: both cover almost the whole DSSC value chain, from components and cells to entire modules.

At that time a further potential candidate was the pigment and additives division of a large chemical company. The company is a supplier of pastes and dyes and invests in research and development of large-scale processes for the production of components and DSSC technology.

Important steps and tasks for setting up and fortifying the DSSC innovation community were:

- Clarification of goals and shared interests of all partners.
- Identification of further potential partners in the field of components and sales.
- Executing smaller joint development projects (e.g., optimizing and testing chemicals and dyes).
- Drawing up and signing a memorandum of understanding and non-disclosure agreement between the two small chemical companies and Borderstep.

Figure 4.5 represents the innovation community in the practice field as it existed at the end of the InnoCo project. As described in Sect. 3.6, the community consists of five promotors. Three more actors can be considered to be supporters and belong to the wider network in which the innovation community is embedded. As described in Chap. 1, promotors often belong to different levels (framing and linking, value chain and company level) of an innovation system. In the case of the best practice field, the key actors (see also the analysis of promotors and supporters

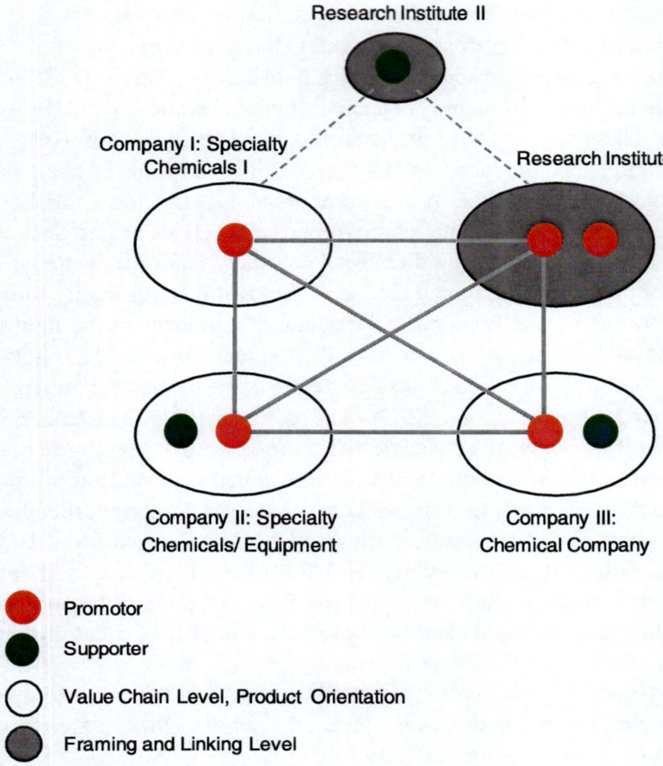

Fig. 4.5 Key actors of the DSSC community and their integration into the innovation system (Source: Author)

in Sect. 3.6) can be attributed to the different levels of the innovation system, as indicated by the white and grey coding in the figure.

The figure also shows that the community involves promotors and supporters from the framing and linking level as well as from the value chain level that in this case is product-oriented. Finally, the company level is especially significant here, as for two of the companies (the specialty chemicals/equipment company and the chemical company), the innovation community had a communicative and unifying effect in their own organization. Furthermore, it is important to emphasize here that the DSSC innovation community maintained a clear entrepreneurial view and strategy throughout the entire innovation process.

4.3.4 Implementing Important Coaching Tasks

The main characteristic of the DSSC practice project is that it constitutes a radical innovation, comprising complex tasks in a highly dynamic environment. It is science and technology intense, and the innovation community is embedded in a larger innovation system.

In view of these facts, the coaching of such an innovation process needs to be systemic. In our understanding, systemic coaching is a solution-oriented procedure that perceives innovation communities as being embedded in an innovation system. It aims at establishing proper framework conditions for innovation projects, effectively supporting innovation communities in improving their performance (see Sect. 3.2) and contributing to the overall success of innovation projects.

As outlined in Sect. 3.2, the contribution of promotor networks to the success of a cooperative innovation project depends on three key factors:

- Situational or task-based composition of the community, in terms of the skills, roles and resources available
- Existence and extent of community identity
- The level of interaction quality

Systemic coaching has to support innovation projects and innovation communities in finding the right composition of the partners to be involved (composition of the community), in developing common goals and team spirit (identity of the community), and in finding a productive and trusting form of collaboration (interaction quality in the community). With these three tasks in mind, coaching has to take the different phases in the development and life-cycle of a community into account. Basically, three phases can be distinguished: initiation, development, and reconfiguration (see Sects. 4.1 and 5.3). In view of the complexity of the whole undertaking, this kind of systemic coaching should ideally be carried out by a neutral external scientific institution.

Clarifying Converging and Diverging Interests and Attitudes and Identifying New Partners

Due to the transition and reconfiguration of the former ColorSol research project to the new innovation community within the InnoCo practice field, a clarification of converging and diverging interests was of particular importance. To that end, interviews were conducted with every single partner. A structured questionnaire was devised for that purpose. A first evaluation of the interviews was presented to the partners for examination and possible amendments. Afterwards, a "map" of the common and diverging interests of all the DSSC innovation community partners was designed and discussed in a strategy meeting. In view of the fact that especially the two small chemical companies have overlapping and therefore competing business interests, this was an important first step in building confidence and managing potential conflicts.

Moreover, several criteria for the assessment and overall evaluation of further strategic partners for the DSSC innovation community (see the table below) were worked out and discussed (Table 4.3).

Table 4.3 Criteria for the assessment and evaluation of partners

Criteria for the assessment and overall evaluation of further strategic partners for the DSSC-innovation community			
Criteria	**Partner**		
	A	**B**	**C**
1. Technological competence			
2. Market position of the product			
3. Proximity to DSSC-markets			
4. Extension and density of the sales network			
5. Quality of theservices			
6. Soundness of the long-term financial basis			
7. Degree of independence of the company			
8. Basis of trust			
9. "feel-good factor"			
General company informations (turnover, cash-flow, patents, licenses etc.)			

Source: Author

Implementing Joint Development Projects

In a next step, several common projects for the two chemical companies to jointly develop were devised. The projects pursued two goals:

- The systematic development of new chemicals that lead to higher efficiency by increasing the light-current yield, absorption spectrum, or long-term stability of DSSCs.
- The optimization of selected chemicals regarding their processing properties, e.g. chemicals with reduced viscosity, printable chemicals.

Within these projects, one of the chemical companies was responsible for the development of optimized chemicals for the use in DSSCs, and the other one for the integration into monolithic DSSC blocks and testing.

This step had two purposes: to improve specific components for the production of pilot DSSC modules and to set the cooperation of the community on a continuously productive and stable basis. Especially this last point required trust as well as a certain orientation on value chains that could create advantages for all members of the community. The next step was therefore to extend the community and to integrate future users.

Setting Up a Producer–User Innovation Community

After establishing the core of the value chain innovation community, the next step was to set up a producer-user innovation community. In order to get a complex technology like DSSCs on the market with the specifications that users really want, it is essential to prepare the corresponding markets early on in the innovation process.

The community was convinced that the DSSC technology could open up other applications and entirely new products that cannot be realized with conventional photovoltaics. For this reason, during the planning of a user innovation workshop, the partners were discussing which potential users might be interested in this new technology. Further aspects examined were the main technical and optical requirements for monolithic DSSCs regarding the various projected applications.

Following the discussion, the following tasks for the preparation of a user innovation workshop were articulated:

- Identification, selection and description of potential applications and producers/ designers of monolithic DSSCs.
- First selection of potential test users for the planned workshop.
- Coordination of the test production of DSSC modules in Switzerland with the coinciding workshop.
- Developing a first concept and programme for the user innovation workshop.
- Monitoring of module production and schedules in view of the planned workshop, and delivery of the first monolithic DSSC modules.

The selected potential applications were grouped into different categories according to their specific requirements, taking into account aspects such as the number of pieces, cost pressure, lifetime, performance, design and processing quality:

1. Consumer and design products (e.g. Sun table)
2. Stationary and mobile applications (e.g. bike taxis)
3. Industrial applications (e.g. stand-alone traffic lights)
4. Architectural applications

The user innovation workshop had three underlying objectives: First, the test and pilot applications for monolithic DSSCs needed to be identified. Then development partners needed to be found for the most practicable applications. Finally, a network of interested users (companies, developers, industrial designers etc.) needed to be set up.

The results of the workshop were encouraging: It was agreed that three test and pilot applications would be realized by a consortium of companies, and the according technical specifications and schedules were discussed and settled.

Expanding the DSSC Innovation Community

The next challenge for the DSSC innovation community was that the Ecole Polytechnique Federale in Lausanne (EPFL) commissioned the Swiss chemical company with assessing the feasibility of a bigger DSSC installation, a storefront for a new congress hall of about 300 m^2. The main tasks that had to be dealt with here were the long-term stability and reproducibility of the DSSC modules, as well as design and colour varieties. Also, a completely new production line with higher capacities had to be laid out and installed, with the aim of mastering the entire production process.

To deal with these tasks, the skills and experience of various partners had to be brought together and combined. Therefore, quite a substantial number of players in the field of DSSC technology were invited to participate in a multiplicator innovation workshop. The aims of this workshop were:

- To familiarize participants with the concept of innovation communities.
- To agree on a plan how to achieve long-term stability of DSSC modules.
- To validate concepts for improving design quality of the modules.
- To discuss how to master the whole production process for pilot production (avoiding rejects, reducing faults in production).

The participants formulated all necessary tasks and schedules in their working groups and appointed community members to accomplish these tasks.

4.3.5 Main Methods Used in Coaching the DSSC Innovation Community

The following methods have successfully been used in coaching the DSSC innovation community (also see Sect. 5.4):

- Strategy workshops for the clarification of converging and diverging interests, aims and attitudes of all partners.
- Strategy workshops for the systematic identification of further partners and their potential contributions to the DSSC innovation community.
- Continuous monitoring and analysis of the technological development of DSSCs using technology roadmapping and other methods, the results of which were also used to identify further development areas and market sectors as well as suitable partners and users.
- Collaborative projects for the development and optimization of electrolytes and dyes, especially in the early stages of the project, between the two small chemical companies.
- Business innovation workshops (see Bierter and Fichter 2005) with the aim of systematically and quickly finding innovative ideas for new products, services or business processes in areas where value creation and competitive advantages are essential.

All the methods used by the Borderstep Institute in this project in fact achieved what they were intended to. In a relatively short time, the DSSC innovation community managed to coordinate the interests and goals of the participating partners, and avoid conflict. The joint projects were implemented for the benefit of all partners, which also had positive effects on the group identity, role typecasting, and trust between partners.

4.3.6 Lessons Learned

The practice field Renewable Energies shows that the development and reconfiguration of the innovation community can successfully be supported and fostered by systemic coaching in the sense that it clarifies interests, systematically identifies and integrates strategic partners and organizes the process of community development. Conclusions that can be drawn from the experience gained in this practice field include:

1. One main focus and therefore one central task is looking for new and suitable partners who have the know-how and competencies required at their disposal. This is necessary to reach the new objectives as established in the transition from the ColorSol project to the new DSSC project.
2. Accordingly, a second focus was on clarifying the strategic interests of all partners involved. This is indispensable for the creation of the appropriate climate and trust, improving communication and common understanding of the respective roles and responsibilities among the members of the innovation community.
3. It was important to set up a producer-user innovation community at a relatively early stage. Only thus was it possible to learn about the desired specifications of potential DSSC products. This is an obvious prerequisite for bringing such a complex technology to market. It is essential that, in addition to the technology in question, the corresponding market(s) are prepared early on in the innovation process.
4. Expanding the DSSC innovation community was the ultimate litmus test, proving that the community can tackle challenging new and complex projects aimed at successfully marketing DSSC products.

Altogether, it was acknowledged by all members of the DSSC innovation community that coaching by an external and neutral coach had an important supportive function in the initiation and development of the innovation community.

4.4 New Materials: Managing the Carbon Nanotubes Innovation Community

Stefanie Bunzel

Advances within the field of new materials such as carbon nanotubes are characterized by very intensive, fast and international research and development efforts. To be able to actively participate in such a complex technology field, it is necessary – even for large companies – to establish innovation networks with the objective of opening new markets with advanced products. Yet the inter-organizational cooperation and communication along extensive value chains proves to be difficult and therefore requires a set of methods and approaches to support the management of such an innovation network. The following chapters give an overview on how the management of the carbon nanotubes innovation community is accomplished with the goal to develop and further improving the network.

4.4.1 Why Carbon Nanotubes?

Carbon nanotubes (CNT) represent a new class of nanoscale materials; they are currently the most extensively researched and well-known nanostructures. Although the first publications reporting cylindrical carbon-based structures appeared as early as the 1950s, the breakthrough for CNTs did not occur until 1991, when the Japanese microscopist Sumio Iijima published transmission electron microscopic images of nanostructures. The great interest in CNTs has mainly been due to their broad functionality: CNTs have special mechanical, optical, magnetic, electrical, biological, thermal, actuatory and sensory properties which are dependent on their geometry (aspect ratio, surface-to-volume ratio, chirality, number of tubes) and chemical structure. CNTs are single-walled (SWCNT) or multi-walled (MWCNT) tubes made of carbon, with diameters of several nanometres and aspect ratios between 1:100 and 1:10,000. They are produced by arc discharge between carbon electrodes, by catalytic decomposition of carbon gas in a furnace, by chemical vapour deposition, or by laser vaporization of graphite (see Fig. 4.6). A great economic potential is anticipated for CNTs in future because

Fig. 4.6 Carbon nano tubes (CNT) (Source: Bayer Material Science)

of their unique structural features. The range of CNT applications is extremely wide, due to their diverse material properties. CNTs are used in areas ranging from field-effect transistors, actuators and sensors to material reinforcement and medical applications (ZTC 2009).

The properties of CNTs are heavily dependent on their quality (defect density). Since CNTs are elongated structures, the biggest differences in properties can be found along the tube axis and not in the radial dimension. In the following, particular properties of CNTs or CNT bundles are described in greater detail (Koker 2006).

- *Mechanical*: Since the C atoms of CNTs are covalently linked, and covalent bonds are among the strongest chemical bonds, very high modulus and tensile strength values are achieved, resulting in exceptional hardness, fracture strength and toughness at low temperatures, and in super-plasticity at high temperatures. CNTs thus have a density of 1.3 (SWCNT) to 1.8 (MWCNT) g/cm and an elasticity (E modulus) of 542 (SWCNT) to 1,200 (MWCNT) GPa; by comparison, steel has a density of 7.8 g/cm and a tensile strength of 207 GPa (ZTC 2009).
- *Electric*: Depending on their structure and diameter, SWCNT can be metallic (conductive) or semi-conductive, with the size of the band gap depending on the average diameter. The structure of the CNT greatly affects its electrical properties. Around two-thirds of all CNT types can be considered semiconductors at room temperature, and one-third metallic conductors.
- *Thermal*: The thermal conductivity of CNTs is 6,000 W/(mK), considerably higher than that of diamonds (2,300 W/(mK) or silver (430 W/(mK) at room temperature. This very good thermal conductivity is effective only along the tubular axis; in perpendicular direction, carbon nanotubes are thermal insulators. For this reason, MWCNTs are, for example, used to cool computer chips. Compared to chips with normal cooling, chips with CNT cooling have an 11% higher heat transfer rate.
- *Optical*: Defect-free CNTs, especially SWCNTs with their discrete and well-defined bandgap and sub-band structures, are ideal for optical and optoelectronic applications. CNTs are found in applications such as scanning probe tips or even field emission displays (FEDs). The latter is particularly attractive because the entire screen market has an annual turnover of several billion U.S. dollars and is continuing to grow.
- *Chemical*: Due to their chemical properties, CNTs can serve as catalyst carriers or act as catalysts themselves. In addition, applications as molecular filters or membranes and in "nano reactors" are being considered.
- *Biological*: This mainly involves interactions with complex biological systems, such as cells or organisms. CNTs are used in the field of bio-nanostructures, for example in the production of structures for the detection of bio-molecules (bio-imprinting).
- *Actuation*: Through their direct conversion of electrical energy into mechanical actuation, micro-actuators provide systems with the possibility to respond, to

generate force and perform movements; they are therefore essential components for many industrial systems, particularly in the fields of robotics and circuitry. Applications for micro-actuators made of CNTs can be found anywhere where actions or responses are required in extremely small spaces, the prime example being minimally invasive medical procedures. One advantage of the CNT-based actuator is that, compared to piezoelectric actuators, the longitudinal extension capacity of CNTs is much more pronounced (by a factor of about 30).

4.4.2 *Motivation for the Formation of a CNT Community*

The many useful properties, geometries, raw materials, manufacturing processes and other parameters such as quality, productivity, purity and dimensions associated with industrial CNT applications show the complexity of bringing together a technological "push" and a market "pull". So far, however, the consistent combination of material properties and product requirements has not been a great problem, neither for nanotechnology in general or for carbon nanotubes in particular. The main challenges involve:

- *Acquisition of knowledge about possible and potential uses on the application side:* Details of nanotechnology solutions, and carbon nanotubes in particular, are in many cases still unknown, due to the complex scientific foundations and the great dynamics and novelty of their development. Therefore, these aspects are only considered in product planning to a limited degree.
- *The development of market-oriented value chains with relevant partners, without creating competitive situations:* As can be seen in Fig. 4.7, the CNT value chain begins with the basic field of material development and production and then includes compounders and formulators, system manufacturers and, finally, end users. Within this value-added chain, in order to align "push" and "pull" factors, CNT manufacturers need to develop an innovative component in cooperation with an end product manufacturer. The end product manufacturer develops these components with a system supplier, who obtains the material from a processor who in turn gets it from a CNT manufacturer. This means that the CNT producer must, even at the beginning of the value-added chain, obtain detailed information about the value-adding functions of his technology for the consumer at the end of the chain – i.e., three to five steps down the line –, and from that information derive critical success factors for his CNTs. To date, the process is characterized by a sequential development of relationships between structures, processing methods,

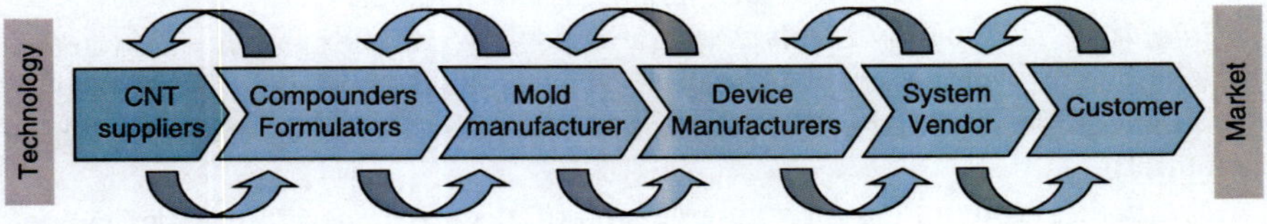

Fig. 4.7 CNT value-added chain (Source: Author)

properties and functions along the value-added chain. Due to the series of interfaces within the value-added chain to be mastered along the way, this process is very time-consuming and often leads to information loss. It is therefore necessary to support the construction of a far-reaching and comprehensively connected community in order to ensure an efficient, quick and targeted performance.

- *The assessment of opportunities and risks with respect to relevant markets, technological developments and social acceptance:* The current development of nanotechnology is affected by the opposing forces of the prospect of innovative products with improved properties made from new and sustainable nanomaterials on the one hand, and the risks involved on the other hand. These risks include potentially toxic effects, lack of acceptance on the part of consumers, costly, complex and expensive manufacturing procedures or material reformulations, or unclear market situations. Scenarios for the use of carbon nanotubes must be defined and developed at an early date, and an exchange organized with all relevant societal sub-sectors.

Significant advances in basic research and increased competition from Japan and the United States have recently led to the decision in Germany to jointly promote CNT development between the industry and the research community in an application-oriented manner, so as not to be left behind in the international race. A major corporation in the chemical industry assumed a key role here, which, together with several other companies and research institutions, in 2006 took the lead in developing the first ideas for a German nationwide CNT network, and then also entered into negotiations with the Federal Ministry of Education and Research (BMBF). The basic idea was to cover all basic steps in technology, material manufacturing, chemical functionalization and dispersion, and to simultaneously fuse promising fields of application with the appropriate components, systems and final processors.

The underlying innovation community of this practice field was analyzed in detail in a case study in Sect. 3.9. In addition to the findings described in this study, the following chapters focus on the experiences and results achieved from the companies and research institutes of the practice field and the active involvement into the evolution of the innovation community. A special interest will be on factors that helped the carbon nanotubes community to evolve and to continuously develop further. Different levels of the innovation system (see Chap. 1) and phases of the community development can serve as parameters for this analysis.

4.4.3 The Formation of the CNT Innovation Community

Between the initial idea and the actual start of the first projects, a 2-year planning, application and acquisition phase took place during which the primary issues were strategic orientation and identification of appropriate community members. Because of the novel, R&D status of the technology and the long and interdisciplinary value chains, the success of the project notably depended on the intensive

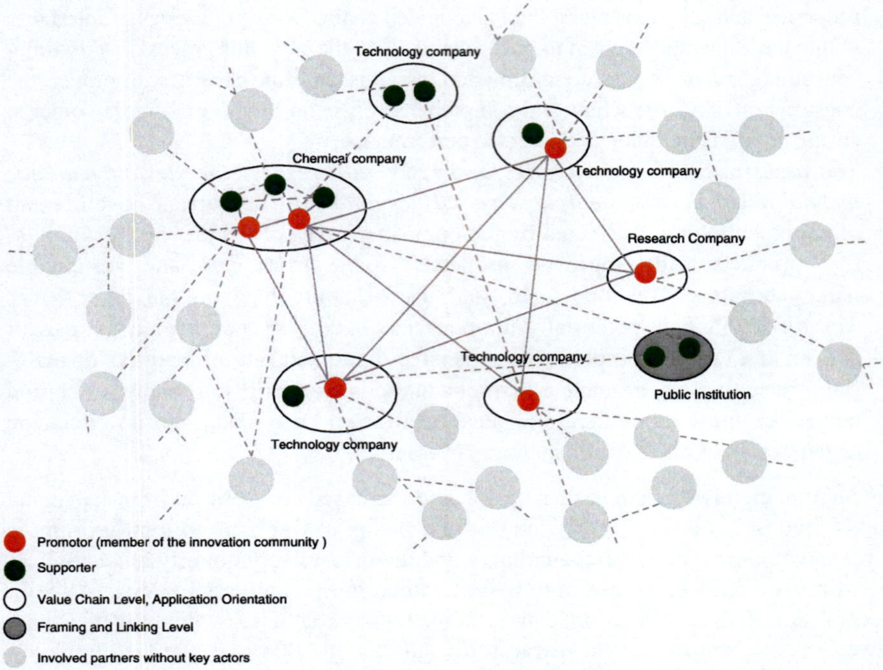

Fig. 4.8 Key actors of the carbon nanotubes community and their integration in the innovation system (Source: Author)

cooperation of several relevant key individuals, who in the present case constitute the innovation community, and who took responsibility for recruiting additional partners. Using the "snow ball" method, it was possible to attract additional potential partners for the project along the entire value-added chain. Figure 4.8 shows the close integration of the promotors (red dots) and supporters (green dots), the various participating companies (grey circles), and the resulting CNT network.

As described in Sect. 3.9, the innovation community consists of six promotors. Nine additional supporters belong to the greater network which the innovation community is embedded in. As described in Chap. 1, promotors often belong to different levels (framing and linking, value chain and company level) of an innovation system. In the case of the best practice field, the key actors (see the analysis of promotors and supporters in Sect. 3.9) can be attributed to the levels of the innovation system, as indicated by the white and grey coding in the figure below.

The figure shows that the community involves promotors and supporters from the framing and linking level as well as from the value chain level that in this case is application oriented. From the accompanying work in the best practice field it is known that there are more actors on the respective company level due to the own interests those partners have within their specific fields of competences.

Based on this sustainable strategic project planning process and the great potential for CNT technology identified in the studies, in 2008, the BMBF approved the research network 80 million euros of funding. Half of that sum was to be provided by the BMBF, the other half by the project partners. This large proportion of self-financing illustrates the network members' motivation to actually carry out the project successfully.

By the time of the official kick-off in April 2008, the initiators of the idea were able to recruit 45 industrial partners and 25 other partners from the research community for the project. Within the project, 18 interconnected, simultaneous R&D projects were established, which addressed both basic issues and practical applications. In the basic research-oriented interdisciplinary areas, the following four topics are being addressed:

- Production and catalysis
- Functionalization and modification
- Dispersion, conditioning and semi-finished products
- Health and safety aspects of CNTs

The results from these single projects are continuously being transferred to the application projects and vice versa (see Fig. 4.9). The projects are grouped into four main areas, in which the focus is placed on the following exemplary applications:

- *Energy and environment*: Fuel cell and battery technology, energy-efficient membranes for desalination and gas separation.
- *Mobility*: Composite materials for motor vehicles and aerospace.

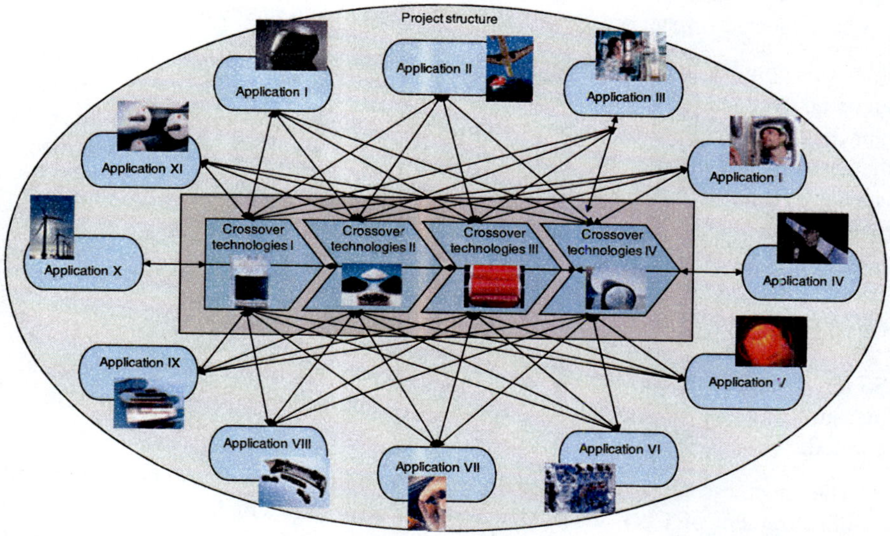

Fig. 4.9 Project structure of the CNT innovation project (Source: Illustration by Author, based on www.inno-cnt.de)

- *Lightweight construction*: Structural elements for automotive engineering and aeronautics applications.
- *Electronics*: Displays, x-ray and microwave generators, photovoltaic applications, high-resolution electron beam instruments.

This matrix project structure, linking basic research to practical industry issues, constitutes one of the strengths of the network. The network members' hopes of creating a relevant framework for the swift economic exploitation of new CNT materials technologies must also be seen in this context.

4.4.4 Potentials for Further Development of the CNT Community

The great challenge facing the established CNT network in Germany is the development and extension of German leadership in CNT technology, establishing global reach and lead markets for CNT applications through the close integration of technology and application development. To achieve these ambitious goals, the innovation community has used a variety of methods and measures which have a supportive effect both for the development of the community and for the innovation project itself. In addition to the continuous monitoring of global developments and potential markets of CNTs, and the identification of additional strategic partners, the following factors are relevant:

- The technical, economic and social importance of innovation sources
- Full coverage of the technology and value-added chain
- Very high scientific and technical quality
- Economic importance
- The expected impact on society, the environment and the economy

In this context, and in view of the dynamics of CNT technologies, a continuing focus on new fields of application is very useful. To that end, the BMBF in 2010 launched a new round of tenders to develop the innovation project further, both thematically and in terms of its partnership structure. As a result, nine new projects were integrated into the overall project, and expanded to include the appropriate partners.

Measures and Methods for Developing the Community

So far, over the entire course of the innovation project, the following methods and measures for the design and support of the innovation community have been applied:

- The most essential tools to support the community in the construction and development of CNT innovation projects were and will in future remain the use of *government funding programmes* of the BMBF and the *formation of strategic alliances*.

- To support efficient and effective activity in the community, regular *project meetings* are held at all levels, including task-specific *workshops, coordination tasks*, and the *transfer of knowledge*. Once a year, all members meet to discuss results from individual projects and exchange information with other network partners. Due to the size of the community, these annual meetings have been shown to be very useful and stimulating for further work.
- Several committees equipped with various tasks and responsibilities have been set up for the *coordination* of the overall project. These include the coordinating committee as a managing body for all general issues, the steering committee which is responsible for performance evaluation and monitoring of the project objectives, and the scientific board which is responsible for the screening, discussion, consultation and communication of scientific results, even beyond the boundaries of the network.
- *External communication* (marketing, interviews, etc.) is seen as an important factor in dealing with the public. In addition to events, the CNT alliance provides news and information on its joint Internet platform. For this purpose, the resources of the participating partners, particularly those of large corporations, are utilized. These measures are designed to dispel possible unfounded fears in the public regarding CNTs, by means of a deliberate and methodical flow of information.
- Within the project InnoCo, the initiative for new product ideas and the evaluation of existing concepts in their application context are facilitated by the formation of a *dynamic scenario* for CNT technology. With this IT-based method, various application cases can be tested in terms of probability, based on the general development of technology, so that market assessments can be made at an early stage. For example, the initial question could be: "What is the probability that CNTs will be successfully introduced into application area X?" The required market characteristics and potential processing technologies can then be directly derived. The big advantage of this method is the reduction of uncertainty along the value chain through the use of systematic assessment approaches and the integration of all key factors in all stages of development.
- Also addressed within the InnoCo-project developing a common understanding on the future development of CNTs is supported by the use of a *technology roadmap* which is continuously updated in the light of available skills, production technologies, products and markets (see Fig. 4.10). From this, it is possible to define strategic orientations in various industries or markets and ascertain general technological lines of development (e.g., which issues occur frequently in the same context, how risk issues are tied to specific applications, CNT formulations, etc.). Furthermore, the need for material properties and processing technologies can be derived from the markets and products. This method proved to be an appropriate network management tool for the continuous development and strategic alignment of a complex innovation network like the carbon nanotube community. The coordination within long value chains – from basic research to consumer applications – often results in misunderstandings regarding needs and requirements due to a different technological understanding. Bringing

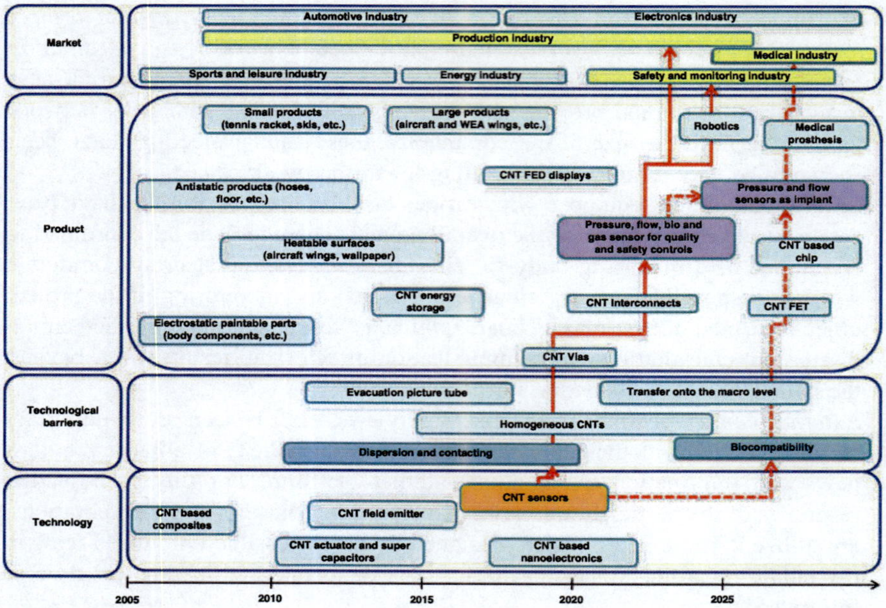

Fig. 4.10 Example of a section of a CNT roadmap (Source: Illustration by Author, based on www.inno-cnt.de)

the technological and market aspects of all involved partners into one roadmap helps to overcome those misunderstandings and allows the innovation process to become more efficient. The technology roadmap supports decisions regarding possible new application fields and thereby also the process of opening the network for new industrial and research partners.

4.4.5 Lessons Learned

Through the constant interlocking of technological push and market pull approaches, development times can be significantly reduced, and success rates in the market increased. Especially in networks that deal with radical innovations and have long value-added chains, encouraging cooperation between all key actors in the innovation community is of fundamental importance. With regular project and coordination meetings, the exchange of information in interdisciplinary and inter-project workshops, the creation of interdisciplinary scenarios and technology roadmaps, and the successful establishment of CNTs on the world market by German companies can be assessed as very realistic. The networked project structure minimizes the risk of development failures due to its very narrow market focus (a typical phenomenon in technology-driven networks). The applied methods for the management of the carbon nanotubes network differed depending on the phase of the focal innovation process. While the main focus in the maturation and initial phase was on setting up an appropriate network of CNT partners throughout Germany, in the subsequent phase, the focus changed to a more market- and application-driven view. The methods presented in this practice field reflect those different focuses.

Due to the methodologically supported procedure, it is to be expected that the CNT community will implement efficient and targeted applications in the next few years, and thus achieve successful market penetration with novel nanotechnological systems and materials with innovative features in a number of different market segments.

4.5 Conclusions for the Management of Innovation Communities

Severin Beucker

In Sect. 4.1, it was concluded that the managerial support of innovation communities depends on the level of the innovation system the community is involved in and on the development phase of the community. To assess these dependencies and to draw conclusions from similarities in the three practice fields, first, the major characteristics and differences of the innovation communities in the practice fields are compiled. In a second step, these results are assessed and structured according to their potential for managerial support of innovation communities on different levels of the innovation system and in different phases of the community development.

As described in Sect. 4.1, the practice fields were selected to cover various types of innovation projects and networks. Over the course of the project InnoCo and the work in the practice fields, the originally intended objectives were often adapted and redefined more precisely according to the needs of the community. Table 4.4 summarizes the redefined objectives in the practice fields, the applied measures of community development and the effects those measures have on different levels of the innovation system. The redefined objectives also indicate the phase of community development the practice field was in at the time the analysis took place. All three phases of community development (initiation, development and reconfiguration) are contained in these three practice fields (Table 4.4).

The table shows that the practice fields differ in terms of the objectives of innovation and the measures applied. All practice fields and their innovation communities are embedded in larger, more complex innovation networks and systems that relate to various levels of the innovation system, ranging from the company level over the value chain to the framing and linking level (see Fig. 1.2, Chap. 1).

From the comparison of applied measures in the different phases of community development, it can be concluded that:

- The development of new innovation communities (ICT: Connected Energy) strongly profits from measures that enhance and provide spaces for communication and organizational structures such as working groups with regular meetings and workshops, and a defined process for the development of project ideas and proposals. At the same time, these communities require methods and tools for the development of strategies and objectives that support the evolution of a joint understanding and shared goals and visions of the new community (e.g., scenario and roadmapping techniques).
- Already existing communities (New materials: CNT) seem to have greater need for measures that support the management and assessment of knowledge, skills and options for further development, such as technology roadmapping, the identification of value chains, or risk and reward assessment.

Table 4.4 Comparison of objectives, applied measures and their effects on the innovation system of the practice fields

InnoCo practice field and project	Objective of the project	Applied measures of community development	Effects on different levels of the innovation system
ICT: Connected Energy	Initiating and establishing a new innovation community	Selection of the Connected Living association (CLA) as network platform CLA General Assembly and 2-day strategy workshop to identify "hot" topics Setting up an "energy" working group within the CLA Development of common understanding on topics and objectives Definition of scenarios for distributed energy management architecture Working group meetings and workshops Joint development of a project idea, based on scenarios in a "living document" (compile knowledge and develop a common understanding)	*Framing and linking level*: Integration in Connected Living association, exchange with industrial and governmental institutions; acquisition of government funding *Value chain level*: Initiation of an innovation community and development of a common understanding of topics and objectives; Setting up a formal project for developing products and services for energy management in households; Institutionalization of the innovation community as a project team
		Scenario development and roadmapping for smart home networks Development of project ideas and redefinition of objectives and tasks Drafting a project outline and applying for government funding Realization of the government-funded project "Connected Energy-SHAPE" Identification of business model and value chains	*Company level*: Internal marketing and improved understanding
New materials: CNT	Developing and improving an existing innovation community	Analysis of existing network Acquisition of knowledge on possible applications of material Dynamic Scenario development Technology roadmapping	*Framing and linking level*: Integration in CNT Initiative, improved relationships with key persons in governmental institutions and R&D organizations; improved public relations

(continued)

Table 4.4 (continued)

InnoCo practice field and project	Objective of the project	Applied measures of community development	Effects on different levels of the innovation system
		Identification of value chains	
		Assessment of opportunities and risks in terms of markets, technological development, and social acceptance	*Value chain level*: Integration of competences along the value chains and products for CNTs; improved trust and cooperation between key players
			Company level: Improved understanding of technology and value chains
Renewable energies: DSSC	Reconfigurating and extending an innovation community	Clarification of interests Development of common understanding on topics and objectives	*Framing and linking level*: Exchange with R&D network
		Definition of cooperation, reaching a non-disclosure agreement Strategic partner search, Identifying matching partners	*Value chain level*: Connecting relevant partners for developing profitable applications and products of DSSCs; shared objectives of partners
		Definition of development projects, work plans and goals Setting up a producer-user community Identification of value chains and market potentials	*Company level*: Greater awareness of customer and market needs

Source: Author

- Communities that are in the process of reconfiguration (Renewable Energies: DSSC) have higher demand for measures that help to structure and clarify objectives and interests of community members and thereby help to find and integrate the right partners. Also, these communities seem to profit from measures that strengthen trust in each other like non-disclosure agreements or the definition of cooperation structures.

In total, all communities seem to benefit from good project management which implies aligned work plans and structures, reliable schedules and open and trustful communication.

In addition to these findings, the table shows that innovation communities that are more active on the framing and linking level of the innovation system (ICT:

Connected Energy and New materials: CNT) have a greater need for managerial measures that align the communities' objectives with those of other interest groups like policymakers or funding agencies. Such measures typically include scenario development or roadmapping techniques with a strategic focus.

On the value chain level, the three practice fields show similarities. All of them seem to rely on measures to identify and complete value chains and market potentials. Smaller differences between the fields might result from their weaker or stronger orientation on technologies or products.

Finally, the table does not offer any ultimate insights on measures at the company level. All practice fields seem to see the value and the need of the innovation community for convincing and integrating internal promotors and decision makers, and for organizing information and knowledge transfer, although the need for such measures is probably more present in communities that incorporate promotors from bigger organizations (ICT: Connected Energy and New Materials: CNT) These in turn can use the innovation community and its activities to promote the innovation in their own organizations as well.

References

Bierter W, Fichter K (2005) Business-innovations-workshop: Startschuss für Innovationsprojekte. In: Fichter K et al (Hrsg) Nachhaltige Zukunftsmärkte. Orientierungen für unternehmerische Innovationsprozesse im 21. Jahrhundert. Metropolis, Marburg

Chesbrough H (2003) Open innovation. Harvard Business School Press, Boston

Fichter K (2005) Interaktives innovationsmanagement. In: Fichter K et al (eds) Nachhaltige Zukunftsmärkte. Metropolis Publishing Marburg, S. 239–268

Future Technology Consulting (ZTC) of the VDI Technology Centre GmbH (2009) Kohlenstoff-Nanoröhren: Potential einer neuen Materialklasse für Deutschland – Technologieanalyse. Zukünftige Technologien, 79, Düsseldorf

Koker T (2006) Konzeption und Realisierung einer neuen Prozesskette zur Integration von Kohlenstoff-Nanoröhren über Handhabung in technischen Anwendungen. Universitätsverlag Karlsruhe, Karlsruhe

Rohrbeck R, Döhler M, Arnold HM (2007) Combining spin-out and spin-in activities - the spin-along approach. ISPIM 2007 conference, Warsaw

Chapter 5
Hands-On Methods for Developing Innovation Communities

Abstract Based on the empirical findings of a survey of innovation networks (see Chap. 2), the results of the five case studies presented in Chap. 3 and the practical experience in developing innovation communities outlined in the previous chapter, this chapter presents hands-on methods for initiating and developing innovation communities.

Section 5.1 presents methods and instruments that innovation networks use. The findings are based on a sample of 107 innovation networks in Germany. While this first section describes methods and instruments that are used by innovation networks in general, Sect. 5.2 is focussing specifically on the methods and measures employed by innovation communities. The chapter differentiates the methods and measures, which have been used in the five innovation cases (see Sects. 3.5–3.9), according to five basic tasks in the management of cooperative innovation projects: (1) Process Support, (2) Foresight, (3) Idea Generation, (4) Technology and Market Analysis and (5) Stakeholder Interaction.

Section 5.3 brings together empirical and practical knowledge as to which methods and measures can be used by innovation and network managers for the initiation and development of innovation communities. The findings are summarized in a "toolbox".

Within the InnoCo project, many tools have been tried and tested in innovation community development. Special attention has been given to two methodological approaches which are as important for the development of innovation communities as they are methodologically complex. These are: systemic innovation coaching of innovation communities, and Delphi-based scenario development in innovation networks. These methods are presented in Sects. 5.4 and 5.5.

K. Fichter and S. Beucker (eds.), *Innovation Communities*,
DOI 10.1007/978-3-642-22128-6_5,

5.1 What Methods and Instruments Do Innovation Networks Use? Findings From a Survey

Alexandra Rese and Daniel Baier

Acknowledgement The authors would like to thank Nicolai Sand for his support in developing the questionnaire and assembling the data.

5.1.1 Empirical Findings on Methods Used in Product Development

Expectations regarding collaboration between separate companies in new product development (NPD) include the reduction of development time, sharing increasing costs or gaining access to new technologies, knowledge and skills (Bruce et al. 1995). However, empirical evidence suggests that often cooperations fail to meet expectations (Veugelers 1998). Between 30% and 50% of collaborations have been found to be unsuccessful and instable (see the studies listed by Das and Teng 2000). Several problems have been identified, for instance the significant costs of starting, managing and monitoring the collaborative project (Farr and Fischer 1992; Das and Teng 2000). Regarding the success of product development across firm boundaries the absorptive capacity of the partners, that is the ability to acquire, evaluate and apply new (external) information, has been argued to be important (Cohen and Levinthal 1990). Empirical studies have shown that additional market information can enhance NPD decision making and consequently NPD performance (Adams et al. 1998). Due to high levels of market uncertainty and a rapidly changing environment, information becomes even more important when developing innovative products. Their development requires permanent information processing over the entire development process (Teece et al. 1997, 520). Furthermore, in inter-organizational product development, according to Lawson et al. (2009), even more coordination and management is necessary than in internal product development.

Here, methods and tools (e.g. brainstorming, focus groups, or more advanced design tools such as quality function deployment, or QFD) which have been introduced to improve a company's performance in NPD could help innovation networks to become more successful. These methods and tools have been classified according to underlying NPD problems in the four areas idea generation, product optimization, marketing mix optimization and prediction of success (Nijssen and Lieshout 1995). They may support the creation and management of (new) information, the early identification of problems, the maintenance of targeted product development efforts, or the collection of data for sales force management (Nijssen and Lieshout 1995; Ittner and Larcker 1997; Nijssen and Frambach 1998). Apart from the creation and provisioning of (new) information, specific methods and tools can facilitate and accelerate communication within an innovating company. For example, for intra-company product development, Griffin (1992) showed that the use of QFD helped to increase the communication within a development team. Despite the potential loss of proprietary information, an efficient flow of information is seen as a requirement for product development with different partners (Bidault and Cummings 1994). In addition, some methods and techniques (e.g. joint workshops, focus groups) could also function as socialization mechanism helping to connect key actors of different companies. This may result in closer and more intensive interaction based on mutual trust and respect (Lawson et al. 2009). Again, Griffin (1992) could show that the use of QFD also encouraged teambuilding in internal product development, in terms of the integration, motivation, commitment, and cooperation of staff members (see also Griffin and Hauser 1992). Altogether, empirical studies have confirmed that NPD methods and tools can have a positive impact on companies' performance (Nijssen and Lieshout 1995; Ittner and Larcker 1997; González and Palacios 2002; Kaynak 2003).

Despite the benefits, except for brainstorming, NPD methods and techniques are rarely used in companies, with an average usage ratio of about 30%, according to empirical studies (Mahajan and Wind 1992; Nijssen and Lieshout 1995; Gaul and Volkmann 2000). Regarding the frequency of employment of single methods and instruments there is no clear picture. One explanation could be the different classifications and hierarchies of methods and tools for improving product development processes that have been developed and empirically investigated (González and Palacios 2002). In the survey of Mahajan and Wind (1992) on US "Fortune 500" companies, focus groups (68%), limited rollouts (42%), and concept testing (26%) were the most frequently used measures in NPD. In contrast, in Dutch industrial enterprises, Nijssen and Lieshout (1995) found brainstorming and concept testing (each 61%) to be widely used, closely followed by in-home use tests (60%). In a study of predominantly British firms (Driva et al. 2000), computer-aided techniques (CAD, CAM, CAE), cross-functional teams and brainstorming were most popular. The study of Yeh et al. (2010) of Taiwanese high-tech firms yielded similar results, except in the case of brainstorming. Instead, project management, specific design software and product data management (PDM) were frequently used when developing new products. In the PDMA 2003 best practice study, Barczak et al. (2009) found that in about half of the investigated radical innovation projects, enabling tools and techniques like teambuilding or IT tools had been used. Collaborative partnerships for NPD existed in 25% of the incremental projects and in 50% of the radical projects.

5.1.2 Implications for the Use of Methods and Instruments in Innovation Networks

Regarding innovation networks and the special focus on innovation communities as groups of innovators that promote specific innovation projects (see Chaps. 1 and 2 for details), methods and instruments were selected that could support the formation and development of innovation communities. The methods and instruments should of course benefit the NPD process, but also address teambuilding issues. If the method or instrument is applied in a teamwork setting, this can be a way of innovation network members getting to know each other and developing a common understanding. Because it is assumed that innovation communities are particularly important in the unstructured early phases of the innovation process ("fuzzy front end"), and in emerging technology fields (Fichter 2009), we especially concentrated on methods for search field analysis (forecasting), idea generation and process support.

In the following, the empirical findings of the survey described in Chap. 2 are presented. The data is evaluated in terms of categories for successful and less successful innovation networks, for radical and incremental networks, and for networks with and without an innovation community.

5.1.3 Empirical Findings in Innovation Networks

Altogether, an average of three methods or instruments was used in the examined networks (mean value: 3.02). The number of methods or instruments used ranged from 1 to 11. In 61.7% of the networks, up to three methods or instruments were used. Only a small proportion of networks (9.6%) used six or more methods. Seven networks did not use any of the prompted methods or instruments at all.

As regards methods and instruments used in the innovation networks, expert workshops (51.1%) were the most popular, followed by innovation workshops (39.4%) and creativity and idea workshops (36.2%). The focus seems to be on rather simple, easy-to-use methods. While creativity techniques are employed to some extent (24.5%), more profound methods (e.g. functional analysis, TRIZ or SIL) are seldom used. Outside opinions are integrated by expert and customer workshops (e.g. focus groups), but the lead user method that helps to identify and integrate lead users into the NPD process also rarely comes into action. For planning and coordinating technological developments in a broader framework, technology roadmapping (30.9%) as well as trend analysis and trend monitoring (24.5%) are utilized to some extent in the networks. Again, the use of more complex methods like scenario analysis and Delphi techniques was infrequent.

As regards process support, about a quarter of the networks are trying to support a common understanding by defining key terms and using semantic networks and ontologies. In contrast, an innovation coach giving advice in the product development process is only found in 12 networks (see Table 5.1). Furthermore, the "stage gate model" that divides development processes into a number of distinct stages separated by quality control checkpoints or decision "gates" (Cooper 1990, 2008) was hardly ever used (4.3%). This usage rate is even lower than it is assumed to be in individual firms: In 2002, González and Palacios found usage rates of 22.2% for the stage gate model in a survey of Spanish companies (González and Palacios 2002).

Compared to less successful innovation networks, successful networks generally tend to use specific methods and instruments to a greater extent and amount (see Table 5.2). In successful innovation networks, expert workshops and innovation

Table 5.1 Descriptive statistics of methods and instruments used in innovation networks (multiple answers)

Forecasting methods		Idea generation	
Technology roadmapping	30.9%	Expert workshop	51.1%
Trend analysis and trend monitoring	24.5%	Innovation workshop	39.4%
Scenario analysis	7.4%	Creativity/idea workshop	36.2%
Delphi method	1.1%	Creativity techniques	24.5%
		Customer workshop	22.3%
Process support		Problem solving methods, e.g. TRIZ	9.6%
Definition of key terms, semantic networks/ontologies	25.5%	Functional analysis/object models (Hubka)	6.4%
Support/process consulting by innovation coach	12.8%	Lead user method	3.2%
Stage gate model	4.3%	SIL method (systematic integration of elements of solutions)	2.1%

Source: Empirical study described in Chap. 2

Table 5.2 Comparison of methods in successful and less successful innovation networks (multiple answers)

Methods	Successful innovation networks ($n = 57$)	Less successful innovation networks ($n = 37$)
Forecasting methods		
Technology roadmapping	31.6%	19.7%
Trend analysis and trend monitoring	29.8%	16.2%
Scenario analysis	7.0%	9.1%
Delphi method	1.8%	0.0%
Idea generation		
Expert workshop	56.1%	43.2%
Innovation workshop	43.9%	32.4%
Creativity/idea workshop	31.6%	43.2%
Creativity techniques	22.8%	27.0%
Customer workshop	22.8%	21.6%
Problem solving methods, e.g. TRIZ	8.8%	10.8%
Functional analysis/object models (Hubka)	7.0%	5.4%
Lead user method	1.8%	5.4%
SIL method (systematic integration of elements of solutions)	1.8%	2.7%
Process support		
Definition of key terms, semantic networks/ontologies	26.3%	24.3%
Support/process consulting by innovation coach	12.3%	13.5%
Stage gate model	7.0%	0.0%
Sum of used methods	178	105
Mean value of used methods	3.12	2.86

Source: Empirical study described in Chap. 2

workshops are carried out more often than in less successful innovation networks. The stage gate model is only used in successful innovation networks. In addition, in successful innovation networks, more use is made of trend analysis and trend monitoring.

By dividing innovation networks according to the degree of innovation and analyzing their use of tools and methods, it becomes evident that a significantly larger number of methods is used in radical innovation projects than in incremental projects. This particularly applies to methods of idea generation like innovation workshops, but also to creativity and idea workshops, expert workshops and customer workshops. Of the more complex methods for gathering new product ideas, exclusively radical innovation networks use the lead user method to identify needs, trends and lead users in an extended process. Furthermore, as was the case for successful networks (see Table 5.3), only the radical innovation networks are making use of the stage gate model.

Table 5.3 Comparison of method use in radical and incremental innovation networks (multiple answers)

Methods	Radical innovation networks (n = 38)	Incremental innovation networks (n = 52)
Forecasting methods		
Technology roadmapping	35.7%	16.9%
Trend analysis and trend monitoring	28.6%	21.2%
Scenario analysis	9.5%	5.8%
Delphi method	0.0%	1.9%
Idea generation		
Expert workshop	61.9%*	42.3%
Innovation workshop	61.9%**	21.2%
Creativity/idea workshop	47.6%*	26.9%
Creativity techniques	26.2%	23.1%
Customer workshop	31.0%*	15.4%
Problem solving methods, e.g. TRIZ	9.5%	9.6%
Functional analysis/object models (Hubka)	7.1%	5.8%
Lead user method	7.1%*	0.0%
SIL method (systematic integration of elements of solutions)	4.8%	0.0%
Process support		
Definition of key terms, semantic networks/ontologies	31.0%	21.2%
Support/process consulting by innovation coach	9.5%	15.4%
Stage gate model	9.5%***	0.0%
Sum of used methods	160	123
Mean value of used methods	3.81^{+}	2.38

Source: Empirical study described in Chap. 2

***Significant with $p = 0.00$; **Significant with $p < 0.01$; *Significant with $p < 0.05$; $^{+}$Significant with $p < 0.1$ (two-sided test)

As was the case with radical innovation networks, innovation networks with an innovation community attached had a significantly higher usage of methods and instruments compared to networks without an innovation community. That is hardly a surprising result because 17 of the 23 innovation networks with an innovation community (73.9%) were working on radical innovation projects. Forecasting methods in particular, as well as trend analysis, trend monitoring and technological roadmapping, were used significantly more frequently in innovation networks with an innovation community (see Table 5.4). In addition, innovation networks with an innovation community were supported much more often by innovation coaches than those without an innovation community (see Sect. 5.4 for more details).

The use of methods also reflects the different perspectives that most likely exist as soon as members of different firms and industries are present. For instance, in those networks with an innovation community, external knowledge in the form of workshops was integrated significantly more often than in those without (see

Table 5.4 Comparison of method use in innovation networks with and without innovation community (multiple answers)

Methods	Innovation networks with innovation community (n = 20)	Innovation networks without innovation community (n = 71)
Forecasting methods		
Technology roadmapping	50.0%*	26.8%
Trend analysis and trend monitoring	55.0%**	15.5%
Scenario analysis	10.0%	7.0%
Delphi method	0.0%	1.4%
Idea generation		
Expert workshop	65.0%	47.9%
Innovation workshop	55.0%	35.2%
Creativity/idea workshop	45.0%	33.8%
Creativity techniques	15.0%	16.8%
Customer workshop	50.0%**	15.5%
Problem solving methods, e.g. TRIZ	10.0%	8.5%
Functional analysis/object models (Hubka)	5.0%	5.6%
Lead user method	5.0%	2.8%
SIL method (systematic integration of elements of solutions)	5.0%	1.4%
Process support		
Definition of key terms, semantic networks/ontologies	25.0%	26.8%
Support/process consulting by innovation coach	25.0%	9.9%
Stage gate model	10.0%	2.8%
Sum of used methods	86	190
Mean value of used methods	4.30	2.69

Source: Empirical study described in Chap. 2
**Significant with $p < 0.01$; *Significant with $p < 0.05$ (two-sided test)

Table 5.4). This indicates that innovation communities favour methods that increase their boundary permeability.

5.1.4 Discussion

The empirical results have shown that in innovation networks in particular, methods and instruments that allowed face-to-face encounters with the group and with external experts (e.g. expert workshops, innovation workshops, and creativity and idea workshops) are widely used. In addition, the systematic search for innovation was supported by methods such as technology roadmapping, trend analysis and trend monitoring. About a quarter of the networks reinforced their common understanding by defining key terms and using semantic networks and ontologies.

When comparing the different classes of innovation networks, one notable point was that successful, radical projects with a supportive innovation community (at least by tendency) generally employed more methods and instruments. It could not be shown that the use of one specific method had significant effects on the performance of an innovation network, but the reasons should be investigated in more detail. One explanation is difficulties arising in information processing and distribution (Griffin and Hauser 1996; Adams et al. 1998); other explanations might be that in high-tech markets in particular, information is frequently "fuzzy" and of a qualitative nature, and future markets are difficult to predict (see Veldhuizen 2008 for more details).

In radical innovation networks in particular, the use of methods for idea generation was significantly higher than in incremental projects. As regards process support, the "stage gate model" that provides a roadmap for the development process and recommends activities for each stage (Cooper 1990) was found in successful and radical innovation networks in particular, and to some extent in innovation networks with an innovation community. Innovation networks with an innovation community in particular made use of methods that integrate different (external) perspectives, such as trend analysis, trend monitoring or customer workshops.

It was not investigated in the empirical study whether the use of methods or instruments had a positive effect on group interaction and group identity. This was done in the case studies which are explained in more detail in the following.

5.2 What Methods Do Innovation Communities Use? Case Studies Results

Klaus Fichter

While Sect. 5.1 presented the empirical findings in terms of methods and instruments used by innovation networks in general, this section focuses specifically on the methods and measures used by innovation communities.

5.2.1 *Questions and Methodology*

The findings presented below are based on the five case studies presented in Sects. 3.5–3.9. For each of the five case studies involved, interviews were conducted with four innovation community members, in which they were asked to answer seven questions about the use of methods and measures. Four of the questions concern the measures and methods used for the management of innovation projects, as follows:

- Which measures and methods were used in the overall project?
- What role did the ascertainment and structuring of technological, market-oriented and social trends play in the overall project?
- How were you able to develop a common understanding of the object of the technology and innovation?
- Where do you see further need for action or improvement in the management of the joint innovation project?

Three additional questions were asked related to the measures and methods applied in innovation community development:

- Which methods particularly support cooperation between the people you have previously identified as key players?
- Which methods particularly augment the group composition of the community?
- Which methods particularly reinforce group cohesion and team spirit?

The answers to these open questions were transcribed and analyzed using MaxQDA text analysis software. It helped identify, for instance, how many and what types of methods and measures had been used in each case. Identical answers submitted by more than one interviewee were counted only once.

The following aspects and their relevance for the development of innovation communities were considered in the assessment of types of methods and measures:

What is the functional focus of the method or measure? Here, four categories are distinguished:

- Information dissemination and training
- Information gathering and analysis

- Organization of the innovation project
- Interaction and personal encounters

To what extent does the method or measure contribute to the development of the innovation community?

Here, the impact of each methods was categorized as "high", "medium" or "low".

To which aspect of the performance of an innovation community does the method or measure contribute?

Here, we refer to the concept of teamwork quality of innovation communities introduced in Sect. 3.2. This concept distinguishes three key aspects that influence the performance of an innovation community:

- The composition
- Identity
- Quality of interaction of the innovation community

Does the use of the method or measure require methodological expertise or knowledge?

One key result in the findings presented in Sect. 5.1 is that easy-to-use methods are more frequently applied in innovation networks, while complex methods are applied less. The complexity of a method is generally correlated with the methodological expert knowledge that is required to use the method effectively.

Which tasks of innovation management do the methods and measures support?

Finally, the analysis differentiates the methods and measures that have been used in the five innovation cases, according to basic tasks in the management of cooperative innovation projects:

(a) Process support: This task refers to the organization and management of an innovation project. It comprises tasks like the identification and selection of partners, the development of joint agreements with cooperation partners, the initiation and management of joint projects, establishing new networks, coaching of the innovation team, and various project-internal forms of communication and collaboration (e.g. meetings, telecommunication, working groups, workshops and excursions).
(b) Foresight: This task comprises activities like trend analysis, trend monitoring, scenario development and roadmapping activities.
(c) Idea generation: This task includes various forms of generating and collecting innovation ideas.
(d) Technology and market analysis: This task covers the analysis of relevant company environments. This can comprise the analysis of emerging technologies, intellectual property rights, markets or competitors.
(e) Stakeholder interaction: This task relates to the communication and cooperation with relevant stakeholders in the innovation process, comprising issues of user integration and communication with stakeholders like venture capitalists, funding agencies, or politicians in particular.

5.2.2 Empirical Findings

If we compile the methods and measures of the cases presented in Sects. 3.5–3.9, and arrange them according to the five task areas of innovation process management discussed above, we obtain the overview presented in Table 5.5. The number of methods and measures mentioned by the interviewees ranges from 11 to 17 in the different innovation cases – significantly higher than in the findings from the large-scale survey of innovation networks for successful and radical innovation projects (see Sect. 5.1). As Table 5.6 shows, the variety of methods and measures implemented is also considerably larger than the values determined in the survey findings. In addition to the task areas Process Support, Foresight and Idea Generation described in Sect. 5.1, a number of methods could be identified in the case studies that can be assigned to the task fields Technology and Market Analysis, and Stakeholder Interaction in particular.

As shown in Table 5.5, the majority of methods and measures were applied in the task area Process Support. Thirty-eight of a total of seventy mentioned applications could be assigned to this area. Twelve of the method applications belonged to the area of Stakeholders Interaction. Thus, some two thirds of all method applications were utilized in these two central task areas.

No correlation between the duration of the innovation process and the scope of the methods applied was identified for the five innovation cases analyzed. By the same token, no correlation could be observed between the number of key persons involved or community size, and the number of methods and measures applied.

Table 5.6 shows the methods and measures as stated by the interviewees. The task Process Support represents methods and measures for the organization and management of innovation projects and innovation networks, wherein "easy-to-use" methods and measures such as meetings, working groups and internal workshops, predominate. Strategic partner selection and innovation coaching were both rarely applied, even though they are extremely important for the development and performance of innovation communities. Excursions were also utilized in two of five case studies. This method can be highly significant for the development of an innovation community, as it can grant access to potential new partners (see the Wood Energy case in Sect. 3.8). It also strengthens community identity, for instance, by improving cohesion and enhancing exchange. Excursions have, however, virtually been ignored in the current discussion of innovation management methodology.

In Process Support methodologies, the single area requiring medium to high expertise is the initiation, adoption and implementation of formal cooperative innovation projects. In all the cases examined, formal, oftentimes funded innovation projects are used as a means of advancing innovative ideas.

All the innovation cases studied applied methods and measures in the area Foresight. The focus here is clearly on trend analysis. Two communities also used the roadmapping method. The scenarios method was only applied in the case of Connected Energy. The comparatively small number of methods

Table 5.5 Number of methods and measures used in fives cases with an innovation community

Focal innovation process		Number of key actors		Number of methods and measures used for …					
Case	Period analyzed	Key persons[a]	Members of the innovation community	Process support	Foresight	Idea generation	Technology/market analysis	Stakeholder interaction	Total
Connected Energy	2009–2011	8	4	7	3	1	3	0	14
Dye-Sensitized Solar Cells	2008–2011	8	5	7	2	0	0	3	12
Tilt Sensors	2006–2010	6	4	6	1	1	3	3	14
Wood Energy	2004–2010	15	7	10	1	0	0	6	17
Carbon Nano Tubes	2006–2011	14	6	8	1	1	3	0	13
Total number				38	8	3	9	12	70

Source: Author

[a]"Key persons" comprises supporters and promotors

Table 5.6 Methods and measures used by innovation communities in five innovation cases[a]

	Functional focus				Contribution to network and community development		Prerequisites
	Info. dissemination & training	Info. collection & analysis	Organization	Interaction and encounter	Degree of contribution to community development	Contribution to which quality aspect?	Methodological expertise required?
A. Process support							
Partner search/selection							
Identification of potential partners (CNT)		X	X	X	High	Composition, identity	Partially
Strategy workshop for partner selection (DSSC)		X	X	X	High	Composition, identity	Yes
Agreements/joint understanding							
Memorandum of understanding/ code of conduct (DSSC)			X		Medium	Identity	Partially
Cooperation agreement (TS)			X		High	Interaction	Yes
Contracts (subcontract, cooperation agreement, R&D contract etc.) (CNT)			X		High	Interaction	Yes
Joint projects							
Definition of joint development projects (DSSC)			X	X	Medium	Identity, interaction	No
Project outline (based on first "Living Document") (CE)	X	X	X	X	High	Identity, Interaction	Yes
Demonstration projects (e.g., pilot plots, model energy farm) (WE)	X		X	X	Medium	Identity, interaction	Yes
Projects (e.g. EU-funded projects) (WE)			X	X	High	Identity, interaction	Yes
Development of software and hardware architecture for the technology (CE)	X	X		X	High	Interaction	Yes

(continued)

Table 5.6 (continued)

	Functional focus				Contribution to network and community development		Prerequisites
	Info. dissemination & training	Info. collection & analysis	Organization	Interaction and encounter	Degree of contribution to community development	Contribution to which quality aspect?	Methodological expertise required?
Exchange of material and mock-ups (DSSC)	X	X		X	Medium	Interaction	Partially
Learning together in projects (WE)	X				High	Identity, interaction	No
Meetings							
Formal kick-off meeting (CE)			X	X	Medium	Interaction	Partially
Meeting with/reporting to project sponsor (CE)	X		X	X	Medium	Identity, interaction	Partially
Project meetings (CE)	X	X	X	X	High	Identity, interaction	No
Project meetings, communication (DSSC)	X	X	X	X	High	Identity, interaction	No
Project meeting (WE)	X	X	X	X	High	Identity, interaction	No
Regular status/project meetings and voting sessions (TS)	X	X	X	X	High	Identity, interaction	No
Annual meeting for the overall project (CNT)	X			X	High	Identity, interaction	No
Regular status and project meetings, coordinating sessions (CNT)	X	X	X	X	High	Composition, identity, interaction	No
Personal meeting (DSSC)	X	X		X	High	Interaction	No
Bilateral personal communication (CE)	X	X		X	High	Interaction	No
Personal or "back room" conversations (WE)	X	X		X	High	Interaction	No

Bilateral talks (TS)	X	X		X	High	Interaction	No
Telecommunication							
Internet communication (TS)	X	X		X	Low	Interaction	No
Internet platform (WE)	X	X		X	Medium	Interaction	Partially
Internet platform and Wiki (CNT)	X	X		X	Medium	Interaction	Partially
Newsletter (CNT)	X			X	Low	Interaction	No
Innovation coaching							
Process consulting, innovation coaching (DSSC)			X	X	High	Identity, interaction	Yes
Working groups							
Internal working groups (TS)	X	X	X	X	High	Identity, interaction	No
Working groups ("Renewable Resources", "Producer's Guide") (WE)	X	X	X	X	High	Identity, interaction	No
Task force on communication and external representation (CNT)	X			X	High	Identity, interaction	Partially
Workshops							
Strategy workshop of the association (CE)	X	X	X	X	High	Composition, identity, interaction	Partially
Internal innovation workshops (TS)	X	X		X	High	Interaction	Partially
Project-internal and cross-innovation workshops (CNT)	X	X		X	High	Interaction	Partially
Excursions							
Project-related excursions (CE)	X	X		X	High	Identity, interaction	No

(continued)

Table 5.6 (continued)

	Functional focus				Contribution to network and community development		Prerequisites
	Info. dissemination & training	Info. collection & analysis	Organization	Interaction and encounter	Degree of contribution to community development	Contribution to which quality aspect?	Methodological expertise required?
Excursions (WE)	X	X		X	High	Composition, identity, interaction	No
Establishing new networks							
Networks & associations (WE)			X	X	High	Identity, interaction	Partially
Establishing new producer groups (WE)			X		High	Identity	Yes
B. Foresight							
Trend analysis							
Trend analysis (DSSC)		X			Medium	Composition, interaction	Yes
Trend analysis/Trend monitoring (TS)		X			Medium	Composition, interaction	Yes
Trend analysis (WE)		X			Medium	Composition, interaction	Yes
Trend detection and analysis (CNT)		X			Medium	Composition, interaction	Yes
Scenarios							
Scenario development (CE)		X		X	High	Identity, interaction	Yes
Roadmapping							
Roadmap (CE)	X	X	X		Medium	Identity, interaction	Yes
Technology roadmap (DSSC)	X	X	X		Medium	Identity, interaction	Yes

C. Idea generation							
Workshops							
Lead user workshops (TS)		X		X	High	Composition, interaction	Yes
Scientific community with external experts (CNT)		X		X	High	Interaction	No
D. Technology and market analysis							
Markets - competitors							
Market research, competition analysis (TS)		X			Low		Yes
Market research (CNT)		X			Low		Yes
Intellectual property							
Literature and patent watch (CNT)	X	X			Low		Partially
Intellectual property rights analysis (TS)		X			Low		Yes
IPR coordination for patent applications (CNT)			X		Medium	Interaction	Yes
Technology assessment							
Technology potential assessments (TS)		X			Low		Yes
E. User integration and stakeholder management							
Workshops							
Innovation workshop with users (DSSC)	X	X		X	High	Composition, interaction	Yes
Multiplicator workshop with added-value partners (DSSC)	X	X		X	High	Composition, interaction	Yes
Workshops with external partners (CE)	X	X		X	Medium	Composition, interaction	Yes

(continued)

Table 5.6 (continued)

	Functional focus				Contribution to network and community development		Prerequisites
	Info. dissemination & training	Info. collection & analysis	Organization	Interaction and encounter	Degree of contribution to community development	Contribution to which quality aspect?	Methodological expertise required?
Industry presentations as a group (CE)	X	X		X	Medium	Composition, interaction	No
Presentation series in the field of micro-systems technology (TS)	X	X		X	Medium	Composition, interaction	No
Discussions with experts from other industries (TS)	X	X		X	Medium	Composition, interaction	No
Information documents							
Information materials (e.g., "Wood energy cultivation") (WE)	X				Medium	Identity (for authors)	Yes
Wood energy producer manual (WE)	X				Medium	Identity (for authors)	Yes
Events and fairs							
Trade fair presentation (CE)	X	X		X	High	Interaction	Partially
Trade fair visits, presence in micro-systems community (TS)	X	X		X	High	Composition, interaction	No
Events (e.g. Wood energy day) (WE)	X	X		X	High	Interaction	No
Presentations ("travelling preacher", "word-of-mouth") (WE)	X			X	High	Composition, interaction	No
Experience, information, and infotainment (e.g. Energy bike path) (WE)	X			X	Medium	Interaction	Partially
Photo contest and exhibition (WE)	X				Medium	Interaction	No

Source: Author, based on results of the case studies (see Sects. 3.5–3.9)

[a]Which method or measure was applied in which of the five innovation cases is indicated in parentheses in the first column. The cases are abbreviated as follows: *CE* Connected energy, *DSSC* Dye-sensitized solar cells, *TS* Tilt sensors, *WE* Wood Energy, *CNT* Carbon Nanotubes

implemented in the Foresight task area might be due, among other factors, to the level of expert knowledge this instrument requires.

The task area Idea Generation merely plays a minor role in the innovation processes in question. This can be explained by the fact that the innovation projects and collaborations examined here already rely on concrete innovative ideas and technologies, implying that the extensive search for ideas is in little demand. The low significance could also be an indication of an imbalance in the appreciation of idea generation and creativity methods, at least compared to the more established Process Support and Stakeholder Interaction methodologies.

In the innovation cases at hand, the application of methods in the area of Technology and Market Analysis is relatively low. In addition to market analyses, the area includes intellectual property rights instruments. In the case of Tilt Sensors, technology potential assessment was also applied as a method.

The use of tools and measures in the Stakeholder Interaction task is significantly higher than the use of methods in Foresight, Idea Generation and Technology/ Market Analysis. The main focus here is on workshops with users and other relevant external innovation partners, and on events that facilitate communication and interaction with potential customers and decision makers. Both tasks have a medium to high significance in network and community development.

As the overview of methods and measures in Table 5.6 shows, it is not only in the Process Support area (e.g. identification and selection of partners, joint projects, meetings, coaching, working groups, workshops, excursions, etc.) that methods and measures contribute to the development and performance of innovation communities, but also in the areas Foresight and Stakeholder Interaction. Thus, many of the methods indicated in the survey are both innovation management instruments in general, and instruments than can be specifically used for innovation community development.

5.3 A Toolbox for Innovation Communities

Klaus Fichter

Both the evaluation of the network survey in Sect. 5.1 and the analysis of methodological application in the innovation communities examined in Sect. 5.2 show that for the initiation and development of promotor networks, a wide range of efficient methods and measures are available. Moreover, the testing of various instruments in the InnoCo practice projects has yielded valuable information about which methods and measures can be used to strengthen and improve the performance of innovation communities (see Chap. 4). In the following, the empirical and practical knowledge as to which methods and measures can be used by innovation and network managers for the initiation and development of innovation communities are brought together. The findings are summarized here in a "toolbox" (Fig. 5.1).

The toolbox is structured as follows:

Tools: Here, methods and measures are listed which can be used to initiate innovation communities, and/or to increase their performance. The tools are classified according to the following categories:

- Process support: Partner search and selection, joint projects, innovation coaching, establishing new networks
- Foresight

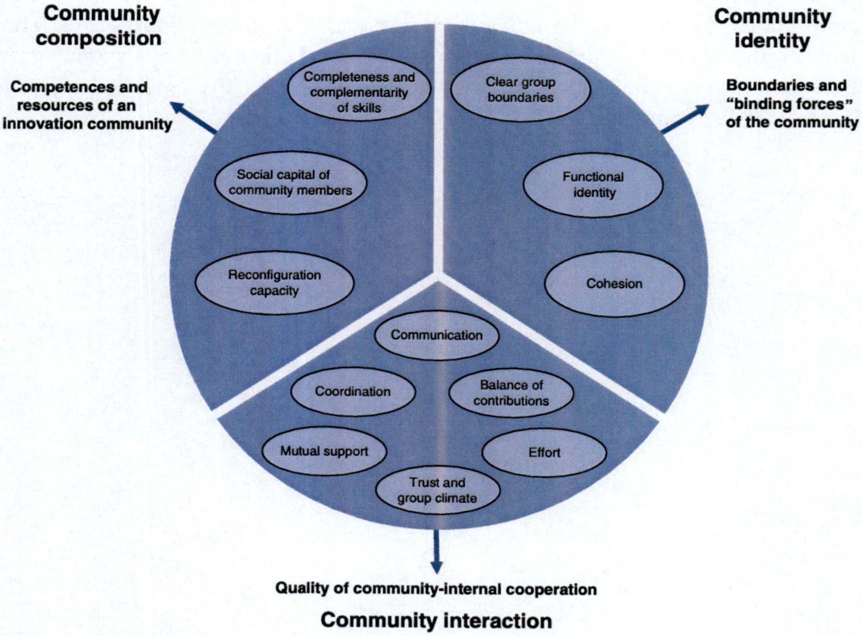

Fig. 5.1 Influencing factors on the performance of innovation communities (Source: Author)

- Idea generation
- Technology and market analysis
- Stakeholder interaction (workshops, information documents, events)

As both the practice projects (see Chap. 4) and the case studies (see Sect. 5.2) have shown, many methods and measures of innovation and project management have a positive effect on the development of the innovation community. Many of the methods and measures listed in the toolbox are already well known from innovation and project management. The advantage of these multifunctional measures is that they can be used both for the management of innovation projects and at the same time for the development of innovation communities. The various tools are therefore classified according to the categories above.

The toolbox also shows the phase of community development for which a particular method or measure is suitable. A distinction is made between three life stages of innovation communities, as developed from the case studies:

- Initiation: Initiation and establishment of a new innovation community.
- Development: Improvement of the performance of an existing innovation community.
- Reconfiguration: Basic modification or expansion of an innovation community.

Provided a method is suitable for a particular phase of community development, it is marked with an "X" in the toolbox.

The toolbox also shows which methods and measures may contribute to which quality aspects of an innovation community. Please refer to the concept developed in Sect. 3.2 on the performance of innovation communities, and the 12 influencing factors or quality aspects developed there (Fig. 5.1 and Table 5.7).

The toolbox also indicates whether the use of a specific tool requires methodological expertise.

Most of the methods and measures listed above are easy to use and self-explanatory, some require some basic methodological knowledge, and others actually require in-depth methodological expertise and experience. In the InnoCo project, many of the tools listed above have been tried and tested in the development of innovation communities. Special attention has been given to the development of two methodological approaches, which are most important to the development of innovation communities and methodologically complex. These are:

- Systemic innovation coaching of innovation communities
- Delphi-based scenario development in innovation networks

Systemic innovation coaching and the method of Delphi-based scenario management and the corresponding software tool "SEMASCORE" are major developments in the InnoCo project and will be presented in Sects. 5.4 and 5.5.

Table 5.7 Toolbox for the development of innovation communities

	Phase of community development			Contribution to network and community development	Prerequisites
Tools (methods and measures)	Initiation	Development	Reconfiguration	Contribution to which quality aspect of the community?	Methodological expertise required?
A. Process support ***Partner search and selection***					
Checklist with criteria for selecting partners	X		X	• Completeness and complementarity of competencies	No
Analysis of existing networks and organizations	X		X	• Social capital	Partially
Excursions for identifying pioneers and experts	X		X	• Completeness and complementarity of competencies • Cohesion • Communication, trust and group climate	No
Identification of suitable partners	X		X	• Completeness and complementarity of competencies • Social capital	No
Strategy workshop for partner selection	X		X	• Completeness and complementarity of competencies • Social capital • Functional identity	Yes
Contacting partners and clarification of interests	X		X	• Clear group boundaries • Functional identity • Communication, trust and group climate	Partially
Memorandum of understanding	X		X	• Clear group boundaries • Functional identity	Partially
Non-disclosure agreement	X		X		

Cooperation agreement	X	X	• Clear group boundaries • Functional identity • Coordination • Balance of contributions	Yes
Joint projects				
Definition of joint development goals	X	X	• Functional identity • Coordination	No
Working group for developing a joint project	X	X	• Functional identity • Clear group boundaries • Communication and coordination • Trust and group climate	No
Definition of key terms, semantic networks, ontologies	X		• Functional identity • Communication	Yes
Developing an outline for a joint project ("living document")	X	X	• Functional identity • Clear group boundaries • Balance of contributions • Communication and coordination	Yes
Applying for funding (government programmes etc.)	X	X	• Functional identity • Clear group boundaries • Cohesion	Yes
Cooperation contract	X	X	• Clear group boundaries • Functional identity • Coordination • Balance of contributions	Yes
Formal kick-off meeting	X	X	• Communication and coordination	Partially

(continued)

Table 5.7 (continued)

	Phase of community development			Contribution to network and community development	Prerequisites
Tools (methods and measures)	Initiation	Development	Reconfiguration	Contribution to which quality aspect of the community?	Methodological expertise required?
Project meetings	X	X	X	• Communication and coordination • Mutual support and conflict resolution • Effort • Trust and group climate	No
Project-related excursions		X		• Cohesion	No
Internal working groups		X		• Communication	No
Task force, e.g. for public relations		X		• Trust and group climate	Partially
Strategy workshop		X		• Functional identity and cohesion • Communication and cooperation • Reconfiguration capacity	Partially
Meeting and reporting to project sponsors		X		• Communication and cooperation	Partially
Annual meeting for the overall project		X		• Functional identity and cohesion • Communication and cooperation	No
Personal meetings	X	X	X	• Communication	No
Bilateral talks	X	X	X	• Mutual support and conflict resolution • Trust and group climate	No
Stage gate model		X		• Coordination	Yes
Exchange of material/mock-ups		X		• Communication • Balance of contributions • Effort	Partially
Learning together in projects	X	X	X	• Communication • Mutual support and conflict resolution	No

E-communication: E-mails, tele- or video conferences	X	X	X	• Communication	Partially
E-collaboration platforms		X		• Coordination	Partially
Project- or community-related Wikis	X	X	X	• Communication • Mutual support	Partially
Innovation coaching					
External facilitator for workshops and meetings	X	X	X	• Communication • Balance of contributions • Mutual support and conflict resolution • Trust and group climate	Yes
Systemic innovation coaching	X	X	X	• Communication and cooperation • Balance of contributions • Mutual support and conflict resolution • Trust and group climate • Reconfiguration capacity	Yes
Establishing new networks					
Initiating and establishing networks and associations		X	X	• Social capital • Reconfiguration capacity	Partially
Establishing new producer groups		X	X	• Completeness and complementarity of competencies • Social capital • Reconfiguration capacity	Yes
B. Foresight					
Trend analysis		X	X	• Functional identity • Coordination	Yes
Expert Delphi for identifying and assessing trends and their impacts		X	X	• Completeness and complementarity of competencies • Functional identity	Yes

(continued)

Table 5.7 (continued)

Tools (methods and measures)	Phase of community development			Contribution to network and community development	Prerequisites
	Initiation	Development	Reconfiguration	Contribution to which quality aspect of the community?	Methodological expertise required?
Trend monitoring		X	X		Yes
Joint scenario development	X	X	X		Yes
Sector roadmapping		X	X	• Functional identity	Yes
Technology roadmapping		X	X	• Coordination	Yes
Product roadmapping		X	X		Yes
C. Idea generation					
Internal innovation workshop		X	X	• Reconfiguration capacity • Functional identity • Coordination	Partially
Lead user method and workshops		X	X	• Completeness and complementarity of competencies	Yes
Meetings with scientists and external experts		X	X	• Social capital • Reconfiguration capacity • Coordination	No
D. Technology/market analysis					
Market research	X	X	X		Yes
Competitor analysis	X	X	X		Yes
Literature/patent watch	X	X	X	• Functional identity	Yes
Intellectual property rights analysis	X	X		• Coordination	Yes
IPR coordination for patent applications		X			Yes
Technology potential assessments	X	X	X		Yes

E. Stakeholder interaction					
Workshops					
Innovation workshop with pilot customers/users		X	X	• Completeness and complementarity of competencies	Yes
Multiplicator workshop with added-value partners		X	X	• Social capital	Yes
Workshops with stakeholders or external partners		X	X	• Reconfiguration capacity	Yes
Setting up a producer-user community		X	X	• Communication and coordination	Yes
Industry presentations as a group	X	X	X		No
Discussions with experts from other industries	X	X	X		No
Information documents					
Information materials for customers or users		X		• Functional identity	Yes
Manuals for partners in the value chain		X		• Coordination	Yes
				• Effort	
Events					
Trade fair visits	X	X	X	• Completeness and complementarity of competencies	No
Trade fair appearances		X	X	• Social capital	Partially
Conferences and events		X	X	• Reconfiguration capacity	No
Presentations ("travelling preacher", "word-of-mouth")	X	X	X	• Communication and coordination	No
Infotainment, e.g. Energy bike path		X			Partially
Contests and exhibitions		X			No

Source: Author, based on results presented in Chaps. 3, 4, Sects. 5.1 and 5.2

5.4 Coaching Innovation Communities

Willy Bierter

5.4.1 *Why Coaching?*

The complexity of an innovative technology or a new solution seems to be a key factor influencing the formation of innovation communities. The complexity obviously makes it necessary that competencies and resources for developing and implementing innovative solutions are distributed among different partners, either within a large firm or among different organizations. The distribution of necessary competencies and resources makes the different organizations and people interdependent.

We have seen in Chaps. 2–4 that innovation communities have a significantly positive effect on project performance, especially in radical innovation projects. For this to occur, the partners of an innovation community, with the support of a coach, have to prepare and create four ingredients that are indispensable to success:

1. The right climate for innovation: This relates to the atmosphere, and the state of mind in which human beings work. This atmosphere is formed and determined by the communication of the members of an innovation community and by the understanding of roles and responsibilities. A management that is under a great pressure to succeed and fears effectively blocks itself. Our urge to give everything a name, because the unknown causes unease, does not take us any further in the innovation process. On the contrary, it is experimenting and learning from mistakes that holds great opportunities here.
2. The intellectual basis of creativity, that is, all sorts of personal characteristics, behaviour and the all-embracing intelligence, is founded in the way of thinking of every human being – this is the second important ingredient. However, there is a multitude of disruptive factors or obstacles that prevent creative and open thinking. The usual mindset in the form of competition and rivalry can inhibit the momentum that innovation processes need. A further factor is our constant search for reasons. At the very moment of starting an innovation, when something new arises, we immediately start looking for arguments (for instance, why this or that does not work), instead of allowing an innovation to take its course. Thereby, we are limiting ourselves in our actions and disrupting the process. However, the greatest disruptive factor is our own identity, the way in which we are regulated by unquestioned assumptions in our minds that inhibit far-sightedness and prevent innovation.
3. The third ingredient refers to the "how" of innovation. It embraces the strategies and processes in an innovative organization, from the idea to the implementation on the market. It goes without saying that market analyses and statistics can provide orientation in order to find out what the customer wants and how he thinks. But this alone does not suffice. Only when we are able to move away

from linear thinking to some degree and open up to lateral thinking, or "thinking around the corner", will we arrive at the sources of inspiration.

4. The fourth ingredient is momentum. It is the engine for innovation, it is active inspiration, it has to do with rituals, surroundings and dialogues which keep creativity and innovation alive and promote them. In companies often only numbers count, and the strict adherence to rules – a serious dealing with ideas and concepts is painfully missing. In fact, innovation needs quite the contrary: imagination, a playful approach, an inspiring environment and if necessary, even the breaking of rules.

Not everything that counts can be counted; and not everything that can be counted counts. (Albert Einstein)

In order for these ingredients to fully take effect, besides the formal structures of cooperation (project leaders, project teams, cooperation agreements etc.), informal aspects and personal relations between innovation partners play a central role. The effectiveness of communities is strongly determined by trust and personal relations between promotors. An essential insight from the research on innovation communities is that the cohesion of groups as well as the stability and performance of the respective community are shaped by ongoing interaction and close communication processes.

Further success factors for innovation communities are that social communication and common understanding must play an important role. Especially in early phases of innovation projects, it is indispensable to work on a common basis of understanding and to exchange complex information, undocumented knowledge and know-how in personal meetings. If situations are coming to a head or if knowledge items or innovation processes are too complex or unstructured, innovation communities can be especially effective if direct personal communication takes place and if the involved members have developed common grounds of understanding and identity.

In the entrepreneurial process of innovation communities, we can distinguish six functions (Fichter 2005, 221):

- Explore the already existing.
- Initiate development and value creation processes.
- Bring together resources and actors.
- Cope with insecurities and risks.
- Build "bridges" (relation function).
- Lead things to success (promotor function).

For these reasons, interaction systems – like innovation communities – will as a rule very quickly communicate with external actors (pioneer customers, competitors, suppliers, R&D institutions, stakeholders etc.), at any rate when they realize that acquiring and using external technological and market know-how is a viable option of expanding internal competencies and resources. This opens the innovation community towards the outside and affords external ideas an equally high status and significance as internal ones. Thereby a change is initiated, which Chesbrough has characterized as a transition from the "closed innovation paradigm" to the "open innovation paradigm" (Chesbrough 2003). With this opening towards the outside, the knowledge and experience of external actors that is not or insufficiently available within the innovation community, can be mobilized and applied early on. Moreover, among these external actors there might also be some that could become important partners for the implementation of a radical innovation.

A consequence of all this is that innovation communities, especially in the case of radical innovations, are subject to a high degree of dynamics as regards time and content, which can result in situations where the boundaries of the community are no longer unambiguously defined. This can – at least in short term – lead to conflicts. A further challenge for the development and efficiency of innovation communities (see Sect. 3.2) is the informal nature of the network relations. "Communities go through an own life cycle and form only in the course of time a more or less stable and fitting ordered structure. . . . Group-dynamic processes are especially critical and the community can easily be thrown off track through power struggles, attempts of dominating, processes of marginalisation, behavior of fare dodger, etc." (Gerybadze 2003, 151) For this reason, sensible external coaching can help to guide innovation communities in productive directions.

5.4.2 What Kind of Coaching?

Before answering the question "What kind of coaching?" we should first differentiate the current methods of coaching, mentoring and cooperative consulting, and briefly characterize them:

- *Coaching* is a form of consulting that aims at increasing the alternatives of personal behaviour of the client or coachee. A good coach steers the consulting process in such a way that clients and coachees can critically examine old patterns of behaviour, test new strategies and integrate them into their behaviour. Voluntariness and the willingness to give up hitherto practised "That's just how it is" mentality are essential preconditions for successful coaching on the part of the client or coachee.
- *Mentoring* describes a consulting process in which a person with considerable experience of life and business accompanies a person who is new to the organization and is supposed to have capacities for development. Good mentors will put their proven knowledge and know-how at the mentee's disposal, discuss tactical procedures of approaching problems and tasks with them, and open up their relation networks for them. Open feedback and critical discussions are essential preconditions for successful mentoring.
- *Cooperative consulting* is a form of consulting in a team where usually five to eight participants work on practical issues to effectively structure them with

regard to time and contents. Participants can present matters of concern from their own field of work and receive ideas, hints, praise, or suggestions for alternative procedures from all other participants. Essential to the success of cooperative consulting is an attitude that accepts alternative solutions and the willingness to deal with arising issues favourably and constructively.

The different and common features of the three methods are summarized in the following table (Table 5.8):

In the case of radical innovations where innovation communities play a key role, coaching is the only possible candidate out of the three methods described above. But we have to go one step further. In view of the fact that

1. Radical innovation projects as a rule comprise complex tasks in a highly dynamic environment (see Sect. 1.1), and that

Table 5.8 Key features of different forms of consulting in innovation management

	Coaching	Mentoring	Cooperative consulting
Target group	As a rule, executives and project managers	New or young employees with capacities for development	Persons with comparable professional background, experience and sphere of activity
Adviser	External or internal coaches as process and solution advisers	Internal executives who advise from experience	Colleagues in the consulting group, possibly supported in the beginning by external or internal moderation
Practice vs. theory	Preoccupation with the field of practice of the client (partly supported by theory)	Preoccupation with practice in the organization of the mentee	Working on practical matters of concern for the participants
Relationship	Hierarchy-free, with equal rights, neutral (easily achieved with external coaching, revision necessary for internal coaching)	Hierarchic, related to common membership in the organization	Ready to help, with equal rights, mostly hierarchy-free
Voluntariness	As a prerequisite	A given if mentor and mentee can "reject" advice	As a prerequisite
Duration	Short to medium term (several months, then coaching can become unnecessary)	Long-term process of promotion and commitment to the enterprise	Medium to long term
Costs	Internal costs or external consultancy fees, and time invested by the client	Internal costs, and time invested by the mentor and mentee	Internal costs, and time invested by the participants, possibly a fee for the external moderator

Source: Author

2. The radical innovation project as well as the innovation community it is carried out in are embedded in a larger innovation system (see Sect. 1.2.2), coaching has to be systemic coaching.

In our understanding, systemic coaching is a solution-oriented procedure that perceives innovation communities as being embedded in an innovation system. It aims at establishing proper framework conditions for innovation projects and at effectively supporting innovation communities in improving their performance (see Sect. 3.2) and in contributing to the success rate of innovation projects.

Before going into the details of systemic coaching, we already mentioned the relations that systemic coaching has to other coaching methods. Systemic innovation coaching has common ground with:

- Individual coaching, in so far as it supports individuals with encouragement and feedback in achieving their goals.
- Team coaching, with its focus on the improvement of team performance.
- Organizational coaching, with its focus on improving skills and know-how of individuals and teams, in order for them to reach their set goals.
- Business coaching, by supporting a distinctive business concept, for instance.
- "Executive coaching", in so far as it supports the executive level in getting the maximum out of their strategies and in implementing them with their teams.

5.4.3 The Essentials of Systemic Innovation Coaching

Systemic coaching aims at creating beneficial conditions for successfully working in innovation communities. Innovation communities operate successfully if all participants bring their competencies into the network, are able to take in external knowledge, and adapt and develop it. The partners of the innovation community must be able to learn and communicate, so that they are prepared to solve all emerging problems. Personal contacts and the adherence to unwritten rules shape the cooperation. Openness, sincerity, trust, agreement on objectives, voluntariness, agreement on diverging interests, and coordination of responsibilities are typical features of successful innovation communities. Minimal contractual regulations (e.g. memorandum of understanding, secrecy agreement) may counteract the uncontrolled outflow of knowledge. As seen in the preceding chapters, one central precondition for the success of innovations is the competent management of the network. Usually, all partners in an innovation community have developed routines of communication and coordination which more often than not differ greatly. Therefore, innovation communities need their own forms of coordination and harmonizing instruments (committees, plans, communication forums etc.) supporting the management of common activities.

Systemic coaching is the art, science and practice of inspiring, setting in motion, and enabling performance, learning and development of all the partners active in an innovation community. The keywords here in fact are:

- Art: Although there are scientific bases for systemic coaching, it is an art in the sense that there is no exaggerated concentration on techniques. As long as it is practised excellently, the coach is fully committed to the participants, and all actions are aimed at the improved cooperation of the involved participants.
- Inspiring and setting in motion: Systemic coaching aims at mobilizing personal energies through effective communication among participants, at providing stimuli and fostering an attitude of "Yes, I can do it".
- Enabling: This implies that the participants have the ability to gain insights, to generate ideas und to think issues through for themselves.
- Performance: Whatever the coach does or says must be guided by the intent to improve performance, that is, to achieve greater effectiveness or efficiency in the innovation project.
- Learning: Broad opportunities for learning can be opened in coaching. Issues like how to tackle a certain task or master a new technology, or how to discover perspectives beyond the immediate goals or the future performance of the organization, greatly depend on learning.
- Development: relates to personal growth and self-confidence.

Systemic coaching

- aims primarily at organizations (networks, innovation communities etc.),
- therefore it is less about *what* should be achieved,
- but more about *how* it can be achieved,
- how inner strengths – creative, emotional, entrepreneurial – of the members of the organization can be developed,
- so that they become tireless innovators, team workers and entrepreneurs,
- accordingly, the focus is on "how" questions that need to be dealt with in organizations.

One key aim of systemic coaching is improving the performance of innovation communities. As outlined in Sect. 3.2, the contribution of promotor networks to the success of a cooperative innovation project depends on three key factors:

- Situational or task-based composition of the community, regarding skills, roles and resources.
- Existence and extent of community identity.
- The level of interaction quality.

Systemic coaching has to support innovation projects and innovation communities in finding the right composition of the partners to be involved (composition of the community), developing common goals and a team identity (identity of the community) as well as establishing productive and trusting forms of cooperation (interaction quality in the community). For these three tasks, coaching has to take the different phases in the life cycle of a community into account. Basically, three phases can be distinguished: initiation, development, and stabilisation or reconfiguration phase of the community (see Table 5.9 with exemplary tasks).

Table 5.9 Key tasks of coaching an innovation community

	Phases in the life-cycle of an innovation community		
Phases	Setting up the community	Developing the community	Reconfiguring the community
Area of responsibility			
Find the right composition	Choosing the right and fitting partners. Etc.	Setting up and carrying out common projects. Etc.	Adapting the community to new condition. Etc.
Strengthen the identity	Reaching agreement on common goals as well as on convergent and divergent interests. Etc.	Effectively developing and implementing a common strategy. Etc.	Re-examine goals and strategies and adapt them if necessary. Etc.
Improve and secure the interaction quality	Reaching agreement on optimal forms of communication and cooperation. Etc.	Settling conflicts effectively; reaching agreement on conflict and risk management. Etc.	Promoting the necessary competencies for all participating actors. Etc.

Source: Author

Important "how" questions that systemic coaching can focus on are:

- How do we organize processes if we want to put new ideas into practice?
- How can we promote the development and application of knowledge in the organization?
- How can we enhance the competencies of all participating actors?
- How do we integrate knowledge into the development of products and services?
- How do we lead our organization?
- How do we successfully design change processes?
- How do we effectively implement our strategies in view of our goals?
- How do we use internal cooperation and how do learn from each other?
- How do we use the cooperation with customers and suppliers for our development and how do learn from and with them?
- How do information and communication barriers arise and how do we conquer them?
- How do we treat mistakes, contradictions, conflicts and growing complexity?

The important thing is not to stop questioning. Curiosity has its own reason for existing. (Albert Einstein)

A characteristic of systemic coaching is the solution-oriented approach in dealing with problems. It tries to shift the focus from the problem itself to new perspectives which could generate viable solutions. That is also why "how" questions are placed in the foreground (see above). Thus, it no longer concentrates on problems, but on goals and better solutions. The aim of systemic thinking is to generate more answers, more hypothetical solutions, more promising prospects – and not to debate more problems, conduct more interrogations and even more problem analyses.

Fig. 5.2 Systemic coaching: discovering interrelations and supporting participants, like in an orchestra (Source: © DeshaCAM - Fotolia.com)

Contrary to non-systemic coaching which places the individual in a prominent position, systemic coaching is a way of sensitively discovering interrelations and encouraging the members of the innovation community to "play" together – like a jazz orchestra (see the image below [Fig. 5.2]).

This is one essential feature of systemic working. The coach targets the overall context in order to find out who is important in which interrelations, which individuals with which competencies are important with respect to clients, and what associations, interactions and networks can be identified that are relevant to the problem at hand. In other words, the coach has to find out who the key persons[1] are for different tasks.

In highly networked systems it hardly makes any difference at what point of the system aspects are changed. Each change will cause other changes and therefore, there is a great freedom of choice in the intervention. With well-targeted interventions, it is possible to bring about amazing changes very quickly. Systemic working rarely deals with the question how a problem has arisen – as it wants to find out how a situation should be. Therefore, it is a very solution- and future-oriented approach. It helps to arrive at solutions relatively quickly, since there is no "long march" through a history of problems.

Systemic thinking and acting implies contextual and holistic thinking and acting. It starts from the premise that events and problems can only be understood with regard to their cultural, social and personal references. For a systemic view, it is central that individuals cannot be understood without this context of interrelations: The whole is more than the sum of its parts. Simple cause-and-effect relations are no longer of that much interest, but as to which meaning a problem has in the

[1] The concept of "key persons" is introduced and discussed in Sect. 1.2.

context of its occurrence and for whom? Contexts and relations are examined from the "feedback" perspective, emphasizing their circular character.

Because human beings construct different realities, there is no right or wrong, merely different perspectives. Values like tolerance, esteem, respect and cooperation can be consistently deduced from this basic idea. As individuals are perceived as unique in their experiences and autonomous in their views, they will never fully be comprehended by others. Therefore, a systemic coach needs skills in promoting processes and relationships in particular.

A further challenge in the coaching process is to dissolve rigid perspectives and to encourage the innovation community partners to generate new perspectives. This creates mobility, a diversity of perspectives and new scopes of action. But, and this is essential, the coach needs to respect the individual reality constructs of the innovation community partners, since they are of the same validity as any other possible construct – albeit not always of the same desirability. Respect is decisive which means that we need to accept all the different presumptions, beliefs, values, or attitudes. The participants and the coach should be guided by curiosity, to explore in detail the implications, conditions, and consequences of every particular construct.

The cooperation in innovation communities requires "interface management" in order to avoid coordination, time and information losses. Different actors work together in innovation communities, and naturally they also have different interests and goals. It is obvious that by the integration into an innovation community, all participants have greater prospects of success, due to system effects and reduced transaction costs. Nevertheless, separate learning processes are necessary in order to let the opportunities arising from the existence of the innovation community positively influence the working process of all participants. Moreover, a great degree of coordination is needed, for experiences to be evaluated and generalized, and consequently, for agreements on common procedures to be reached (Table 5.10).

Essential aspects for a successful coaching process are:

- Neutrality of the coach and fair mediation between the partners of the innovation community.
- Clear goals and results.
- Joint endeavours.
- Continuous efforts, periodically reviewed, in order to make sure that coach and coachees are on course with the agreed goals.

Table 5.10 Activities of a coach

What a coach actually does in practical terms
• Facilitating and moderating community meetings
• Preparation and chairmanship of strategy and innovation workshops
• Evaluating community meetings and workshops
• Mediating in conflict situations
• Promoting regular communication (face-to-face and virtual)

Source: Author

- Regular communication and meetings (face-to-face and virtual).
- No judgements on good or bad, on success or failure.
- An attitude characterized by attentiveness and curiosity.

Basic components of the coaching process are:

- Dealing with the development of skills through practice.
- Analyzing special skill components, techniques and their surroundings, in order to support the coachees effectively.
- Confronting coachees with increasingly challenging tasks.
- Trying to identify problems or weaknesses, in order to address them.
- Discovering potentials, building on strengths and using advantages of talents and opportunities.
- Creating beneficial conditions to allow coachees to organize themselves in order to attain their goals.
- Constructive mediation in case of conflicts.
- Process consulting.

To sum up, one can say that systemic innovation coaching aims at:

1. Improving the effectiveness of the performance of innovation communities, and
2. Enhancing the capabilities for learning and performance of the individuals involved.

5.5 Foresight: Trend Analysis and Scenario Development in Innovation Communities Using the Tool SEMASCORE

Dierk-Oliver Kiehne

5.5.1 Objectives for Developing the Tool SEMASCORE

The objective in developing the SEMASCORE tool was to develop a powerful method for supporting innovation communities to identify and analyze trends that are important for their joint innovation project. SEMASCORE is also meant to help innovation communities to jointly develop scenarios and to establish a common perspective of future developments. The method was developed on the basis of the results of empirical studies into the way that groups of companies and their actors find each other, how their cooperation works in formal and informal ways, what information and interaction needs these actors have, and how they currently meet these needs. The method was tested and further developed in cooperation with companies involved in the InnoCo-project.

5.5.2 SEMASCORE Supports Trend Analysis and Scenario Development

In Sect. 5.3, different methods suitable for innovation communities have been introduced. In this chapter, the method of Delphi-supported trend analysis and scenario development will be briefly introduced in the methodological context of developing the software tool SEMASCORE, which will be presented thereafter.

Based on the scenario technique and Delphi consultation approach, as well as a rating process, an appropriate methodical approach was developed. The Scenario Method (Shoemaker 1995; Mercer 1995; Gausemeier et al. 2009) and the Delphi Method (Sackman 1974; Brown 1968) are commonly known. For this reason, they will not be introduced here in any further detail. The following section will briefly describe the specific approach of Delphi-supported scenario development and monitoring.

According to the first step in the scenario analysis, the examination field has to be defined first. This may be a complete business field, like carbon nanotubes (CNTs) as basic materials; it can also be a specific application, like CNTs used in wind generator parts (power plants), or even a strategic question, like if to make or buy a certain technology. This step is important in order to focus the scenario on relevant factors. Of course, the time horizon for prognosis has to be determined (e.g. 5 years). The describing factors of a future scenario (influence factors) are then determined. In order to give them more structure, they are grouped into different categories (influence fields). For each influence factor there is a certain small amount of specifications (future options). For instance, the price of CNT basic materials may rise, fall or remain as it is. These steps are quite comparable to Scenario Analysis. In our approach, two different validations are necessary. The first is an impact analysis: the user estimates how strong the impact of a certain influence factor on the examination focus is. This rating can be a number, e.g. on a scale of 1–10. The scale doesn't matter as long as it is always used the same way within one project. The second rating step is a "worst/best case" consideration. Here it is to be decided whether a future option would be good (best case), bad (worst case) or neutral for my examined field if it were to occur.

In the literature there are different approaches for determining the incidence rate of a certain future scenario, one for example is cross-impact analysis. Since this method is very time-consuming and failure-prone, we decided for an alternative approach here: The Delphi Questionnaire. Here experts have to be named, for each influence factor at least three to four individuals, while some experts may also vote for different influence factors that are similar in the same influence field. The experts receive a questionnaire where they have to estimate the probability in percent of a certain future option to occur. This estimate reason has to be justified and confirmed, with a literature source for example. The respective answers are then swapped among all experts, so that each expert may reflect his or her own views or compare them to those of other experts. The time required is less than 10 min per expert. In the end, a calculation algorithm presents the most likely future scenario and also its influence on the examination field, or vice versa: How well or badly does the examination field suit this future option? It can also be determined which influence factors are useful and which ones are harmful in terms of the focus of the examination.

Additional methodical, community-specific considerations were also taken into account.

- Megatrends in connection with future scenarios.
- Cluster of assumptions and occurrence probabilities.
- Viability of a software-supported concept with open interfaces (Excel).
- Usability concept for simple applicability of the methods.

5.5.3 *User Requirements for Delphi-Based Scenario Development and Its Software Support*

The following user requirements for the method of Delphi-based scenario development and, accordingly, a software-based instrument, arose from a target group analysis (Table 5.11).[2]

[2] The requirement list was created with some key users of the tool who were involved in the InnoCo project as well as some scientists of the research group who were attending the use cases described above. Therefore three different workshops were held in order to determine user requirements. Furthermore some software engineers of InTraCoM Group were involved in the method and software definition process early on.

Table 5.11 User requirements for Delphi-based scenario development and its software-support

Category	User requirement	Description
Content insight	Insight into a possible future and its determinants	Method is capable of producing a (structured) description of the future through its pertinent characteristics and instances
	Identification of the focus on the technology fields pertinent to the innovation community	Method shows, depending on question, which technology fields are pertinent to the innovation community The fields are, e.g. sorted hierarchically, so that focusing is facilitated
	Potentials for innovation community become identifiable	Method can make qualitative distinctions, which means that it makes different applications or technologies comparable. A potential may depend on a time scale, where, for instance, a technology is not yet ready for the market – but in 5 years it maybe is
	Justification and traceability for future developments	A certain future trend or scenario becomes transparent and reproducible
	Influence of different, relevant characteristics considered	Which characteristics are influential? To what extent?
	Plausibility and consistency	The result must appear non-contradictory and internally conclusive
	Reliable insight	Insights should be reproducible and capable of being documented
Sustainability of results	Time frames for trends (possible entry points for the innovation community and degree of maturity)	Insight regarding the time when a market entry appears reasonable
	"Half-life" of information	Information once gathered loses validity and must be re-collected in certain intervals
	Possibility of anticipating potential problems and barriers	The method forecasts the problems and barriers existing with regard to an innovation project
	Tracking the progress of the trend (deviation from planning or forecasting)	I.e., identifying a "trend line"
"Handling" of the method	Multi-step selection process	When selecting a technology field, I receive detailed information
	Moderate maintenance effort	Effort for maintaining the system is in a healthy relation to total effort
	Self-explanatory, unambiguous	Refers to the insights derived from the method
	Playful effect, inputting information is appealing	It should be fun to interact with

(continued)

Table 5.11 (continued)

Category	User requirement	Description
	Instructions: "How do I proceed when gathering information?" (wizard)	Applies to the entering of information, as well as to the retrieving of information
	Tasks can be run parallel	Information retrieval process must be able to be utilized by different actors
	Detailed information as required	It should be possible for information to be displayed in various forms, from a (comparative) overview for a more detailed view
	Harmonizing of information during information output	The amount of detail in the information should remain consistent within the same view
Documentation	Reproducibility and sustainability of insights and results	The result must be reproducible and must have a certain logic, data must be capable of being verified
	Standardized results output	Graphic visualization of results, status indicator
	Minimal documentation effort	Effort for creating reports from insights is minimal
	Further processing of information possible (avoiding data redundancy)	The information gathered is also available for other uses and does not have to be re-gathered

Source: Author

The Delphi-supported scenario development and monitoring (Jakob et al. 2007) is based on the established method of the scenario technique and Delphi consultation and fulfills the requirements stated above for the method; it is accordingly well suited to serve as the basis for method and software development. Additional methodical elements, such as megatrend analyses, preparation of clusters of assumptions (any scenarios), or a purely textual description of scenarios, should also be considered when developing the software approach.

The central question is: how can the method be software-supported in such a way that it is effective, efficient and sustainable? For this, an appropriate range of solutions was generated from the requirements derived from customer needs (Table 5.12).

The development of the software tool SEMASCORE was based on the requirements and possible implementation options listed above.

5.5.4 *Introduction of the SEMASCORE-Tool and Its Functionalities*

Architecture and the Big Picture

The SEMASCORE system basically consists of two components, a visible and an invisible component. The core of the system is a desktop application that also

Table 5.12 Solutions for software that supports Delphi-based scenario development

Category	Requirements	Specifications	Approach to solution
Future scenario	Structured description of a scenario	For each question, there is a specific structured view of the influence field, influence factor, future option, occurrence probability	Hierarchical tree depiction
	Text description of a scenario	There is a textual view for each structured view. Each view must be capable of being transformed into the other one	Provides a text field for alternative scenario description
	Intuitive operation	No knowledge of the method or tool required for working with it	Use of drag and drop, right mouse button Context-sensitive menus
	Clearly arranged view	You can jump from anywhere in the program to anywhere else in the program. It is clear at all times where you are	- Navigational tree like Windows Explorer on the left-hand side - Show project list as tabs above the navigation tree
	All influence fields and influence factors for all projects can be summarized in a separate view	One list for all influence fields (master list)	A tab "All"; from that view, entire influence fields and/or specific influence factors, including occurrence probabilities (where applicable), can be copied and inserted into another project
	Capturing megatrends	In addition to areas of influence, megatrends can also be assigned to each future scenario	Megatrends treated like an influence factor with one instance and a 100% probability of occurrence
	Depiction of influence factors and influence fields	A maximum of ten category levels can be inserted	Hierarchical view of influence factors and influence fields
	Different future options can be created for each influence factor	There is at least one future option for each influence factor. Future options must be phrased absolutely not coexistent with others	Note when creating future options
	Occurrence probabilities of future options	There is an expert assessment at a specific point in time for each future option	Provides a time axis in the system, time stamp at particular intervals in the survey

(continued)

Table 5.12 (continued)

Category	Requirements	Specifications	Approach to solution
	Trends for occurrence probabilities	The occurrence probabilities are defined at specific times, gathered in intervals to be defined. All expert feedback is "frozen" at one time	Analysis of occurrence probabilities over time
Experts	Setting up lists of experts and assigning them to influence fields, as well as influence factors	List view of experts and influence factors	n2n assignment (multiple assignment)
	Weighting and/or hiding experts	Every expert has a quantifier	Gives each expert in the system a weighting factor, so that an adjusted total product is formed
Surveys	Automatically generate, send and analyze questionnaires from the system	No manual intervention for the preparation, sending and analysis of questionnaires	2-component system - Desktop application - Server application
	Several poll runs per survey	2 (Delphi process)	1. Everyone evaluates the first time 2. Everyone can revise initial evaluation after looking through the other assessments (new evaluation)
	Several surveys possible	Surveys can be repeated easily	In the case of a new survey, save the previous one with a time stamp
	Get expert assessments on instances	For each expert, there is a probability distribution for the initial evaluation, and an additional probability distribution for a second evaluation	For each expert, an initial probability distribution of 100% across the instances is set
Analysis	View of probability of occurrence for any scenario clusters	Depiction in percent	Calculation mode: relate adjusted total product of the selected cluster of instances to the mathematical best case
		Depiction in percent	Calculation mode: relate adjusted total product

(continued)

Table 5.12 (continued)

Category	Requirements	Specifications	Approach to solution
	Probability of occurrence of the most likely scenario		of each instance evaluated highest to the mathematical best case
General	Entire project can be saved	One general file includes all contents	.xml dump file with all contents from other lists
	Interfaces with other systems	Excel	Provides for .csv import and export
	External lists can be imported	Lists of experts, of influence factors, of megatrends	Store standard lists as .csv while saving
	Enables offline work (for workshops, clients)	No constant Internet connection	Desktop application with communications interface

Source: Author

permits working offline, i.e. without direct Internet access, on a local computer. In addition to high speeds for the creation and handling of system functions, this also permits the highest possible data security (safety and security). This part, the application, the front end, is the directly visible portion of the system. The invisible portion of the system is located on a server with a database. This portion is a server application that can process the data from the application, dynamically create individual questionnaires from it, manage access rights, send questionnaires with its own e-mail system, collect them again, and return them as a package to the application. To do so, a separate synchronization process between desktop and server application is established. The questionnaires are dynamic web pages and are accessed and transmitted via a browser (Fig. 5.3).

Description of the Main Functions of SEMASCORE

Creating Projects and Actual Scenario Elements

SEMASCORE has common and useful software methods and structures for creating and managing its scenario projects. Some are shown on the following pages. A project is always determined by a certain (strategic) question or a technology area, which then has to be investigated further in terms of future development.

File System

The tool has both *.csv and *.xml interfaces for creating a clearly arranged project folder structure as needed, where manual changes can easily be accomplished or entire projects with all contents can be managed as single files.

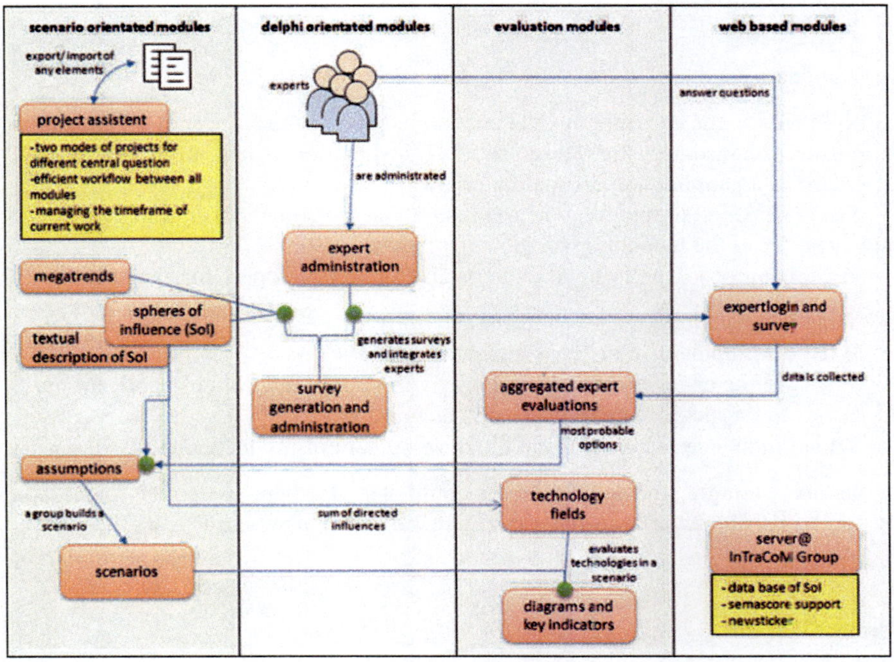

Fig. 5.3 Elements of the system and forms of interaction with the expert community (*above*) and architecture scheme (*below*) (Source: Author)

The possibility to import and save existing data in *.csv format (saved individually with the save wizard) exists for all kinds of project content (influence factors, experts, etc.).

Subjects of Investigations

To process specific questions, two project modes are available: one mode is to compare different technology fields, e.g. applications like carbon nano tubes (CNT) in wind power plants vs. CNT in aircraft engines vs. CNT in human medical prosthesis; the other is to investigate only one single question, the "strategic scenario analysis", e.g. future market developments of CNT material.

With SEMASCORE you can, for example, investigate the following types of issues:

- Technology field
- Technology
- Patent
- Product

- Process
- Material
- Application

Referring to the example of CNT mentioned above, the subject of investigation might be, for instance, the "usage of CNT in functional parts in power plants". It describes an application for the material CNT.

For the sake of simplicity, when talking about the subject of investigations, we always refer to the technology field.

For this purpose, the following questions can be developed, for example:

- What is a technology's level of marketability?
- What development status does a technology have?
- What are the chances that a technology will penetrate a field of applications?
- How does a specific scenario affect a technology?
- Which influence factors impede the development of my technology?

In our example, the key question could be: "Is there a market for carbon nanotube (CNT) material in functional parts in power plants in 5 years?"

Project Wizard

A new project is created using the project wizard. It aids in creating a new project in a clearly arranged form, in entering all necessary metadata, and, if required, in importing scenario elements, such as influence factors, etc.

Project Modes

When creating a new project, a question about the project mode is asked. Each of the two project modes makes available several changes to the graphic user interface (GUI).

- A technology field analysis offers the option to list several technology fields (or products, technologies, etc.) in a project and to directly compare their evaluations. The factors included for the evaluation are: Configuration of the influence factors via the scenarios, ascertainment of an instance's probability via a survey or manual entry and impact of a factor and/or effect of a future option via the setting in a technology field. (This mode is compatible with, e.g., the first, second and fifth of the questions listed above.)
- A strategic scenario analysis avoids the division into several objects (such as the technology fields above) and concentrates on evaluating scenarios by ascertaining the probabilities, impacts and effects directly from the influence factors. This mode refers to the first, third and fourth of the questions listed above.

Table 5.13 Glossary – terms used with the method Delphi-supported scenario analysis

Term	Meaning
Influence field	Category of some influence factors, e.g. the availability of commodities
Influence factor	Factors that influence the technology or the strategic matter that is currently investigated in the scenario project, e.g. the availability of oil
Future option	Different options the influence factor may develop (e.g. the availability of oil may fall, rise or remain as it is)
Technology field	A certain technology that is investigated with the scenario analysis. A question may be how good a technology that is investigated may fit into the future (the scenario that is determined)
Time stamp	The date of determination of the future scenario, e.g. the date when all Delphi Questionnaires were returned and processed by the experts. The time stamp is a point of time when a scenario is determined and is important, considering the fact that a scenario may change within a couple of months, e.g. after a certain incident occurred
Megatrends	Megatrends are influence factors with only one future option, like incidents. E.g. the demand for electric mobility will rise
Scenarios	A bundle of arbitrarily expected future options of different influence factors. This bundle describes an arbitrary future scenario in order to find appropriate strategies for an extreme scenario, for example
Scenario element	All elements mentioned above are elements of a scenario

Source: Author

Navigation

In the left-hand side of the main window, there is an explorer file-based orientation of projects and their content. This view offers the opportunity to comfortably manage project content. Each project has its tab with its own navigation window, so that drag and drop functions between the projects are possible, for example. All editing is possible in navigation form.

Each branch shows an image for quick identification of the element displayed, and on the technology field branches, even the most important values can be seen at first glance.

In addition, common keyboard combinations and extensive context-sensitive menus relating to each element have been implemented, to ensure even faster editing of objects (Fig. 5.4).

Efficient Work Flow

To achieve the highest degree of efficiency during the editing of projects, the most important functions and information have been assembled in the toolbar at the top and the information bar at the bottom.

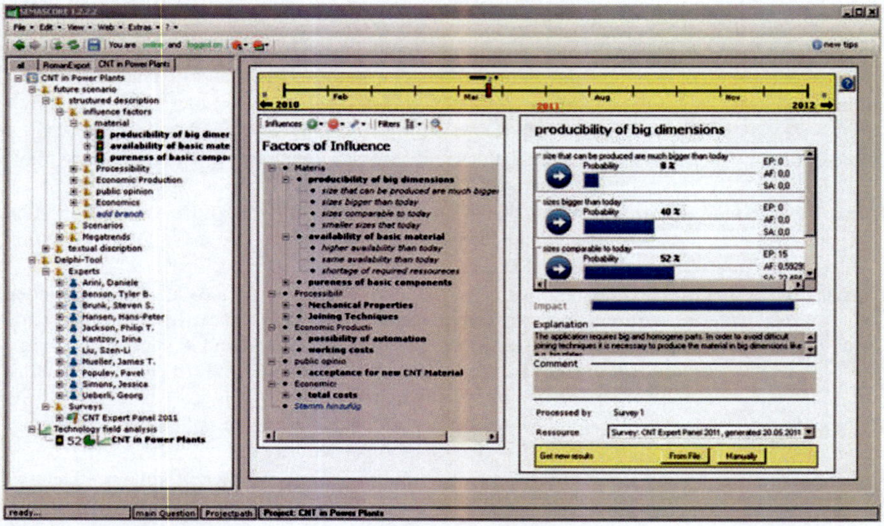

Fig. 5.4 Orientation and navigation in SEMASCORE with context-sensitive menus (Source: Author)

Time Stamp

An innovative function intrinsic to SEMASCORE is the time stamp. Either a small tool window or a shortcut in the toolbar, it permits switching of the time frames that are the focus for the editing of surveys or scenario element values.

Editing of Influence Factors

The influence factor interface consists of the time axis, an overview of project influence factors (incl. a toolbar) and an overview of characteristics of the influence factor currently displayed.

Just as with the orientation, influence factors can be edited in the overview (new creation of factors or fields, deletion, editing of influence fields and factors).

Influence factors are captured and processed by the program in their entirety, including all their characteristics, such as impact, influence field, direction of effect, etc. Here too, care has been taken to ensure that the large number of a project's influence factors can be created and edited quickly and easily.

Time Axis

The time axis clearly shows how many surveys or manual entries of probabilities exist, and when they become available. The time axis can be seen in Fig. 5.4: the ruler above in the right frame. Surveys are all marked with a red bar in the time axis. The beginning and end of the bar show the start and ending point of a survey. By clicking a survey, the estimated probabilities of the future options are shown for the currently displayed factor. In addition, a grey bar shows the time frame for which the survey is intended (when it will be closed), and a white flag shows the time frame to which the survey refers. In the image above the time axis shows a short red bar between May and June. It says that the survey took place within 1 week in the end of May.

Editing of Scenarios and Other Project Elements

For user-friendly editing, a separate mask exists for each scenario element type. Here, the system relates elements to create consistent and coherent projects. For example, scenarios can be created at the push of a button from the most likely future options. If influence factors or their values are changed, the system automatically updates the values and evaluations of the technology fields. A project's elements are, among others:

- Scenarios themselves.
- Megatrends.
- Manual entry of probabilities.
- Filtering influence factors for different perspectives and more complex views.
- Technology fields.
- Assumptions as elements in a scenario.
- Influence factors as text view.

The Delphi Tool

SEMASCORE intelligently combines the scenario technique with the Delphi method. In doing so, it perfectly organizes the modules "manage experts" and "create and execute online surveys".

Preparing and Executing Surveys

Creating Experts

Several experts in a row can be comfortably set up in SEMASCORE via the individual view. Notes for entry, as well as a generator that creates a login and password from the expert's name and a random number, help in this task.

In the overview, all experts are listed with their most important data. There is also an option to import them via SEMASCORE's interfaces, e.g., from Excel (see section above).

Creating Surveys

SEMASCORE has a user-friendly module for the preparation of surveys. They can be planned with all their metadata and assigned to the desired experts and the influence factors to be queried.

Approval and closing of a survey of experts is easily performed at the push of a button, whereby invitation e-mails which have been created with variables to be replaced by the system with expert-specific data, are sent automatically.

The questions about influence factors can be generated automatically or phrased manually.

Experts Answering the Survey

When the survey has been uploaded and the e-mails with the necessary login data have been sent out automatically by the system, an expert can log into the expert portal and answer the questions. Here, there is the option of jumping forward and back to correct answers or to the end of the survey.

Each expert can see only the questions assigned to her or him. The personally generated login data identify the experts and generate the questionnaire that is foreseen for this specific person. If the administrator has closed the survey, no questions can be seen at all (Figs. 5.5–5.7).

Analyzing Surveys

Updating the Survey

With the Delphi tool, we tried to keep the level of automation for the survey as high as possible. The "Update" button performs the following actions:

Fig. 5.5 Expert login. It is a part of the server-based application (Source: Author)

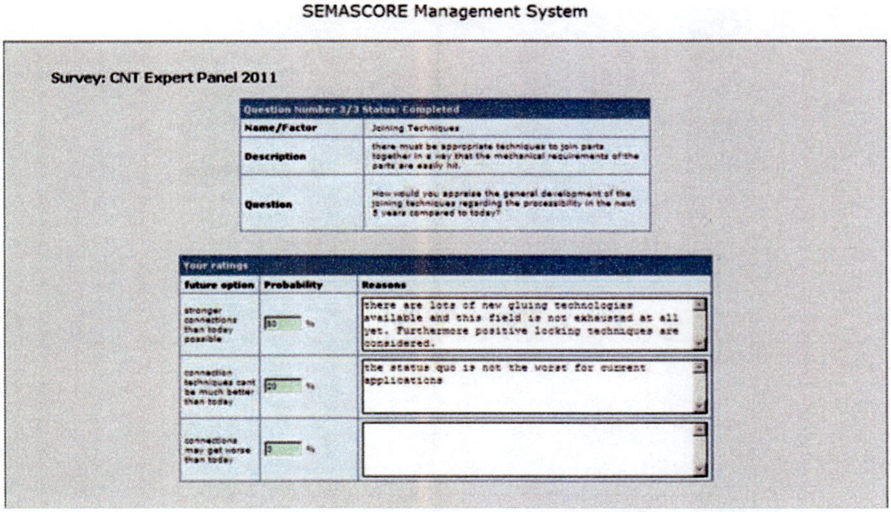

Fig. 5.6 Survey interface: first (*upper*) image Delphi-questionnaire (Source: Author)

- Transferring the status as to whether the survey has been approved.
- Downloading the answers provided by the experts and direct assignment to existing influence factors.
- Updating of the rights of project members.
- A new assignment of questions to individual experts.

Analysis of Technology Fields and Scenarios

As soon as the window with the analysis for the technology fields and/or strategic scenario analysis is called, the existing values are automatically calculated and an evaluation in the form of a traffic light (green = positive degree of maturity) and a number in percent is displayed. The percentage shows the quantitative position of the current configuration (i.e., evaluation of technology fields or scenarios) between the worst case and best case (Fig. 5.8).

Elements of Technology Field and Scenario Analysis

For SEMASCORE, some further values were added to the analysis so that the result can be ideally evaluated. For one, expert points are displayed, which show how many experts on average, and with which weighting, generated this evaluation through their answers. In addition, "SEMASCORE safety" displays percentages of how reliable the evaluation is (Fig. 5.9).

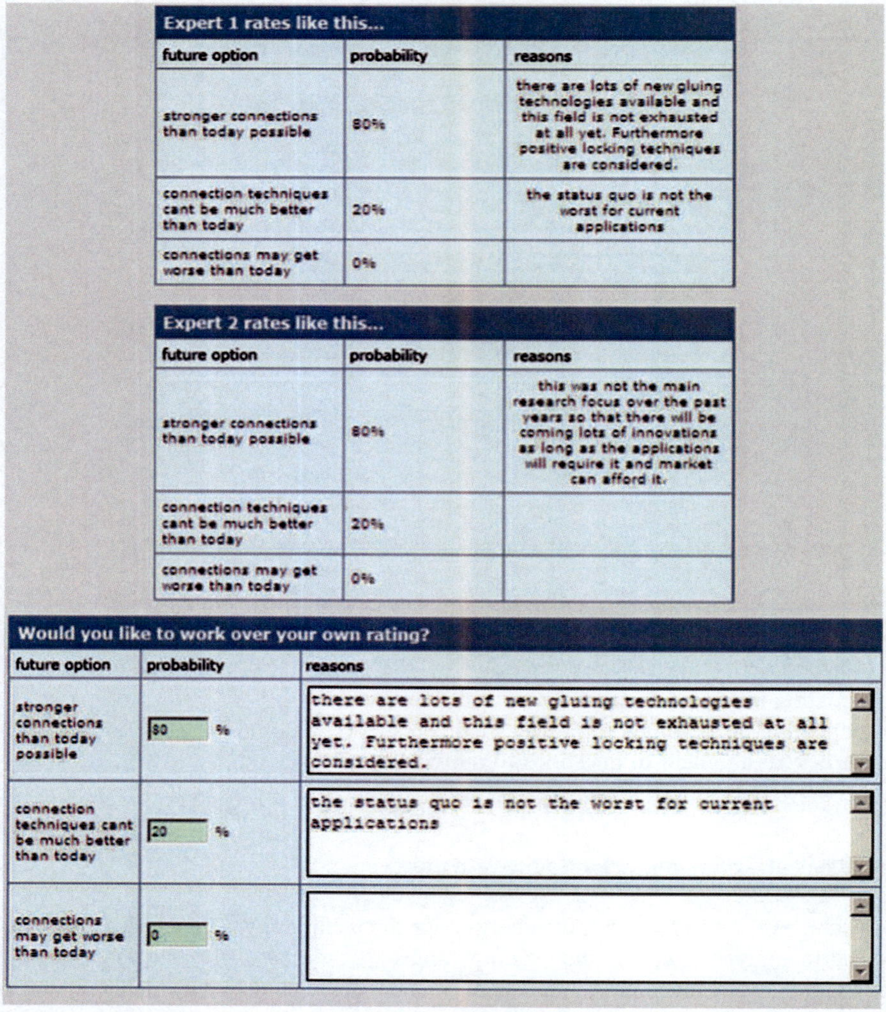

Expert 1 rates like this...

future option	probability	reasons
stronger connections than today possible	80%	there are lots of new gluing technologies available and this field is not exhausted at all yet. Furthermore positive locking techniques are considered.
connection techniques cant be much better than today	20%	the status quo is not the worst for current applications
connections may get worse than today	0%	

Expert 2 rates like this...

future option	probability	reasons
stronger connections than today possible	80%	this was not the main research focus over the past years so that there will be coming lots of innovations as long as the applications will require it and market can afford it.
connection techniques cant be much better than today	20%	
connections may get worse than today	0%	

Would you like to work over your own rating?

future option	probability	reasons
stronger connections than today possible	80 %	there are lots of new gluing technologies available and this field is not exhausted at all yet. Furthermore positive locking techniques are considered.
connection techniques cant be much better than today	20 %	the status quo is not the worst for current applications
connections may get worse than today	0 %	

Fig. 5.7 Survey interface: second (*lower*) image Delphi-questionnaire (Source: Author)

On the level of influence factors, a traffic light symbol indicates if a certain future option is closer to the worst case (future option that may mean a problem for the technology field) or the best case, or neutral. The traffic light displays a red, green or yellow light.

In the example above, the traffic light indicator shows a yellow symbol and the "marketability" says 52%. This means from a current point of view the market and/or technology in 5 years will not be ready so far for marketing the technology, in this example for marketing CNT in wind power plants. The number means that 52% of the best case for this technology will occur, where 100% represent an absolute ideal

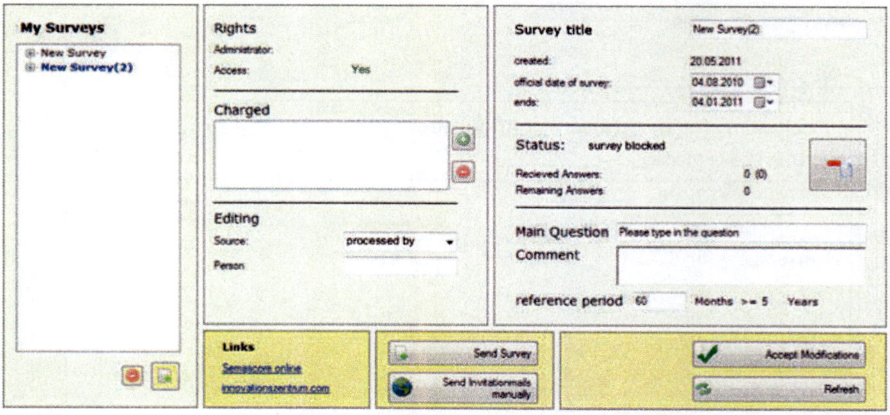

Fig. 5.8 Editing the survey metadata and answer download (Source: Author)

case. The threshold for the traffic lights between green and yellow, yellow and red are adjustable. In the example the chosen threshold were 65% for green and 35% for red.

On this level, the expert feedback is visible, so the reason why they rate a certain future option is also an argument against a certain technology field. This represents a certain threat – SEMASCORE can indicate this at an early stage, e.g. in order to indicate appropriate actions to the user (Fig. 5.10).

Elements of Influence Factors

The average probability is calculated for each future option. The future option with the highest probability is the "most likely scenario". Based on entry data from the SEMASCORE user, whether the most likely scenario will impact the technology field favourably (symbolized by a green traffic light), unfavourably (symbolized by a red traffic light), or not at all (yellow traffic light) is determined. By selecting an influence factor, the experts' answers are displayed in the grey text field, so that an actual-target comparison of the scenario can be quickly analyzed.

Existing Interfaces and Compatibilities

The SEMASCORE tool is a supporting tool for innovation processes as well as innovation communities. Especially in innovation communities, different persons and different processes are involved. Therefore, different methods and environments are used. In order to address this phenomenon, SEMASCORE development was based on existing common standards. The "dump file" of the .sspx scheme contains all necessary project information and has an XML structure; this

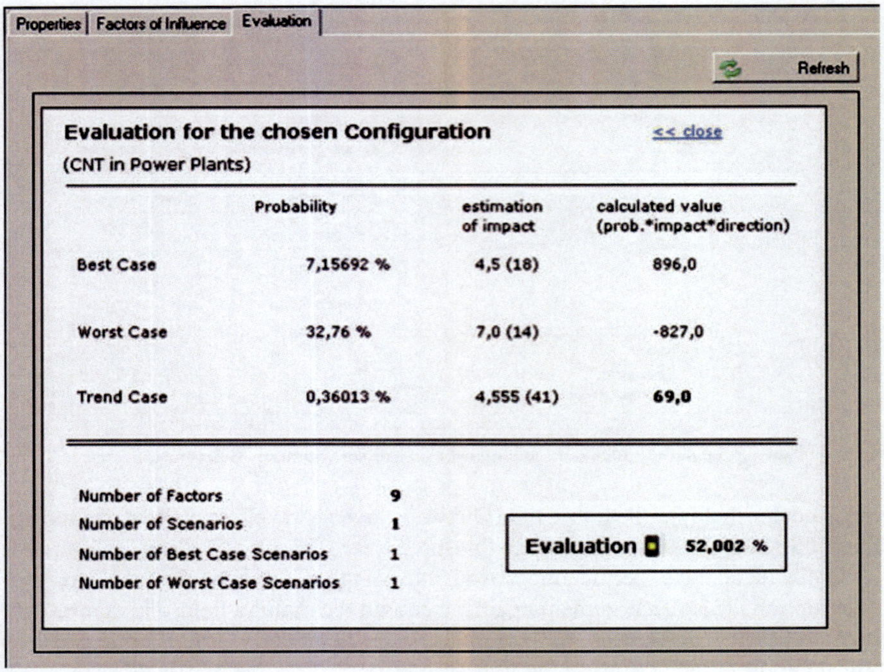

Fig. 5.9 Example of a scenario evaluation analysis[3] (Source: Author)

structure may be extended with further information or any flag in order to import it into other environments or other tools with an appropriate parsing function.

During collaboration with different promotors, experts and other persons involved, information has to be exchanged. During the first testing phase, we found that it is more efficient to let the team directly edit the contents of a project, but without having SEMASCORE installed. Apart from the dump file .sspx, which contains all necessary information, different .csv files are created that contain specific information as already described above, such as the list or catalogue of influence factors. These .csv files can be opened with Microsoft Excel or Open Office Calc, and can be sent to any partner of the community, who may edit the lists and send them back to the administrator working with SEMASCORE. These specific .csv files can be directly created and saved with SEMASCORE.

Other tools may use the open .xml structure in order to import or export SEMASCORE files. Example: InTraCoM developed on behalf of a customer a tool that is used for creating roadmaps, the Roadmap Manager ROMAN. This tool creates,

[3] The example shows the maturity of the examined technology field: Yellow traffic lights and a maturity of 52% means that this technology field is not mature enough in 5 years from the current point of view.

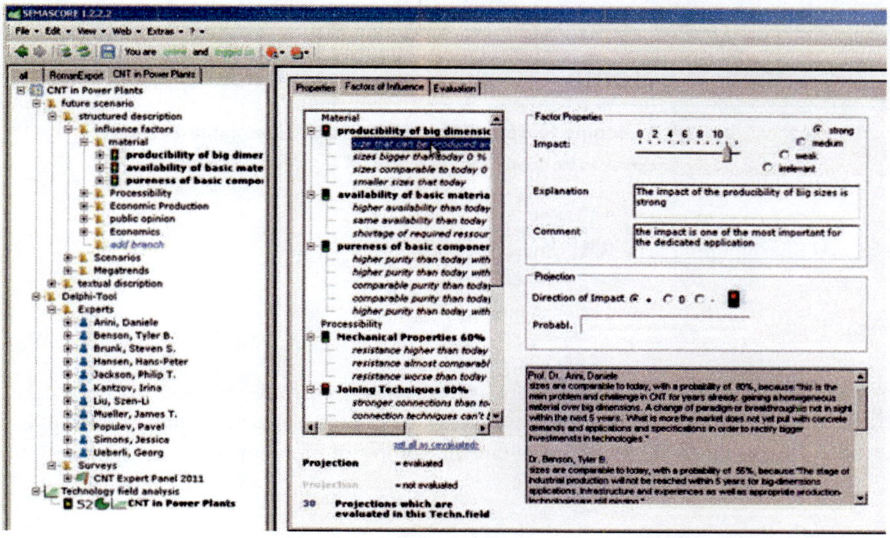

Fig. 5.10 Example for detailed analysis of a scenario[4] (Source: Author)

edits and manages roadmaps. The file structure is also based on xml and uses the same scheme as SEMACORE. The basic idea was to enable the user to start a scenario management project, such as one for a critical path in a roadmap where whether one or another technology should be supported in future must be decided (decision point). On the other hand, a particular future scenario may lead to a certain roadmap, so that the project data of a scenario project has to be imported into the roadmap manager project. Both is possible, based on the same data scheme that SEMASCORE uses.

Support

The SemaWiki

At InTraCoM, a wiki was created for SEMASCORE to ensure permanent communication with the application and to offer support. Notes on the structure of SEMASCORE, on working with the application, and pertinent topics for development, such as "bug report", "additional ideas" and the development roadmap, are located in the SemaWiki (Fig. 5.11).

[4] Besides a probability in percent (e.g. below, on the left tree), it can be analyzed each single influence factor and the potential problems that will occur in future most probably. In the example below the "best case" – bigger sizes than today of a certain material can be produced in 5 years - will not occur after the experts' estimation. It is more probable that the sizes remain like today. The expert feedback in the gray box give the reasons why.

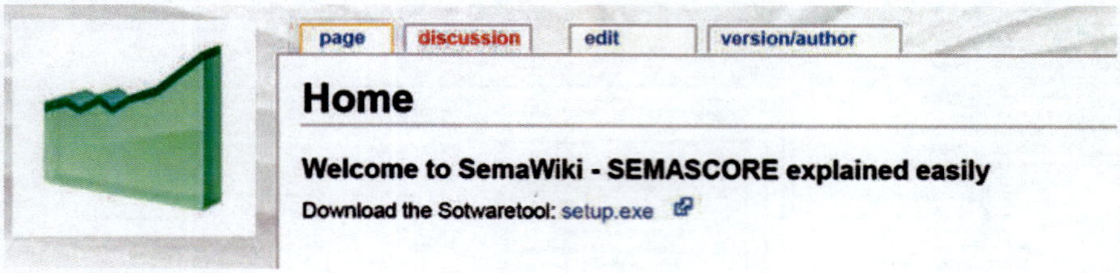

Fig. 5.11 Excerpt from SemaWiki (Source: Author)

SemaTicker

When calling up "Notes", a window with current messages regarding SEMASCORE (availability of new versions, notes regarding the handling of critical bugs such as those that prevent the saving of elements etc.) is displayed.

Error Messages

To simplify error processing, SEMASCORE offers the option to search through known errors that have occurred and are pertinent to programming.

Administrators

SEMASCORE works with information that must be selectively rendered by the database. Project content can only be visible to the person who created it. As an administrator, you are authorized to use SEMASCORE's online component, i.e. upload a survey and edit its project content (in particular, answers provided by experts for a survey). There is also an option here to define further project members who are also authorized to edit the project.

5.5.5 *Outlook and Methods' Scope of Usage*

As shown in this chapter there are many applications for the method and the tool SEMASCORE rendered. It is important to understand that any strategic question or business- or technology-field analysis is suitable for the approach described before. Any question that requires a good structured view into the future may be analyzed with the tool. One of the major requirements was the high performance that means gaining high quality results within a very short period of time. The result with SEMASCORE: there was only one workshop day required to gain all necessary knowledge to feed the system. The survey for the experts is completely automated.

So the results and their analysis are only a few clicks apart in SEMASCORE and never need more than an hour of work.

References

Adams ME, Day GS, Dougherty D (1998) Enhancing new product development performance: an organizational learning perspective. J Prod Innovat Manage 15(5):403–422

Barczak G, Griffin A, Kahn KB (2009) PERSPECTIVE: trends and drivers of success in NPD practices: results of the 2003 PDMA best practices study. J Prod Innovat Manage 26(1):3–23

Bidault F, Cummings T (1994) Innovating through alliances: expectations and limitations. R&D Manage 24(1):33–45

Brown, B.B (1968) Delphi process: a methodology used for the elicitation of opinions of experts. An earlier paper published by RAND (Document No: P-3925, 1968, 15 pp). The RAND Corporation, Santa Monica

Bruce M, Leverick F, Littler D (1995) Complexities of collaborative product development. Technovation 15(9):535–552

Chesbrough H (2003) Open innovation. The new imperative for creating and profiting from technology. Harvard Business School Press, Boston

Cohen WM, Levinthal DA (1990) Absorptive capacity: a new perspective on learning and innovation. Adm Sci Q 35(1):128–152

Cooper RG (1990) Stage-gate systems: a new tool for managing new products. Bus Horiz 33 (3):44–54

Cooper RG (2008) Perspective: the stage-gate® idea-to-launch process - update, what's new, and NexGen systems. J Prod Innovat Manage 25(3):213–232

Das TK, Teng B-S (2000) Instabilities of strategic alliances: an internal tensions perspective. Organ Sci 11(1):77–101

Driva H, Pawar KS, Menon U (2000) Measuring product development performance in manufacturing organisations. Int J Prod Econ 63(2):247–159

Farr CM, Fischer WA (1992) Managing international high technology cooperative projects. R&D Manage 22(1):55–67

Fichter K (2005) Interpreneurship. Nachhaltigkeitsinnovationen in interaktiven Perspektiven unternehmerischen Handelns. Metropolis, Marburg

Fichter K (2009) Innovation communities: the role of networks of promotors in open innovation. R&D Manage 39(4):357–371

Gaul W, Volkmann M (2000) Methodeneinsatz zur Unterstützung erfolgreicher Produktinnovationen. Zeitschrift für Unternehmensentwicklung und Industrial Engineering 49(2):75–78

Gausemeier J, Plass C, Wenzelmann C (2009) Zukunftsorientierte Unternehmensgestaltung [Future-related company design]. Carl Hanser Verlag, München

Gerybadze A (2003) Gruppendynamik und Verstehen in Innovation Communities. In: Herstatt C, Verworn B (eds) Management der frühen Innovationsphasen: Grundlagen, Methoden, Neue Ansätze. Gabler, Wiesbaden, pp 145–160

González MFJ, Palacios TMB (2002) The effect of new product development techniques on new product success in Spanish firms. Ind Market Manage 31(3):261–271

Griffin A (1992) Evaluating QFD's use in us firms as a process for developing products. J Prod Innovat Manage 12(3):171–188

Griffin A, Hauser JR (1992) Patterns of communication among marketing, engineering and manufacturing - a comparison between two new product teams. Manage Sci 38(3):360–373

Griffin A, Hauser JR (1996) Integrating R&D and marketing: a review and analysis of the literature. J Prod Innovat Manage 13(3):191–215

Ittner CD, Larcker DF (1997) Product development cycle time and organizational performance. J Market Res 34:13–23

Jakob M, Kiehne D-O, Schwarz H, Kaiser F, Beucker S (2007) Delphigestütztes Szenario-Management und -Monitoring - Eine Methode zur Beobachtung von Zukunftsentwicklungen und deren Nutzung im unternehmerischen Innovationsprozess. [Delphi-Supported Scenario Management and Monitoring: A method for monitoring future developments and its use in the entrepreneurial innovation process]. Fraunhofer IRB Verlag, Stuttgart

Kaynak H (2003) The relationship between total quality management practices and their effects on firm performance. J Oper Manage 21(4):405–435

Lawson B, Petersen KJ, Cousins PD, Handfield RB (2009) Knowledge sharing in interorganizational product development teams: the effect of formal and informal socialization mechanisms. J Prod Innovat Manage 26(2):156–172

Mahajan V, Wind Y (1992) New product models: practices, shortcomings, and desired improvements. J Prod Innovat Manage 9(1):128–39

Mercer D (1995) Simpler scenarios. Manage Decis 33(4):32–40

Nijssen EJ, Frambach RT (1998) Market research companies and new product development tools. J Prod Brand Manage 7(4):305–318

Nijssen EJ, Lieshout KFM (1995) Awareness, use and effectiveness of models and methods for new product development. Eur J Market 29(10):27–44

Sackman, H (Apr 1974) Delphi assessment: expert opinion, forecasting and group process, R-1283-PR

Shoemaker PJH (1995) Scenario planning: a tool for strategic thinking. Sloan Manage Rev 36:25–40

Teece DJ, Pisano G, Shuen A (1997) Dynamic capabilities and strategic management. Strategic Manage J 18(7):509–533

Veldhuizen HG (2008) Market intelligence for product excellence. Optima Grafische Communicatie, Rotterdam

Veugelers R (1998) Collaboration in R&D: an assessment of theoretical and empirical findings. De Econ 146(3):419–443

Yeh T-M, Pai F-Y, Yang C-C (2010) Performance improvement in new product development with effective tools and techniques adoption for high-tech industries. Qual Quant 44(1):131–152

Chapter 6
Future Perspectives: Innovation Networks Beyond Traditional Management

Klaus Fichter and Severin Beucker

Abstract Taking new challenges of increased complexity, dynamics and division of labour of innovation processes into account, this book has presented the new concept of "innovation communities" as well as first results from the empirical and practical work on innovation communities carried out within the framework of the research and implementation project InnoCo. InnoCo is part of the research programme on "Innovation strategies beyond traditional management", funded by the German Federal Ministry for Education and Research and the European Social Fund.

Section 6.1 summarizes the key insights from the InnoCo project that have been presented in detail throughout the book. The following Sect. 6.2 discusses the implications of the innovation community concept for managerial practice and future implementations. This is based on empirical findings and practical experience from the InnoCo project, as presented in Chap. 2–5.

The research and implementation project InnoCo has demonstrated the relevance of cross-organizational and cross-level networks of promotors for the success of radical innovation endeavours, and has generated many important new insights into the evolution and management of innovation communities. Nevertheless there are important research questions that are still unanswered and call for further investigations. The final Sect. 6.3 discusses the limitations of present research and the resulting pathways for future research.

K. Fichter and S. Beucker (eds.), *Innovation Communities*,
DOI 10.1007/978-3-642-22128-6_6, © Springer-Verlag Berlin Heidelberg 2012

Practical experience as well as empirical research in innovation management has shown that cooperation and efficient networking are key success factors and enablers in (nearly) all innovation processes. At the same time, the role of key persons like entrepreneurs and their importance for the initiation and implementation of innovations has been a major focus of start-up policy makers, venture capitalists, innovation incubators and innovation researchers. However, so far, only little empirical evidence exists regarding the cooperation of key persons across organizational borders and different levels of the innovation system, as well as their roles and significance in the innovation process. A wide array of studies shows that transformational leaders like "champions" and "promotors" are still a key factor for innovation success today, but that their role and contribution to the effectiveness and efficiency of innovation projects needs to be re-examined in the light of changing framework conditions of innovation (see Chap. 1). The increased dynamics and complexity of innovation processes requires disparate information assets to be levelled and various interrelated knowledge and resource bases inside and outside the organization to be integrated. Self-organizing networks have hence become the dominant innovators for complex technological and systems solutions.

Taking the new challenges of increased complexity and dynamics of innovation process into account, this book presents the new concept of "innovation communities" that draws on insights from evolutionary economics, and on the key actor concept from interaction economics. On the basis of extended promotor theory and the three level innovation system concept, the term "innovation community" is here defined as an informal network of likeminded individuals, acting as universal or specialised promotors, who often come from different organizations and companies, team up in a project-like fashion, and jointly promote a specific innovation, either on one or several levels of an innovation system. Innovation communities are therefore characterized as promotor networks or as informal personal networks of innovators (see Chap. 1).

6.1 Key Insights

Innovation communities are a success factor in radical innovation: As the results presented in Chap. 2 have shown, radical innovation projects that include an innovation community are significantly more successful than those that do not. These results confirm the basic assumption of the InnoCo research project, according to which innovation communities represented a key success factor for radical innovation projects.

Key role of universal promotors (champions) and innovation communities: Our empirical results of the large-number survey (see Chap. 2) confirmed the assumption that universal promotors (champions), but not necessarily single-role promotors, are a fundamental part of innovation networks. Indeed, even the presence of one champion occupying all four promotor roles had a positive influence on project performance. Furthermore, larger numbers of champions were positively correlated with better project performance. Champions were also more common in innovation communities. At the same time, we were able to show that group interaction and group identity factors are significantly higher in successful innovation networks. As a test of the quantitative definition of innovation communities, these results indicate that the presence of an innovation community is a success factor for radical innovation networks in particular.

Size of innovation communities: In all of the innovation cases examined, several promotors can be identified that work together closely (interaction quality), perceive themselves as a team (identity) and in this sense, meet the characteristics of an innovation community as defined in Sect. 3.3. The number of members of the innovation community ranges from 4 to 7. This result is supported by the findings from the large-scale survey of innovation networks, described in detail in Chap. 2, and by earlier case studies (Fichter 2009a) that found innovation communities to have a relatively small number of members. This is understood to be due to the fact that innovation communities are informal promotor networks that require personal contacts, team spirit, and a certain degree of trust between community members, all of which can generally only be ensured by a smaller group size.

Universal promotors (champions) play a prominent role in innovation communities: The innovation communities studied in the case studies and in the large-number survey all had at least two "champions", individuals playing two or more professional promotor roles. In four of five cases, the proportion of champions within the entire community was 50% or higher. Only in one case, the share of champions was smaller (two out of five community members – the other three members assumed only one promotor role each).

Not all promotors of an innovation process are members of the according innovation community: The case studies demonstrate that not all promotors in a particular innovation process actually take part in the innovation community. This has become particularly clear in the complex innovation processes associated with Wood Energy and Carbon Nanotubes, where not even half the promotors are in fact members of the innovation community – possibly because some promotors provide professional and technical support, but do not cooperate that closely with other promotors and are therefore not perceived as part of the community. This aspect highlights the need to distinguish between the issue of identifying the promotors of a focal innovation process, and the distinct question whether these promotors collaborate closely, see themselves as part of a team, and can in this sense be ranked as a part of the innovation community.

Key persons comprise promotors and supporters: The large-number survey had shown a rather high percentage of no-role owners (34.7%) in innovation networks. The innovation case studies helped to clarify their "role". An important insight from the innovation case studies is that, in addition to the promotors, other key persons who support the endeavour through their activities, competence or decision-making power also have notable influence on the initiation, progress, and even overall success of the innovation project. Their "drive" in the innovation process may not be as continuous or as straightforward as the promotors', so that they are not defined as promotors. But even so, their selective support is of great relevance for the overall success of the innovation project. To distinguish them from the promotors, they are here referred to as “supporters”. The group of key persons hence comprises promotors and supporters.

Evolution of innovation communities: One of the lead questions of the case studies was: How does the size and structure of the innovation communities change over time? The case studies reveal that the number of community members steadily increases from the initiation or planning phase to the final development phase. This is plausible in the sense that when nearing implementation, the number of organizations involved increases (see Sect. 3.10). The five cases examined further confirm an initial observation from earlier case studies (Beucker 2007; Fichter 2006; Springer 2007): the community composition over time is characterized by the continuity of its personnel on the one hand, and rapid turnover on the other. In all five communities, we can identify promotors that are members from the beginning to the end, or continuously remain members once they have entered the community.

“Core promotors” as continuous members of innovation communities: All innovation communities studied in the framework of the InnoCo-research project comprise at least two or three continuous members. These can be referred to as “core promotors”.

Performance of innovation communities: All the networks of promotors examined in the case studies in fact qualified as innovation communities as defined in Sect. 2.2.3 and 2.4.4, i.e. all analyzed items and constructs were rated with 5 or above (arithmetic mean) by the respondents, on a scale from 1 (very low) to 7 (very high). From the perspective of the respondents (4 per innovation community), each promotor network displayed particularly high values for community identity (5.2 or above) and interaction quality (5.1 or above). Also, the respondents considered their community to be well or very well composed (5.3 or above). This results in high to very high community performance (5.2–6.2) for all the innovation communities examined.

Community performance contributes to innovation success: As the survey presented in Chap. 2 as well as the case studies in Chap. 3 underscore, the efficiency of innovation communities corresponds with good to excellent success in terms of the joint innovation project (project success). The joint endeavour was considered to be successful (5.3 or higher on average) by all respondents. In all cases studied, the contribution of the community to project success is rated as high to very high (5.8 to 6.8). This further supports the findings from the large-scale survey of innovation networks (see Chap. 2) that the community surrounding a radical innovation project is a key factor for its success.

How do innovation communities contribute to innovation success? The case studies support the hypothesis articulated in Sect. 3.1.3 that innovation communities, through their collaboration, contribute to a project’s success by: (1) helping to transfer an innovative idea from an informal stage to a formal organizational stage or project phase, (2) motivating promotors and empowering them to overcome innovation barriers, (3) improving the social capital of innovation networks, through the acquisition of performance-critical resources, such as funding, knowledge, or reputation, and (4) helping solve or mitigate project-threatening conflicts.

In which situations are innovation communities especially important? According to the function of innovation communities described, we can conclude that innovation communities are especially important when: (1) ... an innovative idea needs to be taken from an informal stage to a formal organization or project phase ("midwife function"). (2) ... resources that the individual organizations cannot not or only with considerable effort provide alone (e.g. money, knowledge, skills, or reputation), and (3) ... conflicts or acceptance problems need to be resolved that individual partners cannot or only with considerable effort address.

Innovation communities as cross-organizational and cross-level networks of promotors: The case studies reveal that all innovation communities work across different levels of the innovation system (see Fig. 1.2). They all involve promotors at the company level, the value chain level, and the framing and linking level. Despite the fact that all innovation communities are cross-level promotor networks, most of their activities relate to the value chain level. Here, the activities of an innovation community can either focus on (1) a broader field of a basic technology (e.g. carbon nanotubes), or a comprehensive usage system (e.g. growing wood crops, and subsequently, the entire value chain of using wood crops for energy), or (2) a specific product or application based on a new technology or solution.

Three basic phases of development in innovation communities: Three basic phases of the life cycle of an innovation community can be distinguished: (1) Initiating and establishing a new innovation community, (2) Developing and improving an existing innovation community, (3) Reconfiguring and extending an existing innovation community.

Methods and measures for the development of innovation communities: As the case studies reveal, the development and performance of innovation communities not only requires methods and measures in the Process Support task area (partner identification and selection, joint projects, meetings, coaching, working groups, workshops, excursions, etc.), but also in the areas Foresight and Stakeholder Interaction. Thus, many of the methods applied by innovation communities can be used both as innovation management tools in general, and as instruments specifically suited to the development of innovation communities.

Systemic innovation coaching and Delphi-based scenario development: Many of the methods and measures for innovation communities are easy to use and self-explanatory, some require basic methodological knowledge, and others require thorough methodological expertise and experience. Many of the tools listed in Sec. 5.3 have been tried and tested in the InnoCo project. Special attention was paid to the methods of systemic coaching and Delphi-based scenario development in innovation networks, which are both methodologically complex and at the same time, very important for the development of innovation communities.

6.2 Implications for Managerial Practice and Future Implementation

Practical experience as well as empirical research in innovation management has shown that cooperation and efficient networking are key success factors and enablers in (nearly) all innovation processes. This book has presented scientific evidence and practical examples that demonstrate that, in innovation processes that deal with basic new technologies and are aimed at radical new solutions in particular, the existence of a supporting innovation community is a key success factor. The results of this study indicate that group composition, group interaction, and group identity should not be neglected when developing new products and solutions in networks.

For innovation and network managers as well as innovation intermediaries supporting an innovation network or alliance, the question therefore is which features make innovation communities successful and what aspects have to be taken into account when developing them. Based on the empirical results and practical experiences in implementation and community development within the InnoCo project, the following properties can be described as the specific power of innovation communities, in terms of their contribution to the success of a joint radical innovation project:

6.2.1 Composition of the Innovation Community

- The composition of the innovation community is a key factor of innovation success. Special attention has to be paid to the identification and selection of appropriate partner organizations and promotors. Effective methods and measures for partner identification and selection are available (see Sec. 5.3 for more details).
- The performance of a particular innovation community is shaped and fortified through the involvement of champions.
- In an innovation community, all promotor roles (power promotor, expert promotor, process promotor, relationship promotor) should be represented.
- A core group of promotors and champions that remains part of the community from the beginning to the end of the project seems to be crucial to overall stability and performance over time.
- Since innovation communities are informal networks of promotors that require personal contacts, a team spirit and a sense of trust between all community members, a manageable group size seems to improve performance or enable it in the first place.
- The appropriate size of the innovation community depends on the complexity of the innovation project and the number of involved partner organizations. As a rule, a size of 4 to 7 members of the innovation community seems to work well.
- Regarding the performance and efficiency of innovation communities, it is important that all relevant partner organizations are represented in the community by a promotor from the respective organizations.
- The performance of a innovation community to a large extent depends on the contacts and relationships that individual promotors can provide and are prepared to share with the rest of the promotor network (social capital).

- Reconfiguration capacity: Since project tasks can change during the innovation process and the demand for individual partners may change over time, depending on the stage of the project, not only does the association of innovating organizations regularly need to check its composition, but also the associated innovation community itself.

6.2.2 *Identity of the Innovation Community*

- The performance of a community depends, on its capability to develop common targets and perspectives for the shared project, as well as a common language and basic understanding of project goals and future options.
- Network performance seems to largely depend on the way in which its promotors manage to generate group cohesion.

6.2.3 *Interaction of the Innovation Community*

- Creating trust among community members, which enhances the performance of the group, is one of the key tasks in terms of the functioning of the innovation community.
- Intensive und respectful cooperation: Trust is usually only gained through personal encounters and experience in practical cooperation.
- Conflict management requires the will and ability to resolve conflicts at the individual promotors' level. Resolving personal or professional conflicts within an innovation project can be facilitated if the innovation community is managed or advised by a neutral partner (an innovation coach or technology transfer facility, for instance).
- In addition to their conflict management skills, innovation coaches or intermediaries can also take over the role of independent process or relationship promotors, again significantly improving the performance of the network as a whole.
- The performance of innovation communities ultimately depends on the degree to which it is able to organize an effective exchange of knowledge and experience, e.g. in the form of workshops, conferences or field trips.

6.2.4 *Support by Governments and Innovation Intermediaries*

As this book has demonstrated, radical innovation projects that include an innovation community are significantly more successful than those in which this is not the case. For governments, funding agencies and innovation intermediaries, this implies that they should focus on the initiation and development of networks of promotors, and develop funding and supporting schemes for new technologies and innovation projects. Cooperative innovation projects should incorporate a sound network management concept and allow for the funding of activities for the process support of joint projects (such as systematic partner selection, joint projects, innovation coaching, establishing innovation communities, etc.; see Sect. 5.3 for more details).

The level of complexity of an innovative technology or solution seems to be a key factor influencing the formation of innovation communities. Greater complexity obviously leads to the situation that the necessary skills and resources for developing and implementing innovative solutions are distributed among different partners, either within a large firm or among different organizations. This distribution of resources makes the different organizations and individuals literally interdependent.

One consequence is that innovation communities, especially in the case of radical innovations, are subject to a high degree of dynamics in terms of time and contents, which can itself provoke conflicts. A further challenge for the development and efficiency of innovation communities is the informal nature of network relations. Group-dynamic processes are especially critical, and the community can easily be thrown off track through power struggles, attempts of domination, processes of marginalisation, "fare dodger" behaviour, etc. For this reason, sensible external coaching can help to guide the community in productive directions. Governments, funding agencies and innovation intermediaries should therefore integrate the element of systemic innovation coaching (see Sect. 5.4) into their funding and support programmes for new technologies and radical new solutions.

The fact that innovation communities constitute a success factor in radically new technologies and radical innovation projects should be incorporated into the training of innovation and network managers. Also, the concept of innovation communities should be made an essential component of academic programmes and of extended training courses for technology, innovation, network and start-up management.

6.3 Limitations and Pathways for Future Research

The research and implementation project InnoCo has demonstrated the relevance of cross-organizational and cross-level networks of promotors for the success of radical innovation endeavours and has generated many important new insights into the evolution and management of innovation communities. Nevertheless, there are important research questions that are still unanswered and remain for further investigations. Here are some of the viable pathways for future research:

6.3.1 *Interaction Between Promotors, Supporters and Other Key Persons*

The case studies presented in this book have made clear that innovation communities are a part of a broader network of key persons involved in an innovation process. The innovation community can be considered the "core" of the broader innovation network. The empirical findings underscore the fact that not all promotors of specific innovation processes are members of the innovation community.

The case studies and the large-number survey also demonstrate that besides promotors there is another group of key persons involved, which we have named "supporters". Future research (see also Sect. 2.5.1) could for instance investigate how the interaction between the core of the innovation network – the innovation community –, and the other key persons involved in innovation actually works, and what forms of interaction and cooperation improve the performance of the innovation network and its contribution to project success. Concepts from research on "communities of practice" (Wenger et al. 2002) and from interaction economics (Fichter 2009b) can be helpful for this scientific endeavour. Wenger et al. (2002) distinguish a small, intensively participating core group from active members and peripheral members. Taking an interactional perspective into account, where roles develop from interaction (e.g. Benne and Sheats 1948), it is possible to apply different role concepts altogether. In addition, it should be investigated how roles, activity levels and membership (fluctuation) change over time in the innovation process (for communities of practice see Jahnke 2010). Also, the modelling of different stages of development of innovative communities of practice (Wenger et al. 2002) could be transferred to the innovation community context.

6.3.2 *The Importance of Personality Traits*

In previous studies, important factors were identified that affect the performance of innovation communities (see Sect. 6.2). There is, however, one important aspect that has not been examined so far: the personality traits of the various members of innovation communities. Future investigations could establish how the personal characteristics of promotors affect teambuilding and cooperation in innovation communities, and which personality types or combinations of personality types enhance the performance of innovation communities. Finally an interesting question is if it boosts the performance and quality of an innovation community if the members are supported in the self awareness of the personality characteristics and their promotor roles. Even counteracting characteristics or roles (opponents or inhibitors) can be of interest for the success of an innovation community.

6.3.3 Use of Supporting Methods and Measures in Innovation Communities

The empirical results and practical experiences presented in this book show which methods and measures can be applied to support the creation, development and reconfiguration of innovation communities (see Sect. 5.3 to 5.5). Nevertheless, some questions regarding the specific range of application, design and effects of certain methods and measures remain. These could also be addressed in future research.

6.3.4 Refining the Typology of Innovation Communities

As early as 2003, in his initial work on innovation communities, Gerybadze (2003) defined different types of innovation communities and arranged them according to their respective main tasks and the phases they covered within the innovation process. Moreover, as discussed in Sect. 3.10 and 4.1, innovation communities can also be distinguished according to their own evolutionary stage (initiation, development, or reconfiguration phase), or according to the level of the innovation system they primarily function at. Future studies could further examine whether there are other types or variants of innovation communities, and how they can be differentiated with scientific precision.

6.3.5 Limitations of Success Factor Analysis

Regarding the dimensions of group identity and group interaction, the large-number survey had revealed that we are confronted with the usual methodological limitations of success factor analysis, i.e. that successful networks tend to overestimate factors that explain their own success (Rese and Baier 2011). An in-depth analysis has shown that the factors themselves are affected by multi-collinearity (see Rese et al. 2010 for more details). Further investigations could take into consideration the reciprocal relationship between group identity and group interaction (e.g. cognitive similarity and communication, see Rentsch et al. 2008). The items determining group identity should be further tested and developed in other organizational group contexts. In addition, an external perspective to monitor subjective (internal) perceptions of group members regarding their group boundaries and transactions could be interesting.

6.3.6 *Surveys that Encompass Multiple Perspectives of Community Members*

The case studies presented in this book indicate that the perspectives of innovation community members can differ depending on their role, personality or experience, or on the time they have been involved in the community. This multiple-perspective approach is obviously helpful in analysing the structure and dynamics of innovation communities. Nevertheless, this approach has so far not been used in surveys. In the surveys carried out so far, only one person was questioned as representative of an entire network. The networks are therefore treated here as if they were homogeneous and favoured one single behaviour or strategy. Consequently, it might be interesting to include multiple perspectives (cf. Högl and Gemünden 2001) from each network in future studies.

References

Benne KD, Sheats P (1948) Functional roles of group members. J Soc Issue 4(2):41–49

Beucker S (2007) Die Innovation Solon-Mover der Solon AG. Fraunhofer IRB, Stuttgart

Fichter K (2006) Das "e-place"-Konzept der IBM Deutschland. Fraunhofer IRB, Stuttgart

Fichter K (2009a) Innovation communities: the role of networks of promotors in open innovation. R&D Manag 39(4):357–371

Fichter K (2009b) Interaktionsökonomik [Electronic version]. http://www.borderstep.de/details.php?menue=70&le=en. Accessed April 10, 2011

Gerybadze A (2003) Gruppendynamik und Verstehen in Innovation Communities. In: Herstatt C, Verworn B (eds) Management der frühen Innovationsphasen. Gabler, Wiesbaden, pp 145–160

Högl M, Gemünden HG (2001) Teamwork quality and the success of innovative projects: a theoretical concept and empirical evidence. Organ Sci 12:400–449

Jahnke I (2010) Dynamics of social roles in a knowledge management community. Comput Hum Behav 26(4):533–546

Rentsch JR, Delise LA, Hutchison S (2008) Transferring meaning and developing cognitive similarity in decision making teams: collaboration and meaning analysis process. In: Letsky MP, Warner NW, Fiore SM, Smith CAP (eds) Macrocognition in Teams. Ashgate Publishing, Burlington, VT, pp 127–142

Rese A, Baier D (2011) Success factors for innovation management in networks of small and medium enterprises. R&D Manag 41(2):138–155

Rese A, Sand N, Baier D (2010) Group identity and group interaction in innovation networks. In: 17th international product development management conference, Murcia (Spain), 14-15 June 2010

Springer S (2007) Von der Faser bis zum Druck: Das Coated-Coldset-Netzwerk. Fraunhofer IRB, Stuttgart

Wenger E, McDermott RA, Snyder W (2002) Cultivating communities of practice: a guide to managing knowledge. Harvard Business School Press, Boston, MA